ECOLOGY AND MANAGEMENT OF BLACKBIRDS (ICTERIDAE) IN NORTH AMERICA

Edited by

George M. Linz • Michael L. Avery • Richard A. Dolbeer

CRC Press is an imprint of the
Taylor & Francis Group, an **informa** business

Front cover: Multiple-Species Flock of Blackbirds (Icteridae). Photographed in a sunflower field in central North Dakota. Photo by H. Jeffrey Homan.

CRC Press
Taylor & Francis Group
6000 Broken Sound Parkway NW, Suite 300
Boca Raton, FL 33487-2742

© 2017 by Taylor & Francis Group, LLC
CRC Press is an imprint of Taylor & Francis Group, an Informa business

No claim to original U.S. Government works

Printed on acid-free paper

International Standard Book Number-13: 978-1-498-79961-4 (Hardback)

This book contains information obtained from authentic and highly regarded sources. Reasonable efforts have been made to publish reliable data and information, but the author and publisher cannot assume responsibility for the validity of all materials or the consequences of their use. The authors and publishers have attempted to trace the copyright holders of all material reproduced in this publication and apologize to copyright holders if permission to publish in this form has not been obtained. If any copyright material has not been acknowledged please write and let us know so we may rectify in any future reprint.

Except as permitted under U.S. Copyright Law, no part of this book may be reprinted, reproduced, transmitted, or utilized in any form by any electronic, mechanical, or other means, now known or hereafter invented, including photocopying, microfilming, and recording, or in any information storage or retrieval system, without written permission from the publishers.

For permission to photocopy or use material electronically from this work, please access www.copyright.com (http://www.copyright.com/) or contact the Copyright Clearance Center, Inc. (CCC), 222 Rosewood Drive, Danvers, MA 01923, 978-750-8400. CCC is a not-for-profit organization that provides licenses and registration for a variety of users. For organizations that have been granted a photocopy license by the CCC, a separate system of payment has been arranged.

Trademark Notice: Product or corporate names may be trademarks or registered trademarks, and are used only for identification and explanation without intent to infringe.

Library of Congress Cataloging-in-Publication Data

Names: Linz, George M. (George Michael), editor. | Avery, Michael L., editor. | Dolbeer, Richard A., editor.
Title: Ecology and management of blackbirds (Icteridae) in North America | editors, George M. Linz, Michael L. Avery, and Richard A. Dolbeer.
Description: Boca Raton: CRC Press, 2017. | Includes bibliographical references.
Identifiers: LCCN 20160590991 ISBN 9781498799614 (hardback: alk. paper)
Subjects: LCSH: Blackbirds--North America. | Icteridae--North America.
Classification: LCC QL696.P2475 E25 20171 DDC 598.8/74--dc23
LC record available at https://lccn.loc.gov/2016059099

Visit the Taylor & Francis Web site at
http://www.taylorandfrancis.com

and the CRC Press Web site at
http://www.crcpress.com

Dedications

This book is dedicated to my parents George W. and Theresa C. Linz who successfully raised eight children on an award-winning Pennsylvania dairy farm and my wife Linda Marlene Linz who works and plays beside me on our fantasy farms in Pennsylvania and North Dakota.

George M. Linz

Growing up along the Chesapeake Bay in Maryland, I was forever encouraged by my parents, Bud and Nancy Avery, to explore the outdoors and to appreciate what I found there. This book is dedicated to them, and to my ever-supportive wife, Joy, who brings just that to each day we are together.

Michael L. Avery

This book is dedicated to my wife, Saundra and daughters, Jennie and Cynthia. These remarkable women provided unwavering support, encouragement and excellent advice during my adventures working with blackbirds and other wildlife in North America and abroad. I also dedicate this book to John L. Seubert, my mentor in the early days of my research on blackbirds.

Richard A. Dolbeer

Contents

Preface vii
Editors ix
Contributors xi

Chapter 1
History of Regulations, Policy, and Research Related to Conflicts between Blackbirds and People 1

George M. Linz, Michael L. Avery, and Richard A. Dolbeer

Chapter 2
Ecology and Management of Red-Winged Blackbirds 17

George M. Linz, Page E. Klug, and Richard A. Dolbeer

Chapter 3
Ecology of Yellow-Headed Blackbirds 43

Daniel J. Twedt

Chapter 4
Ecology and Management of the Common Grackle 65

Brian D. Peer and Eric K. Bollinger

Chapter 5
The Brown-Headed Cowbird: Ecology and Management of an Avian Brood Parasite 77

Brian D. Peer and Virginia E. Abernathy

Chapter 6
Effects of Habitat and Climate on Blackbird Populations 101

Greg M. Forcey and Wayne E. Thogmartin

Chapter 7
Dynamics and Management of Blackbird Populations 119

Richard A. Dolbeer

Chapter 8
Chemical Repellents 135

Scott J. Werner and Michael L. Avery

Chapter 9
Frightening Devices 159

Michael L. Avery and Scott J. Werner

Chapter 10
Strategies for Evading Blackbird Damage 175
George M. Linz and Page E. Klug

Chapter 11
Allowable Take of Red-Winged Blackbirds in the Northern Great Plains 191
Michael C. Runge and John R. Sauer

Chapter 12
The Economic Impact of Blackbird Damage to Crops 207
Stephanie A. Shwiff, Karina L. Ernest, Samantha L. Degroot, Aaron M. Anderson, and Steven S. Shwiff

Chapter 13
The Future of Blackbird Management Research 217
Page E. Klug

Index 235

Preface

Urban and rural residents across North America commonly note various species of blackbirds (Icteridae) in their backyards, wetlands, and agricultural fields or simply observe them as large flocks of birds flying overhead. The arrival and departure of blackbirds signify changes in season. In early spring, blackbirds begin calling and displaying from perches to initiate the breeding season, bringing a sense of joy in expectation of warmer days and regeneration. Agriculturalists, however, are aware that newly seeded crops, especially corn and rice, are vulnerable to foraging blackbirds. In the postbreeding fall season, blackbirds gather in flocks augmented by the recently fledged young birds. These congregations, often numbering in the hundreds of thousands, leave their night roosts at daybreak in awesome, synchronous flight displays to forage or to begin southward migration. Although these are spectacular shows of bird life, agriculturalists brace for swarms of blackbirds seeking energy-rich crops to supply the fuel for feather molt and migration. During winter, blackbirds gather in large roosting congregations in the southern United States, often in association with the nonrelated and nonnative European starling (*Sturnus vulgaris*). These large winter roosts, often containing several million birds, can create numerous problems related to public health and agricultural damage.

A.A. Allen's "The Red-Winged Blackbird: A Study in the Ecology of a Cat-Tail Marsh," published in 1914, was arguably the foundation for the more than 1,000 descriptive and experimental studies on the life history and management of blackbirds. There are several reasons for this abundance in publications that continue to this day. First, blackbirds are relatively easy to access and observe. Second, blackbirds show a range of breeding strategies such as polygyny (red-winged blackbirds and yellow-headed blackbirds), monogamy (common grackles), and brood parasitism (brown-headed cowbirds). Third, these species often interact in overlapping breeding habitats and nonbreeding season foraging and roosting habitats, but they differ markedly in breeding strategy and specific habitat and food requirements. Finally, blackbirds are considered economically important because they damage newly planted and ripening crops.

The pioneering publication, F.E.L. Beal's 1900 treatise "The Food of the Bobolink, Blackbirds, and Grackles," stimulated studies on the ecology of blackbirds in relation to the problems of crop damage. Subsequent generations of investigators followed Beal's lead and scrutinized the economic impact of blackbirds on numerous crops, especially corn, rice, and sunflower but also barley, oats, peanuts, sorghum, rye, and wheat. Since the mid-1950s, the overall research effort has expanded in scope and intensity with the injection of millions of dollars to support integrated investigations involving both short- and long-term studies. The wealth of information generated over decades of applied research is found in hundreds of publications in disparate outlets that include government reports, conference proceedings, peer-reviewed journals, monographs, and books.

We sought to summarize and synthesize this vast body of information on the biology and life history of blackbirds and their conflicts with humans into a single volume for researchers, ornithologists, wildlife managers, agriculturists, policy makers, and the general public. Blackbirds are a dominant component of the avifauna in the natural and agricultural ecosystems of North America. Thus, our principal goal is to provide a better understanding of the functional roles of blackbirds in these ecosystems so that improved science-based, integrated management strategies can be developed to resolve conflicts.

We divided the book into 13 chapters. Chapter 1 covers the pertinent history of research and management, policy, and regulations beginning in 1886 when Dr. C. Hart Merriam and Dr. A. K. Fisher established the Division of Economic Ornithology and Mammalogy in the United States Department of Agriculture.

Chapters 2 through 5 are dedicated to the biology and life history of the four most abundant blackbird species: red-winged blackbirds, yellow-headed blackbirds, common grackles, and brown-headed cowbirds.

Effects of habitat modification and climate change on blackbird populations are discussed in Chapter 6. Weather variables have been shown to influence blackbird movements, diet, nest productivity, and so on, whereas effects of climate change are more uncertain, given the limited research available.

Chapter 7 is dedicated to the dynamics and management of blackbird populations, with strong emphasis on historical attempts to manage blackbird populations, most of which failed to the dismay of advocates.

The authors of Chapters 8 and 9 discuss progress on the development of chemical repellents and frightening devices to reduce bird damage to crops. Chapter 8 presents results from cage and field trials of various candidate chemical repellents and discusses the potential utility of these repellents within integrated blackbird management strategies. Chapter 9 focuses on frightening devices to disperse birds from crops. Scientists have found that testing frightening devices is a challenge because controlled and replicated experiments are difficult under field conditions. Thus, with some exceptions, most publications describe limited field trials or demonstrations.

The authors of Chapter 10 discuss strategies to evade damage. Evasion methods should form the base of an integrated pest management strategy because they are not focused on the pest bird itself but are instead intended to manipulate the environment that surrounds the crops that are vulnerable to damage.

Chapter 11 uses the framework of harvest theory and prescribed take level (PTL) to assess allowable take of the red-winged blackbird in the northern Great Plains. The PTL framework has been applied to other species of birds under the Migratory Bird Treaty Act.

In Chapter 12, multiple authors team up to provide a general overview of the economic impact of blackbirds on agriculture and costs associated with management actions.

Finally, the author of Chapter 13 discusses near- and long-term prospects for research on blackbird ecology and management in North America.

The USDA APHIS Wildlife Services IT customer support, especially Pat Anderson, kept GML's computer running at remote locations, and the library staff at the National Wildlife Research Center, especially Cynthia Benton, helped locate obscure publications for various scientists across agencies.

We are grateful to the following scientists who reviewed, commented, and otherwise contributed to the content of this book: Kevin Aagaard, Dave Bergman, Brad Blackwell, Will Bleier, Eric Bollinger, Larry Clark, Dick Curnow, Rick Engeman, Sonia Canavelli Gariboldi, Heath Hagy, Carol Henger, Jeff Homan, Lou Huffman, Larry Igl, Doug Johnson, Fred Johnson, Jim Rivers, Don Snyder, Jessica Stanton, Mark Tobin, Troy Turner, Jack Waide, Mike Ward, Pat Weatherhead, Ken Yasukawa, and Dave Ziolkowski, Jr.

Dr. Larry Clark, Director of the National Wildlife Research Center, provided the impetus for writing a book on the ecology and management of blackbirds that summarizes research conducted over the last 60 years.

George M. Linz
Michael L. Avery
Richard A. Dolbeer
U.S. Department of Agriculture
Animal and Plant Health Inspection Service
Wildlife Services

Editors

George M. Linz was a research wildlife biologist for the U.S. Department of Agriculture, National Wildlife Research Center (NWRC), from 1987 to 2015 and was stationed in Bismarck, North Dakota for much of his career. George earned a BS and an MS in biology from Edinboro University of Pennsylvania and a PhD in zoology from North Dakota State University (NDSU). George was hired as a project leader and was initially stationed at the Denver Wildlife Research Center (now NWRC) headquarters. Two years later, he was transferred to a newly established NWRC field station in North Dakota. George and colleagues authored numerous papers on the ecology and management of blackbirds in relation to sunflower. Over the course of his career, George served as study director and major advisor or co-advisor for 35 graduate students. He was awarded the Professional Wildlife Award by the Central Mountain and Plains Section of the Wildlife Society. George and his wife Linda spend time on their hobby farms in Pennsylvania and North Dakota, where they monitor nearly 100 birdhouses and plant food plots and trees to enhance habitat for wildlife.

Michael L. Avery grew up in Annapolis, Maryland. He attended Johns Hopkins University (BES), North Dakota State University (MS), and the University of California-Davis (PhD) and served as a Peace Corps volunteer in Malaysia (1975–1977). He worked for the National Park Service and U.S. Fish and Wildlife Service before being hired in 1987 as project leader with the Denver Wildlife Research Center (now NWRC) stationed at the Gainesville, Florida field station where he continues to serve as supervisory research wildlife biologist. He is also a courtesy faculty member in the Wildlife Ecology and Conservation Department of the University of Florida. Michael's career embraces a diversity of wildlife management areas, including bird collisions with man-made structures, avian crop depredations, vulture behavior and management, avian contraception, and development of methods to detect and remove large invasive reptiles. Michael and his wife Joy met in the Peace Corps; they have two children and one granddaughter.

Richard A. Dolbeer was a scientist with the U.S. Department of Agriculture from 1972 to 2008, where he led a series of research projects to resolve conflicts between humans and wildlife in North America, Africa, and Asia. With over 200 scientific publications, he is a recipient of the Federal Aviation Administration's Excellence in Aviation Research Award and the inaugural Caesar Kleberg Award for Applied Wildlife Research presented by the Wildlife Society. Richard received degrees from the University of the South, the University of Tennessee, and Colorado State University (PhD, wildlife biology, 1972). Richard currently manages his 56-acre farm, "Bluebird Haven," in Ohio and works as a science advisor to the aviation industry and U.S. Department of Agriculture. He has been married to Saundra for 50 years and has two children and six grandchildren.

Contributors

Virginia E. Abernathy
Research School of Biology
Australian National University
ACT, Australia

Aaron M. Anderson
National Wildlife Research Center
Wildlife Services
Animal and Plant Health Inspection Service
U.S. Department of Agriculture
Fort Collins, Colorado

Michael L. Avery
National Wildlife Research Center
Wildlife Services
Animal and Plant Health Inspection Service
U.S. Department of Agriculture
Gainesville, Florida

Eric K. Bollinger
Department of Biological Sciences
Eastern Illinois University
Charleston, Illinois

Samantha L. Degroot
National Wildlife Research Center
Wildlife Services
Animal and Plant Health Inspection Service
U.S. Department of Agriculture
Fort Collins, Colorado

Richard A. Dolbeer
Retired
Wildlife Services
Animal and Plant Health Inspection Service
U.S. Department of Agriculture
Sandusky, Ohio

Karina L. Ernest
National Wildlife Research Center
Wildlife Services
Animal and Plant Health Inspection Service
U.S. Department of Agriculture
Fort Collins, Colorado

Greg M. Forcey
University of Florida
Gainesville, Florida

Page E. Klug
National Wildlife Research Center
Wildlife Services
Animal and Plant Health Inspection Service
U.S. Department of Agriculture
Bismarck, North Dakota

George M. Linz
Retired
National Wildlife Research Center
Wildlife Services
Animal and Plant Health Inspection Service
U.S. Department of Agriculture
Bismarck, North Dakota

Brian D. Peer
Department of Biological Sciences
Western Illinois University
Moline, Illinois

Michael C. Runge
Patuxent Wildlife Research Center
U.S. Geological Survey
Laurel, Maryland

John R. Sauer
Patuxent Wildlife Research Center
U.S. Geological Survey
Laurel, Maryland

Stephanie A. Shwiff
National Wildlife Research Center
Wildlife Services
Animal and Plant Health Inspection Service
U.S. Department of Agriculture
Fort Collins, Colorado

Steven S. Shwiff
Department of Economics
Texas A&M University
College Station, Texas

Wayne E. Thogmartin
Upper Midwest Environmental Sciences Center
U.S. Geological Survey
Middleton, Wisconsin

Daniel J. Twedt
Patuxent Wildlife Research Center
U.S. Geological Survey
Memphis, Tennessee

Scott J. Werner
National Wildlife Research Center
Wildlife Services
Animal and Plant Health Inspection Service
U.S. Department of Agriculture
Fort Collins, Colorado

CHAPTER 1

History of Regulations, Policy, and Research Related to Conflicts between Blackbirds and People

George M. Linz
National Wildlife Research Center
Bismarck, North Dakota

Michael L. Avery
National Wildlife Research Center
Gainesville, Florida

Richard A. Dolbeer
Wildlife Services
Sandusky, Ohio

CONTENTS

1.1	Migratory Bird Treaty Act	2
	1.1.1 U.S. Depredation Order for Blackbirds	2
1.2	Canadian Wildlife Service	2
1.3	U.S. Department of Agriculture	3
1.4	National Environmental Policy Act	3
1.5	Wildlife Services Decision Model	4
1.6	Blackbird Research in Canada	4
1.7	Blackbird Research in the United States	6
	1.7.1 National Wildlife Research Center Headquarters	6
	1.7.2 Blackbird Research—Headquarters	8
	1.7.3 Blackbird Research—Field Stations	8
	1.7.3.1 Ohio Field Station	8
	1.7.3.2 Florida Field Station	10
	1.7.3.3 North Dakota Field Station	10
	1.7.3.4 Kentucky Field Station	11
	1.7.3.5 California Field Station	12
1.8	Summary	12
References		12

The United States and Canada have invested substantial resources over the past 60 years for developing methods to reduce blackbird (Icteridae) damage to agricultural crops, to manage large winter roosts that create nuisance and public health problems, and to mitigate conflicts with endangered species. It is an indication of the challenging nature of the conflicts with these abundant, highly mobile birds that we are still attempting to improve existing methods and develop new approaches to mitigate the problems. Scientists have tested chemical frightening agents and repellents, mechanical scare devices, bird-resistant sunflowers, decoy crops, habitat management, population management, and cultural modifications in cropping. Methods development proceeds within a framework of federal and state laws and agency policies. Here, we review key laws and policies that guide scientists focused on methods development, and we briefly recount the history of applied blackbird research in the United States and Canada.

1.1 MIGRATORY BIRD TREATY ACT

The Migratory Bird Treaty Act (MBTA) of 1918 is the legal framework governing decisions on management and conservation of native migratory birds in the United States and Canada. The US federal law was first enacted in 1916 in order to implement the convention for the protection of migratory birds between the United States and Great Britain (acting on behalf of Canada). Later amendments implemented treaties between the United States and Mexico (1936), Japan (1972), and the Soviet Union (1976, now Russia).

Blackbirds are native migratory birds and thus come under the jurisdiction of the MBTA. The statute makes it unlawful without a waiver to pursue, hunt, take, capture, kill, or sell birds listed therein ("migratory birds"). The statute does not discriminate between live and dead birds and also grants full protection to any bird parts including feathers, eggs, and nests.

1.1.1 U.S. Depredation Order for Blackbirds

Blackbirds are given federal protection in the United States and Canada under the MBTA. Both countries, however, allow protection of resources and human health compromised by blackbirds, including the use of nonlethal and lethal methods. Blackbirds may be legally killed in the United States under the *Depredation Order for Blackbirds, Cowbirds, Grackles, Crows, and Magpies* (50 CFR 21.43), when found "committing or about to commit depredations upon ornamental or shade trees, agricultural crops, livestock, or wildlife, or when concentrated in such numbers and manner as to constitute a health hazard or other nuisance."

In 2010, rusty blackbirds (*Euphagus carolinus*) were removed from the depredation order and given full protection by the MBTA, as has always been the case for the tri-colored blackbird (*Agelaius tricolor*) (U.S. Department of Interior 2010). Another revision was that nontoxic shot must be used when taking birds by shotgun under the authority of CFR 21.43. Moreover, persons taking blackbirds under CFR 21.43 must provide the U.S. Fish and Wildlife Service (FWS) the following information at the end of each calendar year: name and address, species and number taken, month when birds were taken, state and county where birds were taken, and a general explanation of why the birds were taken. Some states and municipalities have additional restrictions on killing blackbirds. European starlings (*Sturnus vulgaris*), which often associate with blackbirds during the nonbreeding season, are not native to North America and are not protected by the MBTA.

1.2 CANADIAN WILDLIFE SERVICE

The Canadian Wildlife Service is Canada's national wildlife agency. Its core area of responsibility is the protection and management of migratory birds and their nationally important habitats.

Wildlife management in Canada is a constitutionally shared responsibility among the federal, provincial/territorial, and aboriginal governments (Government of Canada 2016a). In Canada, most species of birds are protected under the Migratory Birds Convention Act, 1994 (MBCA). The MBCA was passed in 1917 and updated in 1994 and 2005 to implement the Migratory Birds Convention (Government of Canada 2016b).

A person who owns, leases, or manages land, however, can seek a permit from provincial authorities to scare or kill migratory birds that are causing or are likely to cause damage. Any person may, without a permit, use equipment other than an aircraft or firearms to scare migratory birds. A permit must be acquired to use aircraft or firearms for this purpose.

In situations where scaring migratory birds is not a sufficient deterrent, a permit can be obtained to kill offending birds in a specific time frame and area. A person who controls an area of land may seek a permit to collect and destroy the eggs of migratory birds and to dispose of the eggs in the manner provided in the permit. Where a permit is issued to kill migratory birds that are causing or are likely to cause damage to crops, no person mentioned in the permit shall shoot migratory birds elsewhere than on or over fields containing such crops or shall discharge firearms within 50 m of any water area.

1.3 U.S. DEPARTMENT OF AGRICULTURE

In the United States, the U.S. Department of Agriculture (USDA) Animal and Plant Health Inspection Service (APHIS) Wildlife Services (WS) provides federal leadership and expertise to resolve wildlife conflicts to allow people and wildlife to coexist. The WS program's primary statutory authority is found in the Animal Damage Control Act of 1931. This act gives WS broad authority to investigate, demonstrate, and control mammalian predators, rodents, and bird pests. In 1985, Congress transferred the Animal Damage Control Program (now WS) from the Department of the Interior to the USDA. Another amendment in 1987 gave WS the authority to enter into agreements with public and private entities in the control of mammals and birds that are a nuisance or are reservoirs for zoonotic diseases (Tobin 2012).

The WS Office of the Deputy Administrator, located in Washington, DC, provides national program oversight, with field operations directed from the Eastern Regional Office in Raleigh, North Carolina, and the Western Regional Office in Fort Collins, Colorado. The National Wildlife Research Center (NWRC), headquartered in Fort Collins, is the methods development arm of the WS program. The WS program is aimed at helping to resolve wildlife damage to a wide variety of resources and to reduce threats to human health and safety. Funding for the WS program is a combination of federal appropriations and cooperator-provided funds.

1.4 NATIONAL ENVIRONMENTAL POLICY ACT

In 1969, the National Environmental Policy Act (NEPA) was enacted to establish a national framework for protecting the environment (U.S. Environmental Protection Agency 2015; U.S. Department of Interior 2016). NEPA was intended to assure that all branches of government give proper consideration to the environment prior to undertaking any major federal action that significantly affects the environment. NEPA compliance involves the development of an environmental assessment (EA) to determine if the proposed federal action will have a significant effect on the environment. If the EA shows that the federal action will not have a significant effect on the human environment, then a Finding of No Significant Impact is prepared. If the EA determines that a federal action will have a significant effect on the human environment, then an Environmental Impact Statement (EIS) is prepared. An EA is typically a shorter document than an EIS, and its preparation offers fewer opportunities for public comment or involvement than an EIS. EAs have fewer procedural requirements and therefore take less time and fewer resources to prepare on average than an EIS.

Preparation of an EIS requires public input, and it must be available for 30 days for public review and comment before a final decision is made. Generally, an EIS includes detailed discussions of the following: a statement of the purpose and need for the proposed action, a description of the affected environment, alternatives to the proposed action, and an analysis of environmental impacts and ways to mitigate such impacts. Failure to follow the NEPA process or providing inadequate documentation to support a particular action can result in legal actions to rectify these errors (Cirino 2016).

In 1997, WS completed a national EIS that addressed the need for wildlife damage management (U.S. Department of Agriculture 1997). States with blackbird populations that could impact human endeavors have developed EAs that analyze management options for reducing damage and health hazards (e.g., U.S. Department of Agriculture 2007 and 2015).

1.5 WILDLIFE SERVICES DECISION MODEL

When requests for assistance are received, WS employees are required to use the WS decision model to determine the appropriate damage management strategy (Policy Directive 2.201; Slate et al. 1992; U.S. Department of Agriculture 2014). Requests from the public for assistance include nuisance wildlife, wildlife damage to crops and livestock, and wildlife hazards related to public safety.

If a cursory review of the request is deemed an actionable problem within the purview of WS, the extent and magnitude of the damage is detailed during a site visit (Figure 1.1). The next step is to evaluate available methods for practicality, including legal, administrative, and environmental considerations. This evaluation sometimes occurs during the development of an EA or, less commonly, an EIS. Assuming management options are available, biologists formulate a control strategy that usually includes practical nonlethal methods as a first option. Lethal methods are sometimes an alternative when deemed to be appropriate and to show promise for resolving the conflict (e.g., Dolbeer et al. 1993).

The costs and benefits of using short-term versus long-term solutions and the relative effectiveness of a method or combination of methods are considered. Technical assistance provided includes advice, information, and materials for use in managing the damage problem. Alternately, when funding is available, a wildlife damage specialist can provide direct on-site control, which is particularly appropriate when hazardous materials are used, when endangered species are known to inhabit an area, or when public property is involved. Finally, the wildlife specialist normally monitors the results during site visits to determine effectiveness and whether additional or alternate methods are needed.

1.6 BLACKBIRD RESEARCH IN CANADA

From the mid-1960s to the 1980s, university scientists and graduate students at Carleton University, Ottawa, Ontario, Canada; the Macdonald campus of McGill University, Ste-Anne-de-Bellevue, Quebec, Canada; and University of Guelph, Guelph, Ontario, Canada, conducted the majority of blackbird research related to management of damage to corn and sunflower. This research was largely funded by Agriculture Canada, le Ministère de l'Agriculture du Québec, and the Ontario Department of Agriculture and Food. These scientists published results from numerous studies focused on blackbirds, including roost dynamics, sex-specific food habits, economic and ecological impacts, use of surfactants to manage populations, movement patterns, use of decoy traps, and indirect assessment of damage (e.g., Dyer 1967; Bendell et al. 1981; Weatherhead 1982; Weatherhead et al. 1982). The use of decoy traps and surfactants (wetting agents) to reduce blackbird

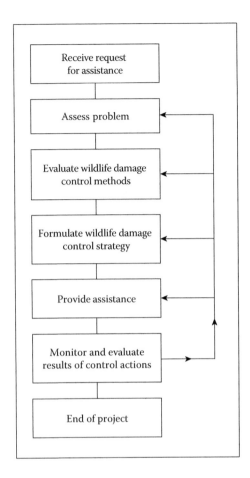

Figure 1.1 The USDA Wildlife Services decision model provides a step-by-step process to address requests for assistance with wildlife damage.

populations were found to be ineffective (Weatherhead et al. 1980a, 1980b). Weatherhead et al. (1982) provided a method of indirectly assessing bird damage and concluded government estimates of bird damage to corn were grossly overestimated. One of the lead researchers, Patrick J. Weatherhead, moved to the University of Illinois at Urbana–Champaign, where he studied the behavior and ecology of birds, including red-winged blackbirds, for many years (e.g., Weatherhead 2005; Weatherhead and Dufour 2000; Weatherhead and Sommerer 2001).

Scientists at the Canada Department of Agriculture Research Station in Winnipeg, Manitoba, contributed information on food habits of red-winged blackbirds in corn and sunflower and provided some of the earliest data showing that blackbirds are attracted to sunflowers (Bird and Smith 1964).

A 5-year project to study the biology and management of blackbirds in relation to corn and sunflower was funded by the Canada/Manitoba Subsidiary Agreement on Value-Added Crop Production (Harris 1983). The project concentrated on testing propane cannons, decoy traps, acoustic devices, pyrotechnics, shotguns, a chemical frightening agent, and bird-resistant sunflower. Harris (1983) reported that acoustic devices were most effective when combined with shotguns and pyrotechnics, whereas decoys traps and a chemical frightening agent were found to be ineffective. Harris participated in an early test of Bird-Resistant Synthetic Sunflower Variety 1 (BRS1), which was developed to thwart blackbird damage (Mah and Neuchterlein 1991). The data showed that blackbirds preferred to eat a commercial oilseed sunflower variety rather than BRS1, which had a lower oil content.

1.7 BLACKBIRD RESEARCH IN THE UNITED STATES

The WS-NWRC is the lead research institution in the United States for developing and evaluating wildlife damage management methods that emphasize practicality, environmental safety, cost-effectiveness, and wildlife stewardship. Scientists study human–wildlife conflicts, wildlife damage, nuisance and pest animals, wildlife disease, invasive species, overabundant wildlife, and overall ecosystem health. To accomplish certain aspects of this research, scientists at headquarters in Fort Collins, Colorado, and field stations throughout the United States collaborate with WS state operational programs, other state and federal agencies, universities, private industries, and nongovernmental organizations. For example, WS cooperated with Utah State University to establish the Jack H. Berryman Institute to enhance education, extension, and research on human–wildlife interactions. This institute was later expanded to include an eastern counterpart at Mississippi State University.

Other university-based scientists contributed important ideas toward our understanding of the impact of blackbirds on crops. For example, Wiens and Dyer (1977) advocated a model-based, indirect approach that included population dynamics, bioenergetics, and diet composition to estimate bird damage to ripening crops. Dyer and Ward (1977) reviewed various bird management strategies and concluded that a single tool approach cannot be used across all bird damage scenarios. Over the last four decades, these concepts were refined and promulgated in numerous publications (e.g., Dolbeer 1980; Dolbeer 1990; Peer et al. 2003; Linz et al. 2011; Linz et al. 2015; Dolbeer and Linz 2016).

1.7.1 National Wildlife Research Center Headquarters

The NWRC is headquartered at the Foothills Campus of Colorado State University (CSU) in Fort Collins (Figure 1.2). Approximately two-thirds of NWRC's 150-person staff is in Fort Collins; the remainder are at field stations in eight states, where they address a range of wildlife damage management issues.

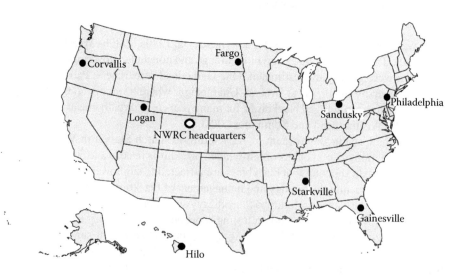

Figure 1.2 The USDA Wildlife Services National Wildlife Research Center (NWRC) headquarters is located on the foothills campus of Colorado State University, Fort Collins, Colorado, USA. The NWRC has eight field stations that conduct research on specific human–wildlife interaction issues.

SIDEBAR 1.1   HISTORY OF THE NATIONAL WILDLIFE RESEARCH CENTER

1886: C. Hart Merriam established the Division of Economic Ornithology and Mammalogy at the United States Department of Agriculture (USDA) and, with A.K. Fisher, pioneered research on methods for controlling damage to agriculture by wildlife.

1905: The USDA Control Methods Research Laboratory in Albuquerque, New Mexico, conducted field and laboratory experiments on various methods for controlling rodent damage to agriculture.

1920: The headquarters were moved from Albuquerque to Denver, Colorado, in 1920 and research was begun on the food habits of wildlife and diseases that affected wildlife. A decade later, the USDA Food Habits Laboratory was established to study the food habits and economic impact of predators, other mammals, and birds in the western United States.

1940: The Fish and Wildlife Service (FWS) was created within the U.S. Department of the Interior (USDI), and the Denver Wildlife Research Laboratory was formed under the FWS. Scientists conducted food habits studies and developed animal control methods.

1948: The Denver Wildlife Research Laboratory initiated a study of blackbird damage to rice in Arkansas. A one-person field station was maintained at Stuttgart from 1950 to 1955 (Meanley 1971).

1956: The FWS was reorganized in 1956 to include the Bureau of Sport Fisheries and Wildlife, which expanded research to include studies of relationships between wildlife populations and their habitats.

1959: The Denver Wildlife Research Laboratory was renamed the Denver Wildlife Research Center (DWRC) and expanded to study the effects of pesticides on wildlife through the Pesticide–Wildlife Ecology Program.

1960: The Section of Animal Damage Control Studies was formed at the FWS Patuxent Wildlife Research Center (PWRC) in Laurel, Maryland, to investigate wildlife damage issues in the eastern United States. Field stations were established in Gainesville, Florida; Newark, Delaware; and Sandusky, Ohio. All personnel and field stations were transferred administratively to the DWRC in 1976.

1967: DWRC scientists took the lead in a long-term international research program in cooperation with the US Department of State's Agency for International Development aimed at discovering, developing, and applying new and better methods to protect world food crops from the ravages of "rats, bats, and noxious birds." Numerous DWRC scientists took assignments lasting 1 to 5 years in various countries in Africa, Asia, and Central and South America. This program ended in 1993.

1970s: The U.S. Environmental Protection Agency's (EPA) registrations for several important chemical tools for managing wildlife damage were canceled, resulting in renewed efforts at the DWRC to develop new, more effective chemical methods for wildlife damage management. Further, the DWRC assumed nationwide leadership for all wildlife damage management research within the FWS.

1980: The DWRC merged with the FWS's National Fish and Wildlife Laboratory. The DWRC's research included a broad array of vertebrate systematic investigations, ecologic and zoogeographic studies, and marine mammal research.

1985: Congress transferred the USDI's Animal Damage Control Program, including part of the DWRC and some of its field stations involved in wildlife damage research, from the FWS to the USDA's Animal and Plant Health Inspection Service (APHIS). The sole focus of DWRC research then became wildlife damage management.

1990 to present: In the 1990s, a state-of-the-art facility was planned and developed on the Colorado State University Foothills Research Campus in Fort Collins. The DWRC was closed and all personnel at DWRC were transferred to Ft. Collins by 1999. Due to the national and international scope of research conducted at this facility, a more fitting name was chosen—the National Wildlife Research Center (NWRC). Facilities on the campus include offices, chemistry laboratories, indoor and outdoor animal research facilities, and a Biosafety Level 3 building for studying diseases transmitted by aerosols or the causes of severe disease.

Source: Miller 2007; U.S. Department of Agriculture 2016.

1.7.2 Blackbird Research—Headquarters

In 1976, all federal animal damage control research was consolidated under the DWRC within the U.S. Department of Interior (USDI) (Sidebar 1.1). Prior to that time, scientists at the Patuxent Wildlife Research Center, Laurel, Maryland; Ohio Field Station, Sandusky, Ohio; and Florida Field Station, Gainesville, Florida, made important scientific contributions in relation to blackbird damage to corn and rice (e.g., Dolbeer 1980, 1990; Meanley 1971). The Ohio Field Station and Florida Field Station continued to research blackbird damage to corn and rice, respectively (e.g., Brugger and Dolbeer 1990; Holler et al. 1982). Concurrently, scientists at DWRC headquarters in Denver focused on corn damage in the Dakotas in the 1950s through the 1970s, and in the late 1970s and 1980s on sunflower damage in North Dakota, South Dakota, and Minnesota (e.g., DeGrazio 1964; Guarino 1984). Scientists during this time spent significant resources over two decades developing the use of 4-aminopyridine, a chemical frightening agent, for protecting ripening corn and sunflower (e.g., Besser and Guarino 1976; Knittle et al. 1988). The product produced inconsistent results due to a variety of reasons, including dense crop canopies obscuring the baits, loss of chemical on baits, poor bait acceptance, and insufficient dosage due to broken bait particles (Knittle et al. 1988). This product is no longer available for protecting growing crops due to extreme toxicity to birds and mammals but is available for other uses (Avitrol Corporation 2013).

Starting in the 1990s and continuing until the present day, scientists from headquarters also conducted research on rice damage in the southeastern United States (e.g., Cummings and Avery 2003; Cummings et al. 2005). Their current research is focused on the development of bird repellents for ripening and sprouting crops (e.g., Werner et al. 2010).

Throughout the history of blackbird research, all of these scientists and their collaborators have conducted both short- and long-term studies that fall into three major research areas: (1) problem definition—defining the extent, magnitude, and frequency distribution of crop losses in relation to roosts, field location, and habitat; (2) ecological studies—estimating breeding and postbreeding populations, investigating food habits, and determining local and migratory movement patterns; and (3) methods development—developing cost-effective and environmentally safe chemical, cultural, and mechanical methods to alleviate damage (Guarino 1984).

1.7.3 Blackbird Research—Field Stations

1.7.3.1 Ohio Field Station

Ohio grows millions of hectares of corn and harbors historically large blackbird breeding populations, including 2.6 million common grackles (*Quiscalus quiscula*) and 2.5 million red-winged blackbirds (Partners in Flight 2013). In the 1960s, blackbird damage levels appeared to be increasing at an alarming rate, and Ohio farmers formed the Bye-Bye Blackbird Committee in 1965 to lobby for government action in reducing crop losses from blackbirds (Figure 1.3). Two years later, this group became the Ohio Coordinating Committee for the Control of Depredating Birds, which attracted congressional attention that resulted in the establishment of a research station in Ohio in 1968. The WS-NWRC Ohio Field Station is located near Sandusky and Lake Erie at Plum Brook Station, a 2,258-ha, fenced facility in Erie County operated by Glenn Research Center, National Aeronautics and Space Administration (NASA). The field station was initially administered from the U.S. Fish and Wildlife Service Patuxent Wildlife Research Center, located in Laurel, Maryland.

The restricted-access facility contains native grassland, reverted farmland, marsh, and woodland adjacent to intensively farmed land and urban settings outside the fence. Field station facilities include indoor and outdoor aviaries, several large bird traps, laboratories and shop space, a 2-ha fenced pond for waterfowl research, and conference rooms. The abundant wildlife populations at the facility allows station scientists to test various wildlife damage methods under controlled

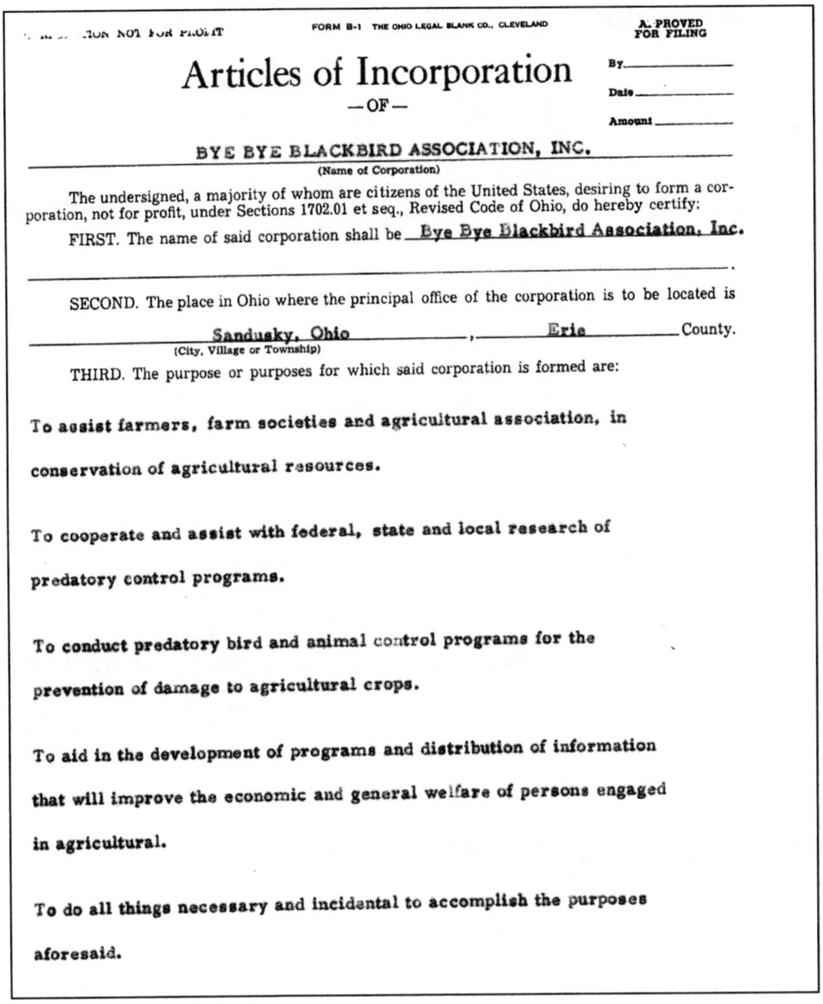

...ION NOT FOR PROFIT FORM B-1 THE OHIO LEGAL BLANK CO., CLEVELAND

APPROVED FOR FILING

By________

Date________

Amount________

Articles of Incorporation

—OF—

BYE BYE BLACKBIRD ASSOCIATION, INC.

(Name of Corporation)

The undersigned, a majority of whom are citizens of the United States, desiring to form a corporation, not for profit, under Sections 1702.01 et seq., Revised Code of Ohio, do hereby certify:

FIRST. The name of said corporation shall be Bye Bye Blackbird Association, Inc.

SECOND. The place in Ohio where the principal office of the corporation is to be located is Sandusky, Ohio (City, Village or Township), Erie County.

THIRD. The purpose or purposes for which said corporation is formed are:

To assist farmers, farm societies and agricultural association, in conservation of agricultural resources.

To cooperate and assist with federal, state and local research of predatory control programs.

To conduct predatory bird and animal control programs for the prevention of damage to agricultural crops.

To aid in the development of programs and distribution of information that will improve the economic and general welfare of persons engaged in agricultural.

To do all things necessary and incidental to accomplish the purposes aforesaid.

Figure 1.3 Articles of Incorporation for the Bye-Bye Blackbird Association in Sandusky, Ohio, 1965.

conditions without incurring costs associated with travel. The field station also leases from NASA 16 ha of farmland immediately outside the facility fence for additional wildlife damage studies. Thus, the Ohio Field Station is ideally located to develop methods for reducing blackbird damage to corn.

In the 1970s, scientists at the Ohio Field Station were primarily concerned with research on agricultural conflicts involving blackbirds and European starlings (e.g., Stickley et al. 1976). These scientists studied the population trends and ecology of these birds and tested the effectiveness of chemical repellents to keep birds from eating crops (e.g., Dolbeer 1978; Dolbeer 1980; Dolbeer 1990). They also assisted the FWS with the management of brown-headed cowbirds (*Molothrus ater*) in Michigan to reduce parasitism of Kirtland's warbler (*Setophaga kirtlandii*) nests (U.S. Fish and Wildlife Service 2012). In the following decade, research continued testing different repellent methods and evaluating various crop hybrids to reduce blackbird feeding without decreasing crop yields (Dolbeer et al. 1986). In the 1990s, research continued to analyze bird depredation problems in agriculture but shifted to a new focus on wildlife hazards to aviation—mainly bird strikes. From the 2000s to the present day, the field station has remained the leading U.S. research facility on wildlife hazards to aviation. Blackbird research moved to other units within the NWRC.

1.7.3.2 Florida Field Station

In 1944, the Florida Field Station was established in a small building in downtown Gainesville as one of the nation's first wildlife research stations. At that time, the field station was under the direction of the USDI Patuxent Research Center in Laurel, Maryland.

By 1961, the original facility was no longer adequate for its expanding wildlife research due to proximity to the growing Gainesville population. Thus, a 10.5-ha tract was acquired on the east side of Gainesville, 4.8 km from the University of Florida. The main office and laboratory building as well as a roofed outdoor aviary were constructed in 1963.

Over the years, additional infrastructure has been added to the facility to maintain research capabilities in light of changing priorities. Significant additions include a pole barn and ATV storage shed, three large outdoor flight pens (1486–2044 m² each), 12 smaller outdoor avian test pens, two roofed aviaries (112 m²) for holding and testing birds, and a dedicated animal care building. The latter is part of a recent (2015–2016) modernization of the facility, which also featured a complete upgrade of the 50-year-old main office and lab building.

The original mission of the field station in the 1940s included the study of rodent damage to Florida sugarcane. This remained a focus of the research program until the 1980s. In the 1950s and 1960s, the research mission broadened to include nuisance birds as well as mammals. In addition, from 1958 through the early 1970s, the field station operated a substation in Stuttgart, Arkansas, where biologists conducted extensive field research on red-winged blackbirds and produced seminal information on the distribution, migration, ecology, and management of blackbirds in relation to damage to rice and other agricultural crops.

Gainesville biologists collaborated with DWRC colleagues in the 1960s and 1970s on blackbird and starling research on development and field testing of surfactants for management and dispersal of large winter roosts in the southeastern United States (e.g., Lefebvre and Seubert 1970). Research during this time also included developing applications for a recently identified avian toxicant, compound DRC-1339 (e.g., Lefebvre et al. 1981). Blackbird research in the 1980s brought more emphasis to nonlethal approaches to reduce crop depredation issues, particularly related to chemical repellents such as methiocarb (e.g., Holler et al. 1982; Avery 1987). Throughout the 1990s the research program continued to investigate potential chemical repellents and other nonlethal methods for controlling bird damage to fruit and grain crops using cage and pen tests and field trials (e.g., Avery et al. 1994, 1997, 1998). Research on repellents expanded to include blackbird damage to wild rice in California (Avery et al. 2000), blackbird (*Agelaius ruficapillus*; also known as *Chrysomus ruficapillus*) damage to rice in Uruguay (Rodriguez and Avery 1996), and dickcissel (*Spiza americana*) damage to sorghum in Venezuela (Avery et al. 2001).

In the 2000s, responsibility for blackbird research on rice shifted to the North Dakota Field Station and headquarters. Concurrently, research at the Florida Field Station began a new phase, which continues today, identifying, evaluating, and developing methods to manage depredation, nuisance, and property damage problems associated with native birds such as vultures and crows, as well as various non-native species such as feral swine (*Sus scrofa*), Burmese pythons (*Python bivittatus*), black spiny-tailed iguanas (*Ctenosaura similis*), and monk parakeets (*Myiopsitta monachus*).

1.7.3.3 North Dakota Field Station

The North Dakota Field Station is located on the campus of North Dakota State University (NDSU), Fargo, where the station began in 1989. However, the station's research on blackbird ecology and behavior patterns in relation to sunflower began in 1979 when the U.S. Congress directed funds for research at NDSU. In 1985, DWRC, CSU, and the FWS's Northern Prairie Wildlife Research Center (NPWRC), located in Jamestown, North Dakota, cooperatively agreed to station a CSU postdoctoral research biologist at the NPWRC facility. The incumbent biologist conducted collaborative studies with DWRC and NDSU scientists on the extent and magnitude of sunflower damage, migration and movement patterns of blackbirds in relation to damage, development of bird-resistant sunflower that featured either chemical or morphological characteristics that were thought to thwart blackbird feeding, and chemical repellents.

The establishment of the field station in 1989 reflected the need for blackbird damage research and the positive benefits of a relationship between NDSU and the NWRC. The field station's primary focus has always been on evaluating, creating, and refining methods used to reduce blackbird damage to sunflowers. From 1996 to 2015, the station was co-located in Bismarck with the North Dakota Wildlife Services operations program. In 1997, both units moved into a new facility that included offices, shops, storage, and bird housing and testing facilities.

After a NWRC field station was formally established at NDSU in 1989, collaborative research across multiple research institutions began with the development of the use of aerial applications of glyphosate herbicide for managing wetland vegetation favored by roosting blackbirds and exploring the use of compound DRC-1339 for population management during spring migration (e.g., Homan et al. 2004; Linz and Homan 2011; Linz et al. 2015). In the 2000s, scientists advanced our understanding of the relationship between blackbird populations, land cover, and climate (e.g., Forcey et al. 2015). Additionally, the use of DRC-1339 for baiting blackbirds feeding in sunflower fields was investigated, bird repellents were tested in the laboratory and in the field, the potential for European starlings to transfer disease within and among feedlots and dairies was elucidated, and the use of wildlife conservation food plots was refined (e.g., Carlson et al. 2011; Hagy et al. 2008; Werner et al. 2011). The addition of Fort Collins personnel in 2008 allowed the expansion of research on the development of bird repellents using test facilities located on the Fort Collins campus. The station's research portfolio expanded to include the movement and migration patterns of European starlings in relation to disease management in the United States. Project biologists later joined a large collaborative group of NWRC and university biologists to discover the role of European starlings in the transfer of disease among feedlots and dairies. Station personnel also were called upon to find methods to deter woodpeckers from damaging wood utility poles and study the movement of American robins (*Turdus migratorius*) and cedar waxwings (*Bombycilla cedrorum*) in relation to fruit damage in Michigan (e.g., Tupper et al. 2010; Eaton et al. 2016).

The North Dakota Field Station is currently charged with testing mechanical and chemical bird repellents, developing strategies to provide alternative food sources for blackbirds repelled from sunflower fields, studying blackbird movement and roosting behavior in relation to sunflower damage, and developing the use of unmanned aerial vehicles for hazing blackbirds and delivering repellents. The field station leader oversees MS and PhD students tasked with specific studies aimed at developing and improving blackbird management tools. In addition, the field station leader collaborates with scientists and graduate students at NDSU and other research institutions while interacting with key stakeholders such as the National Sunflower Association, North Dakota Wildlife Services, and the North Dakota Department of Agriculture to manage the conflict between agriculturalists and blackbirds.

1.7.3.4 Kentucky Field Station

The FWS established the Kentucky Field Station at Bowling Green, Kentucky, in 1977 to conduct research on blackbirds and starlings using winter roosts in the southeastern United States. Studies included the behavior and ecology of winter roosting birds, problem definition, and methods development. In 1988, the Kentucky Field Station staff was transferred to a newly established field station on the campus of Mississippi State University to study fish-eating birds known to prey on farm-raised catfish. During the 11 years of its existence, Kentucky Field Station personnel developed the use of a sprinkler irrigation system for applying PA-14 surfactant to blackbirds and starlings roosting in trees (Heisterberg et al. 1987). The application of PA-14 sometimes reduced roost numbers dramatically. They successfully tested the use of DRC-1339 for reducing starling numbers at feedlots and blackbirds damaging sprouting rice (e.g., Glahn and Wilson 1992). Finally, field station personnel banded 20,000 blackbirds and starlings in Kentucky and Tennessee and discovered that the majority of blackbirds nested in the northeastern United States, whereas nearly 50% of the starlings were hatched in the subject state (Mott 1984).

1.7.3.5 California Field Station

The FWS established the California Field Station at Dixon in the early 1960s to conduct research on blackbirds in California. The station was closed in the mid-1980s when wildlife damage management was moved from the USDI to the USDA. Station personnel collaborated with scientists and graduate students at University of California–Davis. This field station primarily addressed bird depredations on ripening rice and grapes but also conducted studies on blackbird damage to sunflower (Avery and DeHaven 1984). The field station had office space, aviaries, a shop, and a laboratory.

1.8 SUMMARY

The blackbird–agriculture conflict in North America spawned a robust research effort in the 1950s that continues today. We tip our hats to the many scientists who spent countless hours conducting field and laboratory studies, some over several decades. We can learn much from their documented experiences and publications. Classic experimental studies using free-ranging blackbirds in commercial fields were found to be challenging because of the great mobility of foraging bird flocks, changing cropping patterns, and unpredictable precipitation patterns that change the availability of roosting habitat in relation to crops (Stickley et al. 1976; Jaeger et al. 1983; Knittle et al. 1988; Linz et al. 2011). Thus, testing the plethora of mechanical frightening devices, chemical agents, bird-resistant crop hybrids, and lure crops over the years often yielded inconsistent results, partially due to high variability in blackbird foraging behavior between treatments. Scientists are now relying more heavily on cage test designs to screen potential repellents and netted enclosures stocked with blackbirds to simulate replicated field trials. Although these testing strategies are useful, moving from encouraging test results of a particular repellent with captive birds to successful application under field conditions remains difficult when applied to ripening crops.

Budget constraints and shifting research priorities have reduced the number of scientists assigned to this challenging problem. We are cautiously optimistic, however, that progress will be made as a result of new and improved technologies and innovations integrated into an overall pest management strategy.

REFERENCES

Avery, M. L. 1987. Flight pen test of partial treatment with methiocarb to protect rice seed from depredations by flocks of brown-headed cowbirds. *Bird Damage Research Report* 401. U.S. Fish and Wildlife Service Denver Wildlife Research Center, Denver, CO.

Avery, M. L., and R. DeHaven. 1984. Bird damage to sunflowers in the Sacramento Valley, California. *Vertebrate Pest Conference* 10:197–200.

Avery, M. L., J. S. Humphrey, and D. G. Decker. 1997. Feeding deterrence of anthraquinone, anthracene, and anthrone to rice-eating birds. *Journal of Wildlife Management* 61:1359–1365.

Avery, M. L., J. S. Humphrey, T. M. Primus, D. G. Decker, and A. P. McGrane. 1998. Anthraquinone protects rice seed from birds. *Crop Protection* 17:225–230.

Avery, M. L., P. Nol, and J. S. Humphrey. 1994. Responses of three species of captive fruit-eating birds to phosmet-treated food. *Pesticide Science* 41:49–53.

Avery, M. L., E. A. Tillman, and C. C. Laukert. 2001. Evaluation of chemical repellents for reducing crop damage by dickcissels in Venezuela. *International Journal of Pest Management* 47:311–314.

Avery, M. L., D. A. Whisson, and D. B. Marcum. 2000. Responses of blackbirds to mature wild rice treated with Flight Control bird repellent. *Vertebrate Pest Conference* 19:26–30.

Avitrol Corporation. 2013. *Avian Corn Chops Label*. http://www.avitrol.com/pdf/avitrol-us-corn-chops-specimen-label.pdf (accessed June 23, 2016).

Bendell, B. E., P. J. Weatherhead, and R. K. Stewart. 1981. The impact of predation by red-winged blackbirds on European corn borer populations. *Canadian Journal of Zoology* 59:1535–1538.

Besser, J. F., and J. L. Guarino. 1976. Protection of ripening sunflowers from blackbird damage by baiting with Avitrol FC Corn Chops-99S. *Bird Control Seminar* 7:200–203.

Bird, R. D., and L. B. Smith. 1964. The food habits of the red-winged blackbird, *Agelaius phoeniceus*, in Manitoba. *Canadian Field-Naturalist* 78:179–186.

Brugger, K. E., and R. A. Dolbeer. 1990. Geographic origin of red-winged blackbirds relative to rice culture in southwestern and southcentral Louisiana. *Journal of Field Ornithology* 61:90–97.

Carlson, J. C., A. B. Franklin, D. R. Hyatt, S.E. Pettit, and G. M. Linz. 2011. The role of starlings in the spread of Salmonella within concentrated animal feeding operations. *Journal of Applied Ecology* 48:479–486.

Cirino, E. 2016. U.S. courts crack down on feds over mass wildlife culls. *Scientific American*. http://www.scientificamerican.com/article/u-s-courts-crack-down-on-feds-over-mass-wildlife-culls/ (accessed June 28, 2016).

Cummings, J. L, and M. L. Avery. 2003. An overview of current blackbird research in the southern rice growing region of the United States. *Wildlife Damage Management Conference* 10:237–243.

Cummings, J. L., S. A. Shwiff, and S. K. Tupper. 2005. Economic impacts of blackbird damage to the rice industry. *Eastern Wildlife Damage Management Conference* 11:317–322.

DeGrazio, J. W. 1964. Methods of controlling blackbird damage to field corn in South Dakota. *Vertebrate Pest Conference* 2:43–49.

Dolbeer, R. A. 1978. Movement and migration patterns of red-winged blackbirds: A continental overview. *Bird-Banding* 49:17–34.

Dolbeer, R. A. 1980. *Blackbirds and corn in Ohio*. U.S. Fish and Wildlife Service, Resource Publication 38, Washington, DC.

Dolbeer, R. A. 1990. Ornithology and integrated pest management: Red-winged blackbirds *Agelaius phoeniceus* and corn. *Ibis* 132:309–322.

Dolbeer, R. A., J. L. Belant, and J. L. Sillings. 1993. Shooting gulls reduces strikes with aircraft John F. Kennedy International Airport. *Wildlife Society Bulletin* 21:442–450.

Dolbeer, R. A., and G. M. Linz. 2016. *Blackbirds*. Wildlife Damage Management Technical Series, U.S. Department of Agriculture, Animal & Plant Health Inspection Service, Wildlife Services, Washington, DC. https://www.aphis.usda.gov/wildlife_damage/reports/Wildlife%20Damage%20Management%20Technical%20Series/FINAL_Blackbirds_WDM%20Technical%20Series_Aug2016.pdf (accessed October 23, 2016).

Dolbeer, R. A., P. P. Woronecki, and R. A. Stehn. 1986. Blackbird-resistant hybrid corn reduces damage but does not increase yield. *Wildlife Society Bulletin* 14:298–301.

Dyer, M. I. 1967. An analysis of blackbird feeding behavior. *Canadian Journal Zoology* 45:765–772.

Dyer, M. I., and P. Ward. 1977. Management of pest situations. In *Granivorous Birds in Ecosystems*, ed. J. Pinowski, and S. C. Kendeigh, pp. 267–300. Cambridge University Press, New York.

Eaton, R. A., C. A. Lindell, H. J. Homan, G. M. Linz, and B. A. Maurer. 2016. Avian use of cultivated cherry orchards reflects species-specific differences in frugivory. *Wilson Journal of Ornithology* 128:97–107.

Forcey, G. M., W. E. Thogmartin, G. M. Linz, P. C. McKann, and S. M. Crimmins. 2015. Spatially explicit modeling of blackbird abundance in the Prairie Pothole Region. *Journal of Wildlife Management* 79:1022–1033.

Glahn, J. F., and E. A. Wilson. 1992. Effectiveness of DRC-1339 baiting for reducing blackbird damage to sprouting rice. *Eastern Wildlife Damage Control Conference* 5:117–123.

Government of Canada. 2016a. Justice Laws Website. *Migratory Birds Regulations (C.R.C., c. 1035)*. http://lois-laws.justice.gc.ca/eng/regulations/C.R.C.,_c._1035/page-5.html#h-18 (accessed June 16, 2016).

Government of Canada. 2016b. *Birds protected in Canada under the migratory birds convention act, 1994.* https://ec.gc.ca/default.asp?lang=En&n=E826924C-1 (accessed June 23, 2016).

Guarino, J. L. 1984. Current status of research on the blackbird-sunflower problem in North Dakota. *Vertebrate Pest Conference* 11:211–216.

Hagy, H. M., G. M. Linz, and W. J. Bleier. 2008. Optimizing the use of decoy plots for blackbird control in commercial sunflower. *Crop Protection* 27:1442–1447.

Harris, H.A.G. 1983. Blackbird control – an agricultural perspective. *Bird Control Seminar* 9:299–300.

Heisterberg, J. F, A. R. Stickley, K. M. Garner, and P. D. Foster, Jr. 1987. Controlling blackbirds and starlings at winter roosts using PA-14. *Eastern Wildlife Control Conference* 3:177–183.

Holler, N. R., H. P. Naquin, P. W. Lefebvre, D. L. Otis, and D. J. Cunningham. 1982. Mesurol for protecting sprouting rice from blackbird damage in Louisiana. *Wildlife Society Bulletin* 10:165–170.

Homan, H. J., G. M. Linz, R. M. Engeman, and L. B. Penry. 2004. Spring dispersal patterns of red-winged blackbirds, *Agelaius phoeniceus*, staging in eastern South Dakota. *Canadian Field-Naturalist* 118:201–209.

Jaeger, M. M., J. L. Cummings, D. L. Otis, J. L. Guarino, and C. E. Knittle. 1983. Effect of Avitrol baiting on bird damage to ripening sunflower within a 144-section block of North Dakota. *Bird Control Seminar* 9:247–254.

Knittle, C. E., J. L. Cummings, G. M. Linz, and J. F. Besser. 1988. An evaluation of modified 4-aminopyridine baits for protecting sunflower from blackbird damage. *Vertebrate Pest Conference* 13:248–253.

Lefebvre, P. W., N. R. Holler, R. E. Matteson, E. W. Schafer Jr., and D. J. Cunningham. 1981. Developmental status of n-(3-chloro-4-methylphenyl) acetamide as a candidate blackbird/starling roost toxicant. *Bird Control Seminar* 8:65–70.

Lefebvre, P. W., and J. L. Seubert. 1970. Surfactants as blackbird stressing agents. *Vertebrate Pest Conference* 4:156-161.

Linz, G. M., E. H. Bucher, S. B. Canavelli, E. Rodriguez, and M. L. Avery. 2015. Limitations of population suppression for protecting crops from bird depredation: A review. *Crop Protection* 76:46–52.

Linz, G. M., and H. J. Homan. 2011. Use of glyphosate for managing invasive cattail (*Typha* spp.) to protect crops near blackbird (Icteridae) roosts. *Crop Protection* 30:98–104.

Linz, G. M., H. J. Homan, S. W. Werner, H. M. Hagy, and W. J. Bleier. 2011. Assessment of bird management strategies to protect sunflower. *BioScience* 61:960–970.

Mah, J., and G. L. Neuchterlein. 1991. Feeding behavior of red-winged blackbird on bird-resistant sunflowers. *Wildlife Society Bulletin* 19:39–46.

Meanley, B. 1971. *Blackbirds and the southern rice crop.* U.S. Department of Interior, U.S. Fish and Wildlife Service, Resource Publication 100. http://pubs.usgs.gov/rp/100/report.pdf (accessed June 23, 2016).

Miller, J. E. 2007. Evolution of the field of wildlife damage management in the United States and future challenges. *Human-Wildlife Conflict* 1:13–20.

Mott, D. F. 1984. Research on winter roosting blackbirds and starlings in the southeastern United States. *Vertebrate Pest Conference* 11:183–187.

Partners in Flight Science Committee. 2013. *Population Estimates Database, version 2013.* http://rmbo.org/pifpopestimates (accessed June 28, 2016).

Peer, B. D., H. J. Homan, G. M. Linz, and W. J. Bleier. 2003. Impact of blackbird damage to sunflower: Bioenergetic and economic models. *Ecological Applications* 13:248–256.

Rodriguez, E. N., and M. L. Avery. 1996. *Agelaius* blackbirds and rice in Uruguay and the southeastern United States. *Vertebrate Pest Conference* 17:94–98.

Slate, D., R. Owens, G. Connolly, and G. Simmons. 1992. Decision making for wildlife damage management. *North American Wildlife Natural Resources Conference* 57:52–62.

Stickley, A. R., Jr., R. T. Mitchell, J. L. Seubert, C. R. Ingram, and M. I. Dyer. 1976. Large-scale evaluation of blackbird frightening agent 4-aminopyridine in corn. *Journal of Wildlife Management* 40:126–131.

Tobin, M. A. 2012. U.S. Department of Agriculture Wildlife Services: Providing federal leadership in managing conflicts with wildlife. *Wildlife Damage Management Conference* 14:1–2.

Tupper, S. K., W. F. Andelt, J. L. Cummings, C. Weisner, and R. E. Harness. 2010. Polyurea elastomer protects utility pole crossarms from damage by pileated woodpeckers. *Journal of Wildlife Management* 74:605–608.

U.S. Department of Agriculture. 1997. *Animal Damage Control Program: Final Environment Impact Statement (revised).* U.S. Department of Agriculture/APHIS/WS-Operational Support Staff, Riverdale, MD.

U.S. Department of Agriculture. 2007. *Management of blackbird species to reduce damage to sunflower, corn, and other small grain crops in the Prairie Pothole Region of North Dakota and South Dakota: EA monitoring report.* https://www.aphis.U.S. Department of Agriculture.gov/regulations/pdfs/nepa/ND_SD%202007%20Blackbird%20EA%20Amend%20(3-25-08).pdf (accessed June 28, 2016).

U.S. Department of Agriculture. 2014. Wildlife Services Directive 2.201. *Wildlife Services Decision Model.* https://www.aphis.usda.gov/wildlife_damage/directives/2.201_ws_decision_model.pdf (accessed June 16, 2016).

U.S. Department of Agriculture. 2015. *Managing blackbird damage to sprouting rice in southwestern Louisiana.* Environmental Assessment. Animal and Plant Health Inspection Service, Wildlife Services, Baton Rouge, LA. https://www.aphis.U.S. Department of Agriculture.gov/regulations/pdfs/nepa/LA-Blackbird%20EA%20FINAL.pdf (accessed June 23, 2016).

U.S. Department of Agriculture. 2016. *National Wildlife Research Center, History Timeline,* Fort Collins, CO. https://www.aphis.usda.gov/aphis/ourfocus/wildlifedamage/programs/nwrc/sa_history/ct_history (accessed June 16, 2016).

U.S. Department of the Interior. 2010. *Fish and Wildlife Service 50 CFR Part 21. Migratory bird permits; Removal of rusty blackbird and tamaulipas (Mexican) crow from the depredation order for blackbirds, cowbirds, grackles, crows, and magpies, and other changes to the order.* Washington, DC. http://www.fws.gov/policy/library/2010/2010-30288.html (accessed September 9, 2016).

U.S. Department of Interior. 2016. *Bureau of Land Management, NEPA Program,* Washington, DC. http://www.blm.gov/wo/st/en/prog/planning/nepa.html (accessed June 28, 2016).

U.S. Environmental Protection Agency. 2015. *Summary of the National Environmental Policy Act.* Washington, DC. https://www.epa.gov/laws-regulations/summary-national-environmental-policy-act (accessed June 23, 2016).

U.S. Fish and Wildlife Service. 2012. Fact Sheet: *Kirkland's Warbler, (Setophaga kirklandi).* https://www.fws.gov/midwest/endangered/birds/Kirtland/pdf/kiwabiologue.pdf (accessed June 16, 2016).

Weatherhead, P. J. 1982. Assessment, understanding, and management of blackbird agriculture interaction in eastern Canada. *Vertebrate Pest Conference* 10:193–196.

Weatherhead, P. J. 2005. Long-term decline in a red-winged blackbird population: Ecological causes and sexual selection consequences. *The Royal Society B* 272:2313–2317.

Weatherhead, P. J., J. R. Bider, and R. G. Clark. 1980a. Surfactants and management of red-winged blackbirds in Quebec. *Phytoprotection* 61:39–47.

Weatherhead, P. J., and K. W. Dufour. 2000. Fledging success as an index of recruitment in red-winged blackbirds. *Auk* 117:627–633.

Weatherhead, P. J., H. Greenwood, S.H. Tinker, and J. R. Bider. 1980b. Decoy traps and the control of blackbird populations. *Phytoprotection* 61:65–71.

Weatherhead, P. J., and S. J. Sommerer. 2001. Breeding synchrony and nest predation in red-winged blackbirds. *Ecology* 82:1632–1641.

Weatherhead, P. J., S. Tinker, and H. Greenwood. 1982. Indirect assessment of avian damage to agriculture. *Journal of Applied Ecology* 19:773–782.

Werner, S. J., G. M. Linz, J. C. Carlson, S. E. Pettit, S.K. Tupper, and M. M. Santar. 2011. Anthraquinone-based bird repellent for sunflower crops. *Journal Applied Behavioral Science* 129:162–169.

Werner, S. J., G. M. Linz, S. K. Tupper, and J. C. Carlson. 2010. Laboratory efficacy of chemical repellents for reducing blackbird damage in rice and sunflower crops. *Journal of Wildlife Management* 74:1400–1404.

Wiens, J. A., and M. I. Dyer. 1977. Assessing the potential impact of granivorous birds in ecosystems. In *Granivorous Birds in Ecosystems*, ed. J. Pinowski, and S. C. Kendeigh, 205-264. Cambridge University Press, New York.

CHAPTER 2

Ecology and Management of Red-Winged Blackbirds

George M. Linz and Page E. Klug
National Wildlife Research Center
Bismarck, North Dakota

Richard A. Dolbeer
Wildlife Services
Sandusky, Ohio

CONTENTS

2.1	Taxonomy	19
2.2	Breeding Biology	20
	2.2.1 Polygyny and Territoriality	20
	2.2.2 Nesting	21
	2.2.3 Nest Predation	21
	2.2.4 Brood Parasitism	21
2.3	Disease Transmission	22
	2.3.1 Avian Salmonellosis	22
	2.3.2 Chlamydiosis	22
	2.3.3 Johne's Disease	22
	2.3.4 Shiga Toxin-Producing *Escherichia coli*	22
	2.3.5 Encephalitis	23
	2.3.6 Lyme Disease	23
	2.3.7 Histoplasmosis	23
	2.3.8 West Nile Virus	23
2.4	Distribution and Populations	23
2.5	Winter Location	26
2.6	Spring Migration	27
2.7	Fall Migration and Annual Feather Replacement	29
2.8	Food Habits	30
	2.8.1 Southern United States	30
	2.8.2 Midwest United States	30
	2.8.3 Northern Prairie of Canada and United States	31
	2.8.4 Ontario, Canada	31
	2.8.5 Summary	32

2.9 Crop Damage 32
    2.9.1 Rice 33
    2.9.2 Corn 33
    2.9.3 Sweet Corn 34
    2.9.4 Sunflower 34
2.10 Summary 35
References 35

The red-winged blackbird (*Agelaius phoeniceus*) is one of the most abundant bird species in North America, with an estimated spring breeding population of 150 million individuals that nest in emergent wetland vegetation and upland habitats throughout the continent (Yasukawa and Searcy 1995; Forcey et al. 2015; Rosenberg et al. 2016). During the nonbreeding season, red-winged blackbirds are often found in flocks numbering from a few birds to many thousands, sometimes in association with other blackbird species and European starlings (*Sturnus vulgaris*). In winter, red-winged blackbirds and these associated species gather in roosts occasionally numbering over 10 million birds (Meanley and Royal 1976; White et al. 1985).

Migratory male red-winged blackbirds typically arrive at their nesting grounds in early March, a month before the females arrive. At this time, casual bird watchers are apt to notice the robin-sized, male red-winged blackbirds with black feathers and highly conspicuous red and yellow epaulets (definitive plumage), prominently displayed while aggressively confronting intruders approaching their nesting territories (Figure 2.1). Loud singing (*o-ka-leeee*, *konk-a-ree*) by these males from high perches in their chosen territories adds to their aesthetic value. Second-year males returning to their natal area following their hatching year do not have the definitive plumage of adults. Rather, they have a duller black body and light red or orange epaulets (Yasukawa and Searcy 1995). The female, at least 20% smaller and far less noticeable with brownish feathers, is often misidentified as a large streaked sparrow (Figure 2.2) (Yasukawa and Searcy 1995; Jaramillo and Burke 1999).

Published in 1914, A.A. Allen's "The Red-Winged Blackbird: A Study in the Ecology of a Cat-Tail Marsh" was arguably the foundation for all subsequent descriptive and experimental

Figure 2.1 Male red-winged blackbird. (Courtesy of Larry Slomski.)

Figure 2.2 Female red-winged blackbird. (Courtesy of Larry Slomski.)

studies on the biology and management of blackbirds (Icteridae). Red-winged blackbirds' abundance, polygamous breeding system, penchant for sprouting and ripening crops, and propensity to gather in large roosting congregations in the nonbreeding seasons have led scientists to pursue multiple avenues of research, resulting in over 1,000 publications found in peer-reviewed manuscripts, books, monographs, disquisitions, and scientific conference and workshop proceedings (Yasukawa and Searcy 1995). Books focused on the biology of red-winged blackbirds have captured data across many of these studies (Payne 1969; Nero 1984; Searcy and Yasukawa 1995; Yasukawa and Searcy 1995; Beletsky 1996; Beletsky and Orians 1996; Jaramillo and Burke 1999). Here, we highlight key findings of these studies; readers are urged to examine the cited publications for details.

2.1 TAXONOMY

The red-winged blackbird's Latin-derived scientific name *A. phoeniceus* is apt; *Agelaius* means "belonging to a flock" and *phoeniceus* means "deep red." The common name for the red-winged blackbird is taken from the black adult male's distinctive red epaulets, which are visible when the bird is flying or displaying.

Size differences between sexes are substantial, with males and females of the eastern red-winged blackbird (*Agelaius phoeniceus phoeniceus*) subspecies averaging about 68 g and 44 g, respectively, based on birds captured in Ohio and Nebraska (Holcomb and Twiest 1968; Scharf et al. 2008). Individual mass can vary 15%–20% within a population. Additionally, premigratory fattening, often by feeding extensively on ripening crops, can increase body mass by >10% (Linz 1982).

Based on morphological characteristics (e.g., wing and tail length, bill size, and shape), ornithologists suggest that there are perhaps 14 subspecies of red-winged blackbird scattered across North America. Yasukawa and Searcy (1995) encouraged further research using advanced molecular methods to clearly differentiate subspecies.

Regardless of subspecies, males have about 20% longer wing, tail, culmen, and tarsus measurements than do females (Yasukawa and Searcy 1995). For example, the male eastern red-winged blackbird has an average wing length of 121 mm and culmen length of 24 mm, compared to females with 98 mm and 19 mm wing and culmen lengths, respectively (Yasukawa and Searcy 1995).

2.2 BREEDING BIOLOGY

2.2.1 Polygyny and Territoriality

After arriving on the breeding grounds, after-second-year (ASY) male (≥2 years old) red-winged blackbirds search for territories, where they defend exclusive areas in an attempt to attract one or more after-hatching year (AHY) females (≥1 year old), with harems of two to five females common (Searcy and Yasukawa 1995; Beletsky 1996). In a study in upland habitat in Ohio, the ratio of nesting females to male territories remained quite stable in May and June, averaging about 1.9:1.0, but new females constantly moved into territories and nested, as previously established females finished nesting and departed (Dolbeer 1976). For the entire nesting season each year, an average of over 4.0 different females nested per territory. These data suggest that some social mechanism limited the number of nesting females at any one time to a "carrying capacity" of the territories. The total number of females nesting in the territories was maximized by a temporal spacing of nest attempts by different females. During this time of active nesting, males display vigorously from perches within their territory and sing frequently, especially in the morning and evening in response to male song and females encroaching on their territory.

Polygyny results in second-year (SY) and ASY males that do not establish a nesting territory and therefore are referred to as "floaters." ASY males are comparable in size and plumage to territory owners, and virtually all SY males are smaller and have a distinctly duller plumage than ASY males. SY males are sexually mature and physiologically capable of successful mating but rarely successfully defend a territory (Payne 1969). No differences in size, testosterone levels, or reproductive capability have been found between ASY floaters and territorial males (Shutler and Weatherhead 1991; Dufour and Weatherhead 1998), and there is no reason to believe that ASY floaters are incapable of contributing to extra-pair copulation (Moulton et al. 2013). The surplus population of floater males explores an area searching for an opportunity to claim a territory (Shutler and Weatherhead 1992; Yasukawa and Searcy 1995; Sawin et al. 2003a; Moulton et al. 2013). Although a few territories are gained by replacing existing males or inserting themselves between existing territories, the majority of males gain their territories from owners that have disappeared (Picman 1987). Chance probably plays a strong role in initial territory acquisition (Eckert and Weatherhead 1987). Regardless, vacant territories are quickly occupied, sometimes within minutes and commonly within 48 hours (Eckert and Weatherhead 1987; Shutler and Weatherhead 1992; Sawin et al. 2003a).

The annual survival rate of adult red-winged blackbirds ranges from 42% to 62%, with a mean life expectancy of 2.14 years (Beletsky 1996). Based on an average annual mortality rate near 50%, the floater pool should be larger than the population of territorial males. In Washington, Beletsky and Orians (1996) reported that about 56.6% and 26.5% of territorial males initially acquired a territory when they were two and three years old, respectively.

Experienced males typically return to the same territory or nearby habitat the following year. For example, Beletsky and Orians (1996) reported that 52%–65% of males holding a territory in one year also held a territory the next year, with 70% reclaiming their original territories or found in nearby habitat. Similarly, Dufour and Weatherhead (1998) found that 51%–60% of territorial males in Ontario returned to their original or adjacent territory. In British Columbia, Picman (1987) found that 94% of returning males reclaimed their original territories. In comparison, females exhibit weak mate fidelity but strong marsh fidelity and appear to use experience in settlement decisions (Dolbeer 1976; Beletsky and Orians 1991).

2.2.2 Nesting

Emergent wetland vegetation is the preferred breeding habitat of red-winged blackbirds, but they also nest successfully in upland habitats, particularly hay fields, pasture, fallow fields, conservation reserve lands, and even shrubs (Dolbeer 1976; Beletsky 1996). In the Prairie Pothole Region (PPR) of the northern Great Plains, red-winged blackbirds are especially productive because there are about 404,000 ha of emergent wetland vegetation that provide ideal nest substrate through the breeding season (Ralston et al. 2007; Forcey et al. 2015).

Females provide the majority of parental investment; their responsibilities include nest building, incubation, nest defense, and provisioning of nestlings and fledglings. Females typically build a woven open cup nest in 1–3 days in vertical vegetation (Yasukawa and Searcy 1995; Beletsky 1996). Beginning 1–4 days after nest completion, one egg is laid daily, with a mean of three to four eggs, but nests with five eggs are not unusual (Yasukawa and Searcy 1995; Belesky 1996). Females incubate eggs and feed the nestlings from 11–13 and 10–12 days, respectively. Renesting after a nest failure may occur one to two times per season but varies depending on the length of the breeding season and stage of nesting when failure occurs.

In comparison, males limit their direct involvement to include only nest defense and limited provisioning (Searcy and Yasukawa 1995; Beletsky 1996). Regional differences may occur in the amount and quality of parental care provided by males within and among populations (Yasukawa et al. 1990; Linz et al. 2011a). In almost all cases where males feed nestlings, feeding is supplemental to female provisioning and occurs only after nestlings are four days old (Yasukawa et al. 1990). Male feeding has been shown to increase with an increase in brood size, nestling age, proportion of male nestlings, and male experience (Yasukawa et al. 1990; Patterson 1991). Overall, the number of offspring per territorial male has been shown to increase with breeding experience (Orians and Beletsky 1989; Beletsky and Orians 1991). Males with more breeding experience defend nests more intensely than their less experienced counterparts (Knight and Temple 1986; Yasukawa et al. 1987). Yasukawa et al. (1987) and Linz et al. (2014) reported that intensive nest defense did not result in higher nest success, whereas Weatherhead (1990) and Knight and Temple (1986) found that nests defended aggressively were more likely to be successful than nests defended with less vigor.

2.2.3 Nest Predation

Red-winged blackbird nests are frequently lost to a multitude of causes, including abandonment, predation, starvation, weather, brood parasitism, and failure of nest support vegetation (Beletsky 1996; Sawin et al. 2003b). Searcy and Yasukawa (1995) identified predation as the main source of nest failure. Red-winged blackbirds breeding in upland areas and at the edge of wetlands may experience more predation than those breeding in the interior, where deeper water limits access by mammalian predators (Picman et al. 1993). Beletsky (1996) collated data across 14 studies and found that 30%–50% of red-winged blackbird nests were depredated. Likewise, Searcy and Yasukawa (1995) reported that predation ranged from 27% to 50% in 10 studies, with nest predators including avian, mammalian, and reptilian predators. In the northern Great Plains, red-winged blackbirds are excluded from deeper water nesting sites by yellow-headed blackbirds (*Xanthocephalus xanthocephalus*) and thus might be exposed to more predation than counterparts in regions where yellow-headed blackbirds do not breed (Orians and Willson 1964; Twedt and Crawford 1995).

2.2.4 Brood Parasitism

Brood parasitism by brown-headed cowbirds (*Molothrus ater*) can negatively influence reproductive success in red-winged blackbirds (Searcy and Yasukawa 1995; Clotfelter and Yasukawa 1999; Lorenzana and Sealy 1999). For example, Lorenzana and Sealy (1999) found that brown-headed cowbird nestlings compete with host nestlings by consuming food otherwise meant for the host nestlings and can reduce red-winged blackbird productivity by up to 1.5 nestlings. Weatherhead (1989) concluded, however, that brown-headed cowbird parasitism did not reduce reproductive success in red-winged blackbirds, and Ortega and Cruz (1988) found conflicting results among study years. In the northern Great Plains, the populations of both red-winged blackbirds and brown-headed cowbirds are large, and thus nest success is probably affected more by yearly precipitation and predation than brood parasitism (Sawin et al. 2003b; Forcey et al. 2011, 2015; Rosenberg et al. 2016).

2.3 DISEASE TRANSMISSION

Red-winged blackbirds sometimes associate with European starlings (*S. vulgaris*) and other blackbirds in concentrated animal feeding operations and can be found roosting together, often in numbers exceeding 1 million birds, in the nonbreeding season (Meanley and Royall 1976; Dolbeer et al. 1978). At least 65 different diseases transmittable to humans or domestic animals have been reported to occur in pigeons (Columbidae), European starlings, and house sparrows (*Passer domesticus*); a similar level of documentation is not available for red-winged blackbirds (Clark and McLean 2003). It is reasonable, however, that birds roosting and feeding together could share some of the same pathogens. The level of enhanced risk is unknown and warrants additional study before reasoned management options can be developed (Clark and McLean 2003). Here, we list and briefly discuss eight important diseases that have been associated with flocking blackbirds, including red-winged blackbirds. We refer the reader to Conover (2002) and Conover and Vail (2015) for a comprehensive review of diseases that can spread between birds and humans (zoonoses).

2.3.1 Avian Salmonellosis

Avian salmonellosis (primarily *Salmonella* spp.) has been documented in starlings and blackbirds species throughout the United States and is transmissible to humans, poultry, and livestock (Carlson et al. 2010, 2015; Conover and Vail 2015). Poultry production operators, however, have protected their buildings from free-ranging birds and thereby have greatly reduced the threat of an outbreak (Clark and McLean 2003).

2.3.2 Chlamydiosis

Chlamydiosis (also psittacosis, ornithosis, and parrot fever) is carried by starlings and blackbirds and can infect humans and domestic fowl, causing respiratory psittacosis and avian chlamydiosis, respectively. Infections result from inhaling *Chlamydia psittaci* that live in dried feces deposited by birds (Conover 2002; Conover and Vail 2015).

2.3.3 Johne's Disease

Johne's disease (*Mycobacterium avium paratuberculosis*) is a contagious, chronic, and sometimes fatal infection that can be carried by birds and primarily affects the small intestine of ruminants (Clark and McLean 2003; Corn et al. 2005). The bacteria are excreted in feces and milk and annually costs the U.S. dairy industry $200–$250 million in losses (Ott et al. 1999; Beard et al. 2001).

2.3.4 Shiga Toxin-Producing *Escherichia coli*

Shiga toxin-producing *Escherichia coli* (STEC) is another disease that might be transmitted by wild birds to cattle (Swirski et al. 2014; Conover and Vail 2015). In the cattle industry, average annual costs of illnesses related to STEC exceeded US$267 million (National Cattleman's Beef

Association 2004). Humans may get this disease from consuming tainted food products, especially ground beef. Further research is needed to better clarify the role of birds and other factors in the transmission or prevalence of this disease.

2.3.5 Encephalitis

St. Louis encephalitis and western equine encephalitis are zoonotic diseases found primarily in wild vertebrates but transmitted to humans by the bite of a mosquito. The viruses are carried by blackbirds and cause acute inflammation of the brain that leads to illness and sometimes death (Conover 2002; McLean and Ubico 2007; Conover and Vail 2015).

2.3.6 Lyme Disease

Lyme disease is caused by the bacterium *Borreliela burgdorferi* and is classified as a zoonosis because it is transmitted to humans through the bite of infected blacklegged ticks (*Ixodes scapularis*). Lyme disease can be transmitted by ticks that feed on birds (Conover and Vail 2015). Battaly and Fish (1993) established that the American robin (*Turdus migratorius*), common grackle (*Quiscalus quiscula*), and house wren (*Troglodytes aedon*) are hosts for immature ticks and thus are high risk species of concern for human health.

2.3.7 Histoplasmosis

Histoplasmosis (*Histoplasma capsulatum*) is a common and sometimes serious noncommunicable fungal disease that primarily affects the lungs (Conover 2002; Conover and Vail 2015). Humans can become ill with histoplasmosis by inhaling dust at roosts that have large accumulations of bird and bat excreta (Chu et al. 2006). Stickley and Weeks (1985) suggested that only those roosts occupied by birds for three or more years have been shown to be infested with *H. capsulatum*. Accordingly, birds roosting at a site the first winter might be allowed to remain, unless nuisance complaints dictate otherwise.

2.3.8 West Nile Virus

West Nile virus (WNV) is a disease that is life-threatening to humans and wildlife. Confirmed in North America in 1999, WNV rapidly spread across the United States along migratory routes, covering 12 states in one year (Lanciotti et al. 1999; Bernard et al. 2001). Bernard et al. (2001) reported that both red-winged blackbirds and common grackles tested positive for WNV. Sullivan et al. (2006) conducted a serological survey of WNV antibodies in central North Dakota and found the peak WNV antibody prevalence was 22% in August of 2003 and 18% in July of 2004. Their results suggest that migratory red-winged blackbird populations may be an important viral dispersal mechanism with the ability to spread arboviruses such as WNV across North America. Whether WNV or other diseases affects productivity or the speed and distance of a red-winged blackbird's movements and migration are not known.

2.4 DISTRIBUTION AND POPULATIONS

SIDEBAR 2.1   THE NORTH AMERICAN BREEDING BIRD SURVEY

In the 1960s, Chandler S. Robbins led the development of a long-term, large-scale avian survey program known as the *North American Breeding Bird Survey* (NABBS; Ziolkowski et al. 2010). The NABBS, which is jointly coordinated by the U.S. Geological Survey and the Canadian Wildlife Service, provides the best data available for national and regional population estimates and trend analyses on more than 420 bird species. Concerns about the effects of powerful pesticides (e.g., DDT) on bird populations provided the initial impetus for the survey, but habitat loss, changes in land use, contaminants, and climate change are also important threats to birds. The survey, which is conducted during the breeding season, covers most of the United States and Canada and consists of 4,100 routes 39.4 km long, with 3-minute stops at 0.8 km intervals. Skilled observers identify all birds seen and heard along the route. Researchers and statisticians analyze the data for each species and provide population estimates and trends (Sauer et al. 2017). Significant declines in a bird population might result in research to identify the cause and suggest management actions to stop or even reverse the decline (Stanton et al. 2016). The NABBS is often used in conjunction with National Land Cover Data Set and the National Climatic Data Center to model the effects of land use and climate variables on bird abundance (e.g., Forcey et al. 2011, 2015; Bateman et al. 2016).

In 1966, the North American Breeding Bird Survey (NABBS) was initiated to provide data for bird population estimates and trend analyses (Sidebar 2.1). Based on NABBS data, the Partners in Flight Science Committee estimated the breeding population of red-winged blackbirds at 150 million individuals in the United States and Canada (Rosenberg et al. 2016). Although it is one of the most widespread and numerous birds in North America (Figure 2.3), data from the NABBS indicate that the red-winged blackbird population has declined about 0.93% annually from 1966 to 2015 (Figure 2.4) (Sauer et al. 2017). On a continental basis, changing land-use patterns (e.g., wetland drainage, grassland conversion, urbanization) and perhaps climate change are likely to drive a sustained long-term bird population decline (Forcey et al. 2011, 2015). On a statewide basis, Blackwell and Dolbeer (2001) showed that changes in farm practices in Ohio, primarily a reduction in hayfield acreage and earlier spring mowing of hay, caused the red-winged blackbird

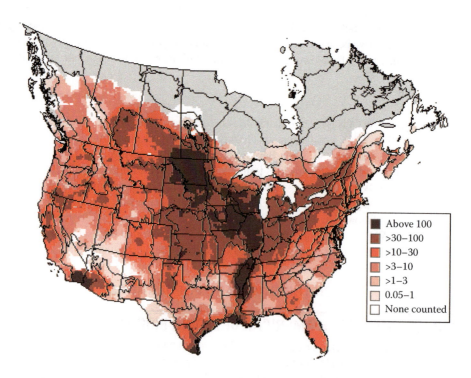

Figure 2.3 Relative abundance of red-winged blackbirds during nesting season (mean number of birds recorded per 39.4 km survey route) based on data from the North American Breeding Bird Survey, 2011–2015. (Sauer et al. 2017.)

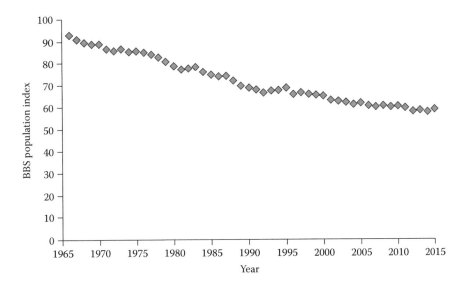

Figure 2.4 Population trend for red-winged blackbirds (mean number of birds recorded per 39.4 km survey route) based on data from the North American Breeding Bird Survey, 1966–2015. (Sauer et al. 2017.)

Table 2.1 Estimated Number of Breeding Pairs of Male Red-Winged Blackbirds Encountered during Surveys Conducted across Multiple Years in North Dakota

Year	Population Size (×1000)	95% CI	Source
1967	2,129	1,745–2,439	Stewart and Kantrud 1972
1981–1982	1,512	1,325–1,570	Besser 1985
1990	1,143	792–1,494	Nelms et al. 1994
1991	1,425	1,382–1,468	Nelms et al. 1999
1992	1,306	1,021–1,591	Igl and Johnson 1997
1993	1,536	1,224–1,848	Igl and Johnson 1997

population to decline by 53% between 1966 and 1996. This scenario is likely repeated throughout North America as agriculturalists make market-oriented decisions that affect land use.

Blackbird breeding populations in the northern Great Plains have received an inordinate amount of attention because their feeding on sunflower (*Helianthus annuus*) causes significant economic harm to growers (Linz et al. 2011b, 2015). In North Dakota, where 40% of U.S. sunflower is grown, both a roadside blackbird index and a general avian density survey were established in 1965 and 1967, respectively. From 1965 to 1981, Besser et al. (1984) conducted a roadside index survey in North Dakota and South Dakota and found that the red-winged blackbird population declined 30% from the first three years (1965–1968) compared to the last three years of the survey (1979–1981). In 1967, Stewart and Kantrud (1972) established 130 plots of 64.75 ha each across North Dakota to estimate the density of all breeding birds. During the inaugural survey, they found two million pairs of red-winged blackbirds. The survey was repeated in 1981–1982 (Besser 1985), 1990 (Nelms et al. 1994), and 1992 and 1993 (Igl and Johnson 1997) (Table 2.1).

In 1991, Nelms et al. (1999) conducted a comprehensive population density survey using a random selection of 10 plots of 64.75 ha within 80 townships (93.2 km^2). They estimated that 1.4 million pairs of red-winged blackbirds were breeding in North Dakota. Overall, these data suggest that the red-winged blackbird population declined >30% between 1967 and the early 1990s. Finally, from 1996 to 1998, Linz et al. (2002) repeated the Stewart and Kantrud (1972) survey on 67 plots located within the PPR and found that the population was 32% higher compared to the average numbers in 1967, 1981–82, and 1990. The years of higher population numbers were marked by above-average precipitation (U.S. Department of Agriculture 1999).

The fluctuation in North Dakota's red-winged blackbird population is likely related to wet–dry precipitation cycles in the state (Besser et al. 1984; Linz et al. 2002; Forcey et al. 2015). In dry years, agriculturalists are able to till the soil in shallow wetlands, destroying emergent vegetation used as nesting substrate. In the 2000s, historically high commodity prices and reduced funding for grassland conservation programs resulted in these lands being converted to grain crops, which resulted in reduced nesting opportunities for birds (Claassen and Hungerford 2014; USDA 2015b). The NABBS data show, however, that between 2003 and 2015 the red-winged blackbird population in the PPR has remained statistically unchanged (Sauer et al. 2017). Future bird surveys are warranted, given ongoing land-use changes due to urbanization, climate change, hydrocarbon extraction, and changes in grassland cover due to the potential increase in grassland acres for biofuels that might offset declining Conservation Reserve Program (CRP) enrollment (Johnson and Igl 1995; Murray and Best 2003; Weatherhead 2005).

2.5 WINTER LOCATION

From 1961 to 1966, DeGrazio et al. (1969) banded 27,000 blackbirds during late summer in eastern South Dakota and recovered red-winged blackbirds, yellow-headed blackbirds, and common grackles from December through February across the wintering range. Most red-winged blackbirds and common grackles were found to overwinter in eastern Texas and western Louisiana, whereas yellow-headed blackbirds were mainly found in central Mexico. Meanley (1964) banded 6,000 red-winged blackbirds in the Patuxent River marsh complex in Maryland and found that most of the banded birds wintered along the eastern coastal states. Mott (1984) reported the majority of blackbirds banded ($n = 20,000$) in Kentucky and Tennessee during the winter were recovered on nesting grounds in the northeastern United States, whereas 50% of the starlings were hatched in the subject states. In a comprehensive analysis of the continental movement and migration patterns of red-winged blackbirds, Dolbeer (1978, 1982) concluded that red-winged blackbirds in the eastern United States tend to winter in the southeastern United States, whereas birds from the northern plains winter in the south-central states of Texas and Louisiana. Adult birds usually returned to the same area to breed each year, but in winter roosts there was a high degree of intermingling of birds from northern areas. Female red-winged blackbirds from a given northern area typically migrated further south than did males (Dolbeer 1982).

During winter, red-winged blackbirds gather nightly in winter roosting congregations, often with other blackbirds and European starlings. The last comprehensive national survey of roosts, in the winter of 1974–1975, recorded 723 roosts containing an estimated 537 million blackbirds and starlings. Red-winged blackbirds comprised 38% (204 million) of the total roosting population (Meanley and Royall 1976).

These large aggregations of blackbirds led to citizen concerns about health issues, crop damage, structural damage due to bird droppings, and safety related to bird–aircraft strikes. As a result, in 1977 the U.S. Fish and Wildlife Service established the Kentucky Field Station at Bowling Green, Kentucky, USA, to research these problems and to find potential management solutions. Scientists developed the use of a sprinkler-irrigation system for applying a wetting agent (PA-14 surfactant) to blackbirds and starlings roosting in trees. The applications of PA-14 caused hypothermia and as a result sometimes killed millions of birds (Heisterberg et al. 1987; Dolbeer et al. 1997). PA-14 is no longer registered for this use due to environmental concerns, and its replacement compound, sodium lauryl sulfate, is not commonly used to manage bird populations (Dolbeer et al. 1997; U.S. Department of Agriculture 2012). They also successfully developed the avicide compound DRC-1339 (a.i., 3-chloro-p-toluidine hydrochloride), a compound still in use for reducing starling numbers at feedlots and blackbirds damaging sprouting rice (e.g., Glahn and Wilson 1992; U.S. Department of Agriculture 2015).

2.6 SPRING MIGRATION

Dolbeer (1978) used band recoveries to conduct a comprehensive analysis of continental movement and migration patterns of red-winged blackbirds. He suggested that red-winged blackbird spring migration occurs from about February 21 to April 24 (pre-reproductive period) in most areas of the United States. Male red-winged blackbirds generally leave the wintering areas earlier than females and arrive in breeding areas to establish territories before females arrive (Yasukawa and Searcy 1995).

Migration from the southern United States begins in February and peaks in March. Wilson (1985) analyzed morphological data (larger birds presumed to be migrants) to conclude that 78% of the red-winged blackbirds in Louisiana during March were local breeding birds, suggesting that most of the damage to newly seeded and sprouting rice was caused by these birds. Using banding data, Bruggers and Dolbeer (1990) found that resident birds constituted 16 of 20 non–banding-station recoveries during spring planting in March and April. This further corroborated Wilson (1985), who suggested that local red-winged blackbirds were responsible for most crop damage during planting. Potential bias associated with collecting data at banding stations prompted Bruggers and Dolbeer (1990) to advocate mass-marking and additional banding to further refine the relationship of migration and rice damage in the Gulf Coast region. To that end, 7 million red-winged blackbirds were aerially tagged with a fluorescent particle marker in Louisiana in 1995 (Sidebar 2.2) (Cummings and Avery 2003). Birds collected the following spring on breeding grounds across North America showed that the marked birds in Louisiana roosts were scattered across 13 states and central Manitoba, Canada.

SIDEBAR 2.2 FORMULATION AND APPLICATION OF AERIAL MASS COLOR-MARKING

Aerial mass color-marking was used to track the local and regional movements of blackbirds using night roosts from 1982 to 2001 (Otis et al. 1986; Knittle et al. 1987; Linz et al. 1991; Homan et al. 2004). The formulation consisted of an acrylic adhesive, food-grade propylene glycol, water, fluorescent-pigmented resin, surfactant, and foam suppressor (Homan and Linz 2005). The formulation was applied to flocks of blackbirds with a fixed-wing aircraft from an altitude of 15–20 m. The pilot usually flew high over the wetland to flush waterfowl and wading birds. The sprays occurred during the 20–30-minute period of twilight following sunset. A 416-liter load was sufficient to mark 100,000 birds in 15 minutes. A coarse droplet size of approximately 400 microns was used because it leaves well-defined splash marks of color on the birds. The spray dries in 3–5 minutes and adheres particularly well to feather surfaces as the birds fly through the descending spray mist. Studies on marked, free-ranging red-winged blackbirds showed that 30% of the initial marks were lost 4–6 weeks after spraying (Knittle and Johns 1986). Resin particles lodged in the barbules, however, can remain much longer, often several months after the date of application (Homan et al. 2004). The marker formulation is nontoxic to freshwater fish and chironomid larvae (Bills and Knittle 1986; Knittle and Johns 1986). Color-marking was coordinated with the U.S. Geological Survey Bird Banding Laboratory.

Source: Homan et al. 2005.

Crop damage in the northern Great Plains has prompted researchers to expend considerable effort to define the spring migration routes of red-winged blackbirds in the central United States, with the assumption that population management might be implemented at key stopover locations (Knittle et al. 1987; Homan et al. 2004). Attempts to reduce the red-winged blackbird population were not undertaken, but detailed information on their spring migratory timing and dispersal patterns was obtained (Blackwell et al. 2003; Homan et al. 2004; Linz et al. 2011b). Migration to breeding areas in the northern Great Plains occurs in April, when large migrating flocks can be found in northeastern South Dakota, until mid-April, when their numbers begin to decrease (Linz et al. 2003; Homan et al. 2004). In some years, flocks of female red-winged blackbirds can be found in northeastern South Dakota in early May (G. Linz, personal observation).

In 1982, 1983, and 1985, Knittle et al. (1987, 1996) used an aerially applied fluorescent pigment to mark millions of male red-winged blackbirds in northwestern Missouri, eastern South Dakota, and western Minnesota. Red-winged blackbirds using spring roosts in northwestern Missouri and southeastern South Dakota migrated northwest to breeding sites in or near sunflower-producing areas. Using these data and bird-banding data, Stehn (1989) proposed that blackbirds responsible for sunflower depredation originated from breeding territories located in a very large area including most of North Dakota (except the southwest corner), the eastern third of South Dakota, far western Minnesota, southern Manitoba, and southeastern Saskatchewan (Figure 2.5).

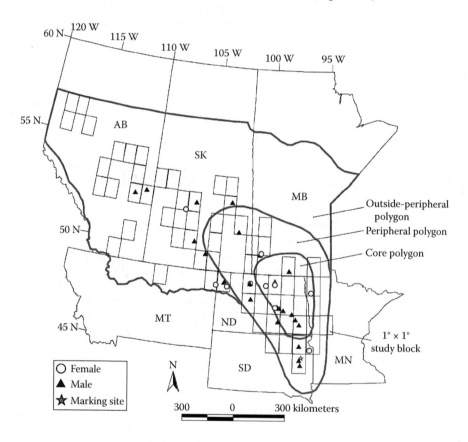

Figure 2.5 In April 2001, three blackbird roosts located in eastern South Dakota (44°48' N, 97°21' W) were aerially marked with a fluorescent pigment. Breeding red-winged blackbirds were randomly collected in the United States and Canada, of which 33 were marked. The polygons were based on an analysis of banding and re-sighting data, physiography, and proximity to the area of concentrated sunflower production. (Homan et al. 2004.)

Homan et al. (2004) refined our knowledge of migration patterns by marking red-winged blackbirds in eastern South Dakota and collecting both males and females in breeding territories. They determined that 82% of marked birds using staging areas in eastern South Dakota during spring migration were breeding in or near the core region of sunflower production in the PPR. Peer et al. (2003) used NABBS data and density estimates to determine that red-winged blackbird population sizes in this region were 27 million in the spring and 39 million post-reproduction. Using banding data, Dolbeer (1978) showed that in the United States local nesting red-winged blackbirds and their offspring stay within 200 km of their breeding area until feather molt and replacement is completed in early October. He found, however, that birds breeding in Alberta and Saskatchewan Canada moved an average of 729 km, suggesting that they were moving south during feather molt (i.e., the process of losing feathers and replacing them).

2.7 FALL MIGRATION AND ANNUAL FEATHER REPLACEMENT

During the peak of annual feather molt, the ability of birds to fly efficiently is compromised and explains why blackbirds are difficult to frighten out of roosts and crop fields during late summer (Smith and Bird 1964). For example, Handegard (1988) reported that hazing blackbirds from sunflower fields with airplanes was particularly difficult during August and early September, which coincides with the peak of molt. Additionally, growing feathers requires extra energy and increases predation risk because of reduced flying capability (Rappole 2013). Thus, a thorough understanding of this critical period of the red-winged blackbird's life cycle is necessary in developing an integrated crop damage management plan.

Meanley and Bond (1970) recognized the importance of the red-winged blackbird's annual molt in relation to fall migration. They found that across Arkansas, Maryland, and Michigan, the majority of red-winged blackbirds initiated molt in late July to early August, with most of the birds completing molt by October 1. At any stage of development, the red-winged blackbird usually had two, but sometimes three and rarely four, nonfunctional primary remiges (wing flight feathers). All feather tracts completed molt at approximately the same time. Further, Meanley and Bond (1970) concluded that most red-winged blackbirds complete feather molt before migrating.

Linz et al. (1983) studied the molt of red-winged blackbirds across five age–sex classes in a migratory population in Cass County, North Dakota. They found that SY males (generally nonbreeding) began molt earlier and were more synchronized than ASY males and AHY females. By the last week of July, the molt of the SY males had advanced one primary remige ahead of ASY males and AHY females. During the latter part of nesting season (July 22 through August 18), a larger percentage of the males (~95%) were molting than were the females (80%–85%). The percentage of nonmolting females may have represented renesting birds and later-nesting SY females. Linz et al. (1983) suggested that differences in molt timing among and within each of the age–sex classes was probably related to the duration of nesting activities.

Red-winged blackbirds normally remain within 200 km of their breeding area until molt is complete (Dolbeer 1978). However, through an analysis of banding returns Dolbeer (1978) showed that birds nesting in western Canada may migrate during August and September. Molt would not be complete during the earlier part of this period. Linz et al. (1983) calculated that individual red-winged blackbirds complete molt in at least nine weeks. Therefore, the first birds should have completed molt by September 9. However they did not collect a red-winged blackbird with complete winter plumage until the week of September 23. This suggested that many birds leave North Dakota upon completion of molt or just before the completion of molt and that birds less advanced in molt arrive from the north during the same period.

2.8 FOOD HABITS

Crase and DeHaven (1975) published an annotated bibliography that lists 233 references on the food habits of blackbirds published from 1831 to 1974. Since 1974, additional food habits studies have been conducted to help identify and quantify what species observed in the crop fields were eating. Food habits studies are labor intensive but are considered a necessary early step in defining a given damage issue. Data from these studies were used to develop economic models (Weatherhead et al. 1982; Peer et al. 2003) and provide insight into potential damage management strategies, especially for the development of alternative foraging sites (Hagy et al. 2008). Here, we review major red-winged blackbird food habits studies conducted in the southern and midwest regions of the United States, the northern prairie region of the United States and Canada, and Ontario, Canada.

2.8.1 Southern United States

Meanley (1961) analyzed the gizzard contents of 130 red-winged blackbirds collected in August and September 1959 in an uncultivated wild rice (*Zizania aquatica*) bed in a tidal marsh in Maryland. Wild rice is a 2-m tall plant that features many tillers and stems, which provide both roosting habitat and a food source for birds. Meanley (1961) found that weed seeds made up 58% of their diet, followed by wild rice (24%), ripening corn (12%), and 5% animal matter (mostly insects). Since Meanley's publication, production of cultivated wild rice has become an established industry in California, Minnesota, Oregon, and Wisconsin, while harvesting of uncultivated rice has continued in numerous northern states (Marcum and Gorenzel 1994; Hauan 2015). To our knowledge, a bird's food habits study related to cultivated wild rice has not been conducted.

Wilson (1985) conducted the most recent food habits study in the rice growing region of the southeastern United States. From October 1979 to August 1980 and August to September 1983, he collected 402 red-winged blackbirds in southwest Louisiana and found that cultivated white rice made up 37% of the birds' annual diet, which was lower than the 54% and 45% reported previously by Kalmbach (1937) and Neff and Meanley (1957), respectively. During March and April, the primary planting months, red-winged blackbird diets averaged 24% white rice, 50% weed seeds, 18% red rice, and 6% insects. During July and August, when the rice crop was maturing, red-winged blackbird diets comprised 37% white rice, 52% weed seeds, 9% animal matter (mostly insects) and 2% red rice. Wilson (1985) suggested that this amount of rice in their diet was a minimum because rice in the milk and early dough stage of development is difficult to quantify.

2.8.2 Midwest United States

Stockdale (1959) analyzed the gizzard contents of 136 red-winged blackbirds collected from February 22 to November 15, 1959, in north-central Ohio. He divided the collections into five periods based on distinctly different activities. During the arrival period (February 22 to March 19) and mating and territory establishment period (March 20 to April 30), their diet was 96% vegetable matter, with waste corn a major component. During the nesting period (May 1–31), the birds contained 74% and 26% vegetable and animal matter, respectively. From June 1 to July 15 (fledging period), their diets comprised 62% animal matter and 38% vegetable matter. Beetles (Scarabaeidae and Curculionidae) were commonly consumed in both the nesting and fledgling periods.

Throughout the flocking period (July 15 to November 15), which coincided with the availability of ripening corn, red-winged blackbirds ate 90% vegetable matter and 10% animal matter. Stockdale (1959) found corn in 76% of the birds during this time, with weed seeds and small grains found in 35% and 16% of the birds, respectively. Across the entire study period, vegetable food and animal matter made up 69% and 31% of their food consumption, respectively.

2.8.3 Northern Prairie of Canada and United States

Bird and Smith (1964) analyzed the gullet and gizzard contents of 183 red-winged blackbirds collected from May to October 1960 in a diverse agriculture area in southern Manitoba, near the North Dakota border. During May, 66% and 26% of the birds' diets were made up of waste grains (sunflower, corn, small grains) and animal matter (largely insects), respectively.

During June and July (nesting season), 67% of their diets consisted of animal matter. From August to mid-October (harvest), red-winged blackbirds ate mostly vegetable matter (72%), with ripening sunflower (38%) dominating this portion of their diet. Animal matter made up 20% of their diet at this time.

Mott et al. (1972) analyzed the gizzard contents of 702 red-winged blackbirds collected from 1959 through 1965 during the spring, summer, and early fall in northeastern South Dakota. The highest use of oats (20%) was between July 1 and August 15 and millet (16%) was used most heavily from September 16 to 25. From August 16 to September 25, use of weed seeds and ripening corn peaked, comprising 39% and 25% of the red-winged blackbird diet, respectively. Overall, they reported that weed seeds (23%), corn (11%), oats (10%), wheat (7%), and millet (3%) comprised the majority of their grain diet, whereas animal matter (mostly insects) contributed 25% of their diet. Finally, Mott et al. (1972) noted that males ate significantly more corn (29%) than did females (9%).

Linz et al. (1984) examined the esophageal contents of 1,182 red-winged blackbirds collected from late July to early November of 1979 and 1980 in sunflower and corn fields in southeastern North Dakota. At that time, this area was unique in North Dakota because agriculturalists regularly planted both field corn and sunflower in juxtaposition. Male diets in corn fields consisted of 45% corn and 41% weed seeds, whereas female diets contained 16% corn and 70% weed seeds. Males collected in sunflower fields contained 69% sunflower and 18% weed seeds, whereas females consumed 57% sunflower and 31% weed seeds. In both crops, males caused significantly more economic damage than did females.

2.8.4 Ontario, Canada

Hintz and Dyer (1970) analyzed the stomach contents of 650 red-winged blackbirds collected in southern Ontario to determine the daily rhythms and seasonal changes in diets. From July 20 to August 14, ripening wheat and oats comprised 58% of the stomach contents and animal matter accounted for 36%; from August 15 to September 10, corn constituted 81% of the stomach contents and animal matter accounted for 11%. During both of these times, weed seeds made up the remainder of their diet. Overall, Hintz and Dyer (1970) found that red-winged blackbirds' caloric intake tended to be higher in the morning than afternoon and higher in the late summer than in the breeding season.

McNicol et al. (1982) collected 440 red-winged blackbirds from March to October 1977 and found that overall their diets were made up of 42% animal matter (mostly insects), followed by 31% corn, 18% weed seeds, and 6% oats. During the postbreeding and fall flocking period (July 11 to October 28), they found that male and female diets contained 56% and 28% corn, respectively. The remainder of their diets were largely made of weed seeds (males—24%; females—44%).

Finally, from August 15 to September 14, 1978, Gartshore et al. (1982) collected 143 red-winged blackbirds (72% males) from four corn fields that contained food in their esophagus and gullet. They analyzed the dry weight of foods for birds collected in three fields and found no difference in the amount of corn consumed by males and females, averaging 94%.

2.8.5 Summary

We note certain commonalities among the results of these food habits studies. Foremost, red-winged blackbird diets are flexible and food selection is dependent on availability. During the nesting season, red-winged blackbird diets consist of a high percentage of insects, whereas they feed insects to nestlings. The birds select insects because they are high in calories and protein and also plentiful in their nesting habitats. After nesting is complete, usually in July, red-winged blackbirds form small flocks and begin to exploit ripening cereal grains near wetland roosts. Their annual feather molt begins in earnest in late July and August, and they also initiate premigratory fattening. Red-winged blackbirds can meet their high calorie demand by exploiting the super-abundant food available in rice, corn, and sunflower fields. We note that as corn and sunflower crops mature and become more difficult to access and handling time increases, weed seeds found in crop fields and waste grain in harvested fields become important food sources, especially for females. Studies comparing the food habits of male and female red-winged blackbirds showed that females, which have smaller bodies and bills, cause less crop damage than do males (McNicol et al. 1982; Mott et al. 1972; Linz et al. 1984).

Transitioning from insects to crops is not complete, as food habits data show that red-winged blackbirds will also take harmful insects in crops (e.g., corn borers and seed weevils). Thus it is reasonable to suggest that red-winged blackbirds provide an ecological service by taking weed seeds and harmful insects that could benefit agriculture productivity (Bendell et al. 1981; McNicol et al. 1982; Bollinger and Caslick 1985; Dolbeer 1990; Okurut-Akol et al. 1990; Kirk et al. 1996). This service might be particularly important for organic growers who must avoid the use of chemical pesticides to maintain their certification for organic products. As a point of emphasis, this fast-growing segment of agriculture has increased from US$3.4 billion in 1997 to US$43.3 billion in 2015 (Hornick 2016). Regardless, it is prudent for agriculturalists and wildlife managers to consider the cost-benefits of blackbird management along with practicality, environmental safety, and wildlife stewardship (Dolbeer 1981; Slate et al. 1992; Blackwell et al. 2003; Linz et al. 2015).

2.9 CROP DAMAGE

For much of the year, red-winged blackbirds forage on insects, waste grain, and weed seeds (Stockdale 1959; White et al. 1985). During late winter, they might also be found feeding on high quality food found in concentrated animal feedlot operations (Dolbeer et al. 1978). However, red-winged blackbirds can sometimes cause significant economic damage to sprouting and ripening crops. Objective assessments of losses to birds is an arduous and expensive task but necessary to obtain scientifically defensible data for documenting economic losses, justifying the use of valuable resources, and assessing efficacy and cost-effectiveness of improved or new damage management techniques (Besser 1985). Shwiff and colleagues detail the economics of bird damage to crops in Chapter 12 of this volume. Here, we review studies that have sought to define the level of economic damage.

The red-winged blackbird prebreeding population (150 million) can sometimes cause significant damage to spring-seeded crops—especially rice but also corn and sunflower (Meanley 1971; Besser 1985; Wilson et al. 1989). Peer et al. (2003) calculated that the red-winged blackbird postbreeding population increases 45% in the northern Great Plains, which provides a reasonable national estimate of 217.5 million. These birds, along with other blackbird species, concentrate in wetland roosts containing up to several million birds. These birds forage in nearby ripening crops, especially field corn, rice, and sunflower, where they sometimes cause economically significant damage (Tyler and Kannenberg 1980; Weatherhead et al. 1982; Linz et al. 1984; Besser 1985; Wilson 1985; Dolbeer et al. 1986; Wilson et al. 1989; Peer et al. 2003).

Blackbird damage is typically 1%–2% of the crop and most of that damage is within 8 km of a roost (Dolbeer 1981; Otis and Kilburn 1988; Wywialowski 1996; Dolbeer and Linz 2016). The uneven distribution of economic damage across producers, however, is the core of this wildlife management problem. That is, a small percentage of producers suffer the majority of the economic loss, which sometimes results in the abandonment of an otherwise profitable crop (Wywialowski 1996; Linz et al. 2011b; Linz and Hanzel 2015). Often blackbirds roost on public land or non–farmer-owned private land but forage on nearby crops. This leads to additional frustration by producers, who feel their hands are tied.

2.9.1 Rice

Rice is considered a minor crop in the United States, although about 1 million ha are planted annually (Cummings et al. 2005). Much of this crop is planted in the southeastern United States, which is also a favored wintering location for blackbirds (Meanley and Royall 1976). A substantial research effort began in the mid-1950s and intensified in the mid-1970s to expedite the development of damage management options (Meanley 1971). Researchers identified red-winged blackbirds as largely responsible for damaging sprouting and ripening rice, but brown-headed cowbirds, common grackles, and to a lesser extent boat-tailed grackles (*Quiscalus major*), great tailed grackles (*Quiscalus mexicanus*), and dickcissels (*Spiza americana*) have also been identified foraging in rice fields (Meanley 1971; Wilson 1985; Glahn and Wilson 1992; Avery et al. 2005).

Rice is readily available to foraging birds after seeding in the spring and while ripening prior to harvest in the fall. Bird damage to rice is not uniformly distributed but is localized and generally proportional to the size of nearby blackbird roosts (Avery et al. 2005). Both early-seeded and late-seeded fields are more likely to receive damage because both resident and spring-migrant blackbirds damage the sprouting rice, whereas resident and fall-migrant blackbirds damage ripening rice (Wilson et al. 1989). In some cases, locally severe blackbird damage to newly planted rice can result in a total loss and require that the crop be replanted (Wilson et al. 1989).

Reliable quantitative bird damage estimates are scarce due, in part, to difficult logistics associated with walking through ripening rice and newly planted, flooded fields. The most recent objective surveys were conducted in the 1980s. At that time, bird damage to newly planted rice cost growers in southwestern Louisiana and east Texas a combined total of about US$8 million (Wilson et al. 1989; Decker et al. 1990). As an alternative to objective surveys, Cummings et al. (2005) surveyed rice growers in the United States and found that between 1996 and 2000 the average annual blackbird damage to newly planted rice ranged from 6% to 15% and the average percent loss to ripening rice ranged from 6% to 14%. Louisiana and Arkansas respondents reported the highest damage.

2.9.2 Corn

Corn is a major crop in the United States, with about 14 million ha planted annually. In 1957, an intense research effort was initiated to alleviate blackbird damage to field corn in Ohio (Stockdale 1959). Scientists recognized that national damage was <1% but local damage near roost sites could be economically significant, which was defined as over 5% (Dolbeer 1980). Although red-winged blackbirds cause the greatest economic loss to ripening field corn, common grackles and yellow-headed blackbirds (in the central United States) also damage ripening corn (Dolbeer 1980; Twedt et al. 1991; Klosterman et al. 2013).

From 1977 to 1979, U.S. Fish and Wildlife Service personnel conducted a statewide survey in Ohio to quantify bird damage to field corn (Dolbeer 1980). The statewide estimates showed that primary damage (the actual corn removed by the birds) averaged 0.6%, and secondary damage (molding or sprouting resulting from moisture entering the opened ear) averaged 0.1%. Dolbeer (1980) reported that <2.5% of cornfields in Ohio had losses >5% and that all of these fields were within 8 km of a major wetland roost of blackbirds.

In 1981, Besser and Brady (1986) conducted an objective survey of bird damage to ripening field corn in 10 major producing states, representing 79% of the planted corn in the United States. The percentages of corn ears and fields with no damage were 98% and 84%, respectively. In 1993, Wywialowski (1996) assessed wildlife damage to corn in the top ten corn-producing states and found that the average loss was 0.19%, with damage 2.4 times higher on the edge of the fields versus the interior.

In North Dakota, corn planting increased threefold from 404,686 ha in the early 2000s to $1,416 \times 10^3$ ha in 2016 (U.S. Department of Agriculture 2016). Growers observed blackbirds in their fields and speculated that significant damage was occurring. Over a 2-year study, however, Klosterman et al. (2013) found that bird damage to cornfields in North Dakota averaged only 0.2% and no fields were found with over 5% damage.

Bird damage to field corn is likely ameliorated because corn is most vulnerable during the milk and dough development stages, a period of 3–4 weeks (Nielsen 2013). After that time, the corn kernel hardens, reducing its attractiveness to blackbirds, particularly females of both the red-winged blackbird and yellow-headed blackbird, which are both smaller than the males. Male blackbirds can damage field corn for several more weeks as the kernels mature and dry down to <20% at harvest. Finally, bird feeding damage predisposes the ears to the development of various ear molds and rots, some of which may subsequently lead to the development of dangerous mycotoxins (Nielsen 2009).

2.9.3 Sweet Corn

Over the last decade, sweet corn was planted on an average of 98,000 ha in the United States (National Agricultural Statistics Service 2016). Systematic national surveys of blackbird damage to sweet corn have not been conducted. However, several countywide surveys conducted across four states in 1974 showed that damage ranged from 4.5% to 23.5% (Dolbeer et al. 1986). In 1965, surveys in Ontario showed sweet corn damage was 2.4% (range 0%–13%). Dolbeer (1990) pointed out that consumers find even a small amount of bird damage unsavory and will not purchase the corn. Damaged sweet corn delivered to canneries must be culled or trimmed, resulting in increased labor costs (Dolbeer 1990).

2.9.4 Sunflower

Sunflower is a minor crop in the United States, with 665,810 ha planted in 2016 (U.S. Department of Agriculture 2016). Blackbird (Icteridae) damage is the most common reason that sunflower producers in North Dakota stop planting sunflower (Linz et al. 2011b; Hulke and Kleingartner 2014). Ripening sunflower is particularly vulnerable to blackbirds because the crop is susceptible from early seed-set in mid-August until harvest in mid-October, a period of 8 weeks (Cummings et al. 1989; Linz et al. 2011b). In the 1970s, sunflower became an economically viable crop in the northern Great Plains but also presented a high-caloric food source for local nesting premigratory and migrating blackbirds. Beginning in 1979 and continuing to this day, scientists from the USDA APHIS Wildlife Services, National Wildlife Research Center, and North Dakota State University are collaborating on research designed to reduce blackbird damage to sunflower.

In 2009 and 2010, Klosterman et al. (2013) assessed bird damage to randomly selected sunflower fields in North Dakota's PPR and found that the average annual blackbird damage was 7.5×10^3 tons (2.7%). Hothem et al. (1988) conducted a statewide (87% of the samples in PPR), bird damage survey in 1979 and 1980 and found that blackbird damage averaged 27.6×10^3 tons (1.8%). Thus, total tonnage lost was three times greater in the Hothem et al. (1988) survey than that reported by Klosterman et al. (2013). Assuming similar numbers of blackbirds were foraging in North Dakota across studies, the birds might have made up the difference in sunflower intake by seeking other food sources such as corn, small grains, waste grains, and weed seeds.

From 2002 to 2013 (except 2004), the National Sunflower Association sponsored comprehensive national surveys of blackbird damage in physiologically mature sunflower fields throughout the foremost sunflower growing states. Kandel and Linz (2016) analyzed and summarized the magnitude of blackbird damage in eight states from 2009 to 2013 and found that among biological production issues, blackbird damage to sunflower ranked third behind disease and weeds. Blackbird damage was substantial, with growers losing 2.6% and 1.7% per year of oilseed sunflower and confectionery production, respectively. Of the eight surveyed states, North Dakota ranked first in bird damage to confectionery and oilseed hybrids.

2.10 SUMMARY

Ubiquitous male red-winged blackbirds, with bright red epaulets framed by black feathers, are easily detected while perched and loudly singing (*o-ka-leeee, konk-a-ree*) on breeding territories, whereas females are more difficult to find because they are smaller and their feathers are cryptic brown with streaking. Most people in North America view "redwings" positively because of their splendor and as a harbinger of spring in northern areas. Their polygynous mating system combined with the conspicuous displays of males on territories and their propensity to conflict with human activities have made this species the subject of hundreds of scientific studies. Their large continental population is a testament to the success of this breeding system and ability to nest successfully in wide array of habitats. Their population is kept in check with a high annual mortality resulting from predation and weather events.

During much of their annual life cycle, they perform an important ecological service by eating waste grain, weed seeds, and insects and by serving as a prey base for many predatory birds and mammals. When grain crops begin to ripen, however, red-winged blackbirds exploit this superabundant food source, causing economic harm to growers. Moreover, nuisance, disease, and crop damage concerns arise in winter when they join with other blackbird species and European starlings in roosts throughout the southern United States. Since the 1950s, a substantial research effort has led to a better understanding of the ecology of red-winged blackbirds in relation to crop damage and the development of management methods designed to reduce the conflict between red-winged blackbirds and people. These management methods will be discussed in following chapters.

REFERENCES

Allen, A. A. 1914. The red-winged blackbird: A study in the ecology of a cattail marsh. *Linnaean Society of New York* 24–25:43–128.

Avery, M. L., S. J. Werner, J. L. Cummings, J. S. Humphrey, M. P. Milleson, J. C. Carlson, T. M. Primus, et al. 2005. Caffeine for reducing bird damage to newly seeded rice. *Crop Protection* 24:651–657.

Bateman, B. L., A. M. Pidgeon, V. C. Radeloff, J. VanDerWal, W. E. Thogmartin, S. J. Vavrus, and P. J. Heglund. 2016. The pace of past climate change vs. potential bird distributions and land use in the United States. *Global Change Biology* 22:1130–1144.

Battaly, G. R., and D. Fish. 1993. Relative importance of bird species as hosts for immature *Ixodes dammini* (Acari: Ixodidae) in a suburban residential landscape of southern New York state. *Journal of Medical Entomology* 30:740–747.

Beard, P. M., M. J. Daniels, D. Henderson, A. Pirie, K. Rudge, D. Buxton, S. Rhind, et al. 2001. Paratuberculosis infection of nonruminant wildlife in Scotland. *Journal of Clinical Microbiology* 39:1517–1521.

Beletsky, L. D. 1996. *The red-winged blackbird: The biology of a strongly polygynous songbird*. Academic Press, San Diego, CA.

Beletsky, L. D., and G. H. Orians. 1991. Effects of breeding experience and familiarity on site fidelity in female red-winged blackbirds. *Ecology* 72:787–796.

Beletsky, L. D., and G. H. Orians. 1996. *Red-winged blackbirds: Decision making and reproductive success.* University of Chicago Press, Chicago, IL.

Bendell, B. E., P. J. Weatherhead, and R. K. Stewart. 1981. The impact of predation by red-winged blackbirds on European corn borer populations. *Canadian Journal of Zoology* 59:1535–1538.

Bernard, K. A., J. G. Maffei, S. A. Jones, E. B. Kauffman, G. Ebel, A. P. Dupuis, 2nd, K. A. Ngo, et al. 2001. West Nile virus infection in birds and mosquitoes, New York State, 2000. *Emerging Infectious Diseases* 7:679–685.

Besser, J. F. 1985. Changes in breeding blackbird numbers in North Dakota from 1967 to 1981–82. *Prairie Naturalist* 17:133–142.

Besser, J. F., and D. J. Brady. 1986. *Bird damage to ripening field corn increases in the United States from 1971 to 1981.* U.S. Fish and Wildlife Service, Fish and Wildlife Leaflet 7, Washington, DC.

Besser, J. F., J. W. DeGrazio, J. L. Guarino, D. F. Mott, D. L. Otis, B. R. Besser, and C. E. Knittle. 1984. Decline in breeding red-winged blackbirds in the Dakotas, 1965–1981. *Journal of Field Ornithology* 55:435–443.

Bills, T. D., and C. E. Knittle. 1986. *Toxicity of DayGlo® fluorescent pigment material to four species of fish.* Bird Damage Research Report 359. U.S. Fish and Wildlife Service, Denver Wildlife Research Center, Denver, CO.

Bird, R. D., and L. B. Smith. 1964. The food habits of the red-winged blackbird, *Agelaius phoeniceus*, in Manitoba. *Canadian Field-Naturalist* 78:179–186.

Blackwell, B. F., and R. A. Dolbeer. 2001. Decline of the red-winged blackbird population in Ohio correlated to changes in agriculture (1965–1996). *Journal of Wildlife Management* 65:661–667.

Blackwell, B. F., E. Huszar, G. M. Linz, and R. A. Dolbeer. 2003. Lethal control of red-winged blackbirds to manage damage to sunflower: An economic evaluation. *Journal of Wildlife Management* 67:818–828.

Bollinger, E. K., and J. W. Caslick. 1985. Red-winged blackbird predation on northern corn rootworm beetles in field corn. *Journal of Applied Ecology* 22:39–48.

Brugger, K. E., and R. A. Dolbeer. 1990. Geographic origin of red-winged blackbirds relative to rice culture in southwestern and southcentral Louisiana. *Journal of Field Ornithology* 61:90–97.

Carlson, J. C., A. B. Franklin, D. R. Hyatt, S. E. Pettit, and G. M. Linz. 2010. The role of starlings in the spread of Salmonella within concentrated animal feeding operations. *Journal of Applied Ecology* 2:479–486.

Carlson, J. C., D. R. Hyatt, J. W. Ellis, D. R. Pipkin, A. M. Mangan, M. Russell, D. S. Bolte, et al. 2015. Mechanisms of antimicrobial resistant *Salmonella enterica* transmission associated with starling-livestock interactions. *Veterinary Microbiology* 179:60–68.

Chu, J. H., C. Feudtner, K. Heydon, T. J. Walsh, and T. E. Zaoutis. 2006. Hospitalizations for endemic mycoses: A population-based national study. *Clinical Infection Diseases* 42:822–825.

Claassen, R., and A. Hungerford. 2014. *2014 Farm act continues most previous trends in conservation.* U.S. Department of Agriculture, Economic Research Service, Washington, DC. http://www.ers.usda.gov/agricultural-act-of-2014-highlights-and-implications/conservation.aspx (accessed June 23, 2016).

Clark, L., and R. G. McLean. 2003. A review of pathogens of agricultural and human health interests found in blackbirds. In *Management of North American Blackbirds: Special Symposium of the Wildlife Society Ninth Annual Conference*, ed. G. M. Linz, pp. 103–108. National Wildlife Research Center, Fort Collins, CO.

Clotfelter, E. D., and K. Yasukawa. 1999. Impact of brood parasitism by brown-headed cowbirds on red-winged blackbird reproductive success. *Condor* 101:105–114.

Conover, M. R. 2002. *Resolving human-wildlife conflicts: The science of wildlife damage management.* CRC Press/Taylor & Francis, Boca Raton, FL.

Conover, M. R., and R. M. Vail. 2015. *Human diseases from wildlife.* CRC Press/Taylor & Francis, Boca Raton, FL.

Corn, J. L., E. J. B. Manning, S. Sreevatsan, and J. R. Fischer. 2005. Isolation of *Mycobacterium avium* subspecies *paratuberculosis* from free-ranging birds and mammals on livestock premises. *Applied and Environmental Microbiology* 71:6963–6967.

Crase, F. T., and R. W. DeHaven. 1975. *Selected bibliography on the food habits of North American blackbirds.* Special Scientific Report Wildlife 192. U.S. Department of Interior, U.S. Fish and Wildlife Service, Washington, DC.

Cummings, J. L., and M. L. Avery. 2003. An overview of current blackbird research in the southern rice growing region of the United States. *Wildlife Damage Management Conference* 10:237–243.

Cummings, J. L., J. L. Guarino, and C. E. Knittle. 1989. Chronology of blackbird damage to sunflowers. *Wildlife Society Bulletin* 17:50–52.

Cummings, J., S. Shwiff, and S. Tupper. 2005. Economic impacts of blackbird damage to the rice industry. *Wildlife Damage Management Conference* 11:317–322.

Decker, D. G., M. L. Avery, and M. O. Way. 1990. Reducing blackbird damage to newly planted rice with a nontoxic clay-based seed coating. *Vertebrate Pest Conference* 14:327–331.

DeGrazio, J. W., J. F. Besser, and J. L. Guarino. 1969. Winter distribution of as related to corn damage control in Brown County, South Dakota. *North America Wildlife Conference* 34:131–136.

Dolbeer, R. A. 1976. Reproductive rate and temporal spacing of nesting of red-winged blackbirds in upland habitat. *Auk* 93:343–355.

Dolbeer, R. A. 1978. Movement and migration patterns of red-winged blackbirds: A continental overview. *Bird-Banding* 49:17–34.

Dolbeer, R. A. 1980. *Blackbirds and corn in Ohio*. U.S. Department of the Interior Fish and Wildlife Service Resource Publication 136, Washington, DC.

Dolbeer, R. A. 1981. Cost-benefit determination of blackbird damage control for cornfields. *Wildlife Society Bulletin* 9:44–51.

Dolbeer, R. A. 1982. Migration patterns for sex and age classes of blackbirds and starlings. *Journal of Field Ornithology* 53:28–46.

Dolbeer, R. A. 1990. Ornithology and integrated pest management: Red-winged blackbirds *Agelaius phoeniceus* and corn. *Ibis* 132:309–322.

Dolbeer, R. A., and G. M. Linz. 2016. *Blackbirds*. Wildlife Damage Management Technical Series, U.S. Department of Agriculture, Animal & Plant Health Inspection Service, Wildlife Services, Washington, DC.

Dolbeer, R. A., D. F. Mott, and J. L. Belant. 1997. Blackbirds and starlings killed at winter roosts from PA-14 applications, 1974–1992: Implication for regional population management. *Eastern Wildlife Damage Control Conference* 7:77–86.

Dolbeer, R. A., P. P. Woronecki, and R. A. Stehn. 1986. Resistance of sweet corn to damage by blackbirds and starlings. *Journal of the American Society for Horticultural Science* 111:306–311.

Dolbeer, R. A., P. P. Woronecki, A. R. Stickley, Jr., and S. B. White. 1978. Agricultural impact of a winter population of blackbirds and starlings. *Wilson Journal of Ornithology* 90:31–44.

Dufour, K. W., and P. J. Weatherhead. 1998. Reproductive consequences of bilateral asymmetry for individual male red-winged blackbird. *Behavioral Ecology* 9:232–242.

Eckert, C. G., and P. J. Weatherhead. 1987. Competition for territories in red-winged blackbirds: Is resource-holding potential realized? *Behavioral Ecology and Sociobiology* 20:369–375.

Forcey, G. M., W. E. Thogmartin, G. M. Linz, W. J. Bleier, and P. C. McKann. 2011. Land use and climate influences on waterbirds in the Prairie Potholes. *Journal of Biogeography* 38:1694–1707.

Forcey, G. M., W. E. Thogmartin, G. M. Linz, P. C. McKann, and S. M. Crimmins. 2015. Spatially explicit modeling of blackbird abundance in the Prairie Pothole Region. *Journal of Wildlife Management* 79:1022–1033.

Gartshore, R. G., R. T. Brooks, J. D. Sommers, and F. F. Gilbert. 1982. Feeding ecology of the red-winged blackbird in field corn in Ontario. *Journal of Wildlife Management* 46:438–452.

Glahn, J. F., and E. A. Wilson. 1992. Effectiveness of DRC-1339 baiting for reducing blackbird damage to sprouting rice. *Eastern Wildlife Damage Control Conference* 5:117–123.

Hagy, H. M., G. M. Linz, and W. J. Bleier. 2008. Optimizing the use of decoy plots for blackbird control in commercial sunflower. *Crop Protection* 27:1442–1447.

Handegard, L. L. 1988. Using aircraft to control blackbird/sunflower depredations. *Vertebrate Pest Conference* 13:293–294.

Hauan, H. 2015. *Zizania aquatica L., Wild Rice; An evaluation of cultivation, domestication, and production for use in the United States*. University of Minnesota Digital Conservancy, Minneapolis, MN.

Heisterberg, J. F, A. R. Stickley, K. M. Garner, and P. D. Foster, Jr. 1987. Controlling blackbirds and starlings at winter roosts using PA-14. *Eastern Wildlife Control Conference* 3:177–183.

Hintz, J. V., and M. I. Dyer. 1970. Daily rhythm and season change in the summer diet of adult red-winged blackbirds. *Journal of Wildlife Management* 34:789–799.

Holcomb, L. C., and G. Twiest. 1968. Red-winged blackbird nestling growth compared to adult size and differential development of structures. *Ohio Journal of Science* 68:277–284.

Homan, H. J., and G. M. Linz. 2005. Aerial mass color- marking of blackbird roosts. *Northern Great Plains Workshop*. South Dakota State University, Brookings, SD.

Homan, H. J., G. M. Linz, R. M. Engeman, and L. B. Penry. 2004. Spring dispersal patterns of red-winged blackbirds, *Agelaius phoeniceus*, staging in eastern South Dakota. *Canadian Field-Naturalist* 118:201–209.

Hornick, M. 2016. Organic sales reach milestones. *The Packer/Farm Journal Media*, Lenexa, KS. http://www.thepacker.com/news/organic-sales-reach-milestones (accessed July 20, 2016).

Hothem, R. L., R. W. DeHaven, and S. D. Fairaizl. 1988. *Bird damage to sunflower in North Dakota, South Dakota, and Minnesota, 1979–1981*. Fish and Wildlife Technical Report 15, U.S. Fish and Wildlife Service, Washington, DC.

Hulke, B. S., and L. W. Kleingartner. 2014. Sunflower. In *Yield gains in major U.S. field crops*, eds. S. Smith, B. Diers, J. Specht, and B. Carver, pp. 433–457. Soil Science Society of America Special Publication 33. American Society of Agronomy, Crop Science Society of America, and Soil Science Society of America, Madison, WI.

Igl, L. D., and D. H. Johnson. 1997. Changes in breeding bird populations in North Dakota: 1967 to 1992–93. *Auk* 114:74–92.

Jaramillo, A., and P. Burke. 1999. *New world blackbirds: The Icterids*. Princeton University Press, Princeton, NJ.

Johnson, D. H., and L. D. Igl. 1995. Contributions of the conservation reserve program to populations of breeding birds in North Dakota. *Wilson Bulletin* 107:709–718.

Kalmbach, E. R. 1937. *Blackbirds of the Gulf Coast in relation to the rice crop with notes on their food habits and life history*. U.S. Bureau of Sport Fisheries and Wildlife, Denver Wildlife Research Center, Denver, CO.

Kandel, H., and G. M. Linz. 2016. Bird damage is an important economic agronomic factor influencing sunflower production. *Wildlife Damage Management Conference* 16:75–82.

Kirk, D. A., M. D. Evenden, and P. Mineau. 1996. Past and current attempts to evaluate the role of birds as predators of insect pests in temperate agriculture. In *Current Ornithology. Vol 13*, eds. V. Nolan, Jr. and E. D. Ketterson, pp. 175–269. Plenum Press, New York.

Klosterman, M. E., G. M. Linz, A. A. Slowik, and H. J. Homan. 2013. Comparisons between blackbird damage to corn and sunflower in North Dakota. *Crop Protection* 53:1–5.

Knight, R. L., and S. A. Temple. 1986. Why does intensity of avian defense increase during the nesting cycle? *Auk* 103:318–327.

Knittle, C. E., and B. E. Johns. 1986. Field-spray comparison of two particle-marker formulations used to mass-mark red-winged blackbirds. *Bird Damage Research Report 371*. U.S. Fish and Wildlife Service Denver Wildlife Research Center, Denver, CO.

Knittle, C. E., G. M. Linz, B. E. Johns, J. L. Cummings, J. E. Davis, Jr., and M. M. Jaeger. 1987. Dispersal of male red-winged blackbirds from two spring roosts in central North America. *Journal of Field Ornithology* 58:490–498.

Lanciotti, R. S., J. T. Roehrig, V. Deubel, J. Smith, M. Parker, K. Steele, B. Crise, et al. 1999. Origin of the West Nile virus responsible for an outbreak of encephalitis in the northeastern United States. *Science* 286:2333–2337.

Lorenzana, J. C., and S. G. Sealy. 1999. A meta-analysis of the impact of parasitism by the brown-headed cowbird on its host. *Studies in Avian Biology* 18:241–253.

Linz, G. M. 1982. Molt, food habits, and brown-headed cowbird parasitism of red-winged blackbirds in Cass County, North Dakota. PhD Dissertation. North Dakota State University, Fargo, ND.

Linz, G. M., S. B. Bolin, and J. F. Cassel. 1983. Postnuptial and postjuvenal molts of red-winged blackbirds in Cass County, North Dakota. *Auk* 100:206–209.

Linz, G. M., E. H. Bucher, S. B. Canavelli, E. Rodriguez, and M. L. Avery. 2015. Limitations of population suppression for protecting crops from bird depredation: A review. *Crop Protection* 76:46–52.

Linz, G. M., and J. J. Hanzel. 2015. Sunflower and bird pests. In *Sunflower: Chemistry, production, processing, and utilization*, eds. E. M. Force, N. T. Dunford, and J. J. Salas, pp. 175–186. AOCS Press, Urbana, IL.

Linz, G. M., H. J. Homan, S. W. Werner, H. M. Hagy, and W. J. Bleier. 2011b. Assessment of bird management strategies to protect sunflower. *BioScience* 61:960–970.

Linz, G. M., C. E. Knittle, J. L. Cummings, J. E. Davis, Jr., D. L. Otis, and D. L. Bergman. 1991. Using aerial marking for assessing population dynamics of late summer roosting red-winged blackbirds. *Prairie Naturalist* 23:117–126.

Linz, G. M., G. A. Knutsen, H. J. Homan, and W. J. Bleier. 2003. Baiting blackbirds (Icteridae) in stubble grain fields during spring migration in South Dakota. *Crop Protection* 22:261–264.

Linz, G. M., B. D. Peer, H. J. Homan, R. L. Wimberly, D.L. Bergman, and W. J. Bleier. 2002. Has an integrated pest management approach reduced blackbird damage to sunflower? In *Human conflicts with wildlife: Economic considerations: Third NWRC Special Symposium,* eds. L. Clark, J. Hone, J. A. Shivik, R. A. Watkins, K. C. Vercauteren, and J. K. Yoder, pp. 132–137. National Wildlife Research Center, Fort Collins, CO.

Linz, G. M., R. S. Sawin, and M. W. Lutman. 2014. The influence of breeding experience on nest success in red-winged blackbird. *Western North American Naturalist* 74:123–129.

Linz, G. M., R. S. Sawin, M. W. Lutman, and W. J. Bleier. 2011a. Modeling parental provisioning by red-winged blackbirds in North Dakota. *Prairie Naturalist* 43:92–99.

Linz, G. M., D. L. Vakoch, J. F. Cassel, and R. B. Carlson. 1984. Food of red-winged blackbirds (*Agelaius phoeniceus*) in sunflower fields and corn fields. *Canadian Field-Naturalist* 98:38–44.

Marcum, D. B., and W. P. Gorenzel. 1994. Grower practices for blackbird control in wild rice in California. *Vertebrate Pest Conference* 16:243–249.

McLean, R. G., and S. R. Ubico. 2007. Arboviruses in birds. In *Infectious diseases of wild birds,* eds. N. J. Thomas, D. B. Hunter, and C. T. Atkinson, pp. 17–62. Blackwell/Wiley, Hoboken, NJ.

McNicol, D. K., R. J. Robertson, and P. J. Weatherhead. 1982. Seasonal, habitat, and sex-specific food habits of red-winged blackbirds: Implications for agriculture. *Canadian Journal of Zoology* 60:3282–3289.

Meanley, B. 1961. Late-summer food of red-winged blackbirds in a fresh tidal-river marsh. *Wilson Bulletin* 73:36–40.

Meanley, B. 1964. Origin, structure, molt, and dispersal of a late summer red-winged blackbird population. *Bird-Banding* 35:32–38.

Meanley, B. 1971. *Blackbirds and the southern rice crop.* U.S. Department of Interior, U.S. Fish and Wildlife Service, Resource Publication 100. http://pubs.usgs.gov/rp/100/report.pdf (accessed June 28, 2016).

Meanley, B., and G. M. Bond. 1970. Molts and plumages of the red-winged blackbird with particular reference to fall migration. *Bird-Banding* 41:22–27.

Meanley, B., and C. W. Royall, Jr. 1976. The 1974–75 winter roost survey for blackbird and starlings. *Bird Control Seminar* 7:39–40.

Mott, D. F. 1984. Research on winter roosting blackbirds and starlings in the southeastern United States. *Vertebrate Pest Conference* 11:183–187.

Mott, D. F., R. R. West, J. W. DeGrazio, and J. L. Guarino. 1972. Foods of the red-winged blackbird in Brown County, South Dakota. *Journal of Wildlife Management* 36:983–987.

Moulton, L. L., G. M. Linz, and W. J. Bleier. 2013. Responses of territorial and floater male red-winged blackbirds to models of receptive females. *Journal of Field Ornithology* 84:160–170.

Murray, L. D., and L. B. Best. 2003. Short-term bird response to harvesting switchgrass for biomass in Iowa. *Journal of Wildlife Management* 67:611–621.

National Cattleman's Beef Association. 2004. *A basic look at E. coli 0157.* National Cattlemen's Beef Association, Centennial, CO.

Neff, J. A., and B. Meanley. 1957. *Blackbirds and the Arkansas rice crop, Bulletin 584.* Agricultural Experiment Station, University of Arkansas, Fayetteville, AR.

Nelms, C. O., W. J. Bleier, D. L. Otis, and G. M. Linz. 1994. Population estimates of breeding blackbirds in North Dakota, 1967, 1981–82, and 1990. *American Midland Naturalist* 132:256–263.

Nelms, C. O., D. L. Otis, G. M. Linz, and W. J. Bleier. 1999. Cluster sampling to estimate breeding blackbird populations in North Dakota. *Wildlife Society Bulletin* 27:931–937.

Nero, R. W. 1984. *Redwings.* Smithsonian Institution Press, Washington, DC.

Nielsen, (B.) R. L. 2009. Corn ear damage caused by bird feeding. *Pest and Crop Newsletter.* Purdue Cooperative Extension Service, Purdue University, Purdue. https://extension.entm.purdue.edu/pestcrop/2009/issue24 (accessed February 5, 2016).

Nielsen, R. L. 2013. Grain fill stages in corn. *Corny News Network,* Purdue University, West Lafayette, IN. https://www.agry.purdue.edu/ext/corn/news/timeless/grainfill.html (accessed June 23, 2016).

Okurut-Akol, F. H., R. A. Dolbeer, and P. P. Woronecki. 1990. Red-winged blackbird and starling feeding responses on corn earworm-infested corn. *Vertebrate Pest Conference* 14:296–301.

Orians, G. H., and L. D. Beletsky. 1989. Red-winged blackbird. In *Lifetime reproduction in birds*, ed. I. Newton, pp. 183–197. Academic Press, New York.

Orians, G. H., and M. F. Willson. 1964. Interspecific territories of birds. *Ecology* 45:736–745.

Ortega, C. P., and A. Cruz. 1988. Mechanisms of egg acceptance by marsh-dwelling blackbirds. *Condor* 90:349–358.

Otis, D. L., and C. M. Kilburn. 1988. *Influence of environmental factors on blackbird damage to sunflower.* U. S. Fish and Wildlife Service, Fish and Wildlife Technical Report 16. Washington, DC.

Otis, D. L., C. E. Knittle, and G. M. Linz. 1986. A method for estimating turnover in spring blackbird roosts. *Journal of Wildlife Management* 50:567–571.

Ott, S. L., S. J. Wells, and B. A. Wagner. 1999. Herd-level economic losses associated with Johne's disease on U.S. dairy operations. *Preventive Veterinary Medicine* 40:179–192.

Ralston, S. T., G. M. Linz, W. J. Bleier, and H. J. Homan. 2007. Cattail distribution and abundance in North Dakota. *Journal of Aquatic Plant Management* 45:21–24.

Rappole, J. H. 2013. *The avian migrant: The biology of bird migration.* Columbia University Press, New York.

Rosenberg, K. V., J. A. Kennedy, R. Dettmers, P. J. Blancher, G.S. Butcher, W.C. Hunter, D. Mehlman, et al. 2016. *Partners in Flight Landbird Conservation Plan: 2016 Revision for Canada and Continental United States.* Partners in Flight Science Committee. http://www.partnersinflight.org/ (accessed September 25, 2016).

Patterson, C. B. 1991. Relative parental investment in the red-winged blackbird. *Journal of Field Ornithology* 62:1–18.

Payne, R. B. 1969. *Breeding seasons and reproductive physiology of tricolored blackbirds and red-winged blackbirds.* Publications in Zoology 90. University of California Press, Oakland, CA.

Peer, B. D., H. J. Homan, G. M. Linz, and W. J. Bleier. 2003. Impact of blackbird damage to sunflower: Bioenergetic and economic models. *Ecological Applications* 13:248–256.

Picman, J. 1987. Territory establishment, size, and tenacity by male red-winged blackbirds. *Auk* 104:405–412.

Picman, J., M. L. Milks, and M. Leptich. 1993. Patterns of predation on passerine nests in marshes: Effects of water depth and distance from edge. *Auk* 110:89–94.

Sauer, J. R., D. K. Niven, J. E. Hines, D. J. Ziolkowski, Jr, K. L. Pardieck, J. E. Fallon, and W. A. Link. 2017. The North American Breeding Bird Survey, results and analysis 1966–2015. Version 2.07.2017 USGS Patuxent Wildlife Research Center, Laurel, MD. https://www.mbr-pwrc.usgs.gov/bbs/bbs.html (accessed April 30, 2017).

Sawin, R. S., G. M. Linz, R. L. Wimberly, M. W. Lutman, and W. J. Bleier. 2003a. Estimating the number of nonbreeding male red-winged blackbirds in central North Dakota. In *Management of North American Blackbirds: Special Symposium of the Wildlife Society Ninth Annual Conference,* ed. G. M. Linz, pp. 97–102. National Wildlife Research Center, Fort Collins, CO.

Sawin, R. S., M. W. Lutman, G. M. Linz, and W. J. Bleier. 2003b. Predators on red-winged blackbird nests in eastern North Dakota. *Journal of Field Ornithology* 74:288–292.

Scharf, W. C., J. Kren, P. A. Johnsgard, and L. R. Brown. 2008. Body weights and species distribution of birds in Nebraska's Central and Western Platte Valley. *Papers in Ornithology.* http://digitalcommons.unl.edu/biosciornithology/43/ (accessed June 16, 2016).

Searcy, W. A., and K. Yasukawa. 1995. *Polygyny and sexual selection in red-winged blackbirds.* Princeton University Press, Princeton, NJ.

Shutler, D., and P. J. Weatherhead. 1991. Owner and floater red-winged blackbirds: determinants of status. *Behavioral Ecology and Sociobiology* 28:235–241.

Shutler, D., and P. J. Weatherhead. 1992. Surplus territory contenders in male red-winged blackbirds: Where are the desperados? *Behavioral Ecology and Sociobiology* 31:97–106.

Slate, D., R. Owens, G. Connolly, and G. Simmons. 1992. Decision making for wildlife damage management. *North American Wildlife Natural Resources Conference* 57:52–62.

Smith, L. B., and R. D. Bird. 1964. Autumn flocking habits of the red-winged blackbird in southern Manitoba. *Canadian Field-Naturalist* 83:40–47.

Stanton, J. C., B. X. Semmens, P. C. McKann, T. Will, and W. E. Thogmartin. 2016. Flexible risk metrics for identifying and monitoring conservation-priority species. *Ecological Indicators* 61:683–692.

Stehn, R. A. 1989. *Population ecology and management strategies for red-winged blackbirds.* Bird Section Research Report No. 432. U.S. Fish and Wildlife Service, Denver Wildlife Research Center, Denver, CO.

Stewart, R. E., and H. A. Kantrud. 1972. Population estimates of breeding birds in North Dakota. *Auk* 89:766–788.

Stickley, A. R., Jr., and R. J. Weeks. 1985. Histoplasmosis and its impact on blackbird/starling roost management. *Eastern Wildlife Damage Control Conference* 2:163–171.

Stockdale, T. M. 1959. Food habits and related activities of red-winged blackbirds (*Agelaius phoeniceus*) of north central Ohio. MS Thesis. The Ohio State University, Columbus, OH.

Sullivan, H., G. M. Linz, L. Clark, and M. Salman. 2006. West Nile virus antibody prevalence in red-winged blackbirds (*Agelaius phoeniceus*) from North Dakota (2003–2004). *Vector-Borne Zoonotic Disease* 6:305–309.

Swirski, A. L., D. L. Pearl, M. L. Williams, H. J. Homan, G. M. Linz, and N. Cernicchiaro. 2014. Spatial epidemiology of *Escherichia coli* O157:H7 in dairy cattle in relation to night roosts of *Sturnus vulgaris* (European starling) in Ohio, USA (2007–2009). *Zoonoses and Public Health* 61:427–435.

Twedt, D. J., W. J. Bleier, and G. M. Linz. 1991. Geographic and temporal variation in the diet of yellow-headed blackbirds. *Condor* 93:975–986.

Twedt, D. J., and R. D. Crawford. 1995. Yellow-headed Blackbird (*Xanthocephalus xanthocephalus*). No. 192. In *The Birds of North America*, ed. A. Poole. Cornell Laboratory of Ornithology, Ithaca, NY.

Tyler, B. M. J., and L. W. Kannenberg. 1980. Blackbird damage to ripening field corn in Ontario. *Canadian Journal of Zoology* 58:469–472.

U.S. Department of Agriculture. 1999. *North Dakota agricultural statistics 1999*. North Dakota Agricultural Statistical Service, Fargo, ND.

U.S. Department of Agriculture. 2012. *Sodium lauryl sulfate: European starling and blackbird wetting agent*. Wildlife Services Tech Note. https://www.aphis.usda.gov/publications/wildlife_damage/content/printable_version/WS_tech_note_sodium.pdf (accessed July 29, 2016).

U.S. Department of Agriculture. 2015. *Environmental Assessment: Managing blackbird damage to sprouting rice in southwestern Louisiana*. U.S. Department of Agriculture Wildlife Services, Washington, DC. https://www.aphis.usda.gov/regulations/pdfs/nepa/LA-Blackbird%20EA%20FINAL.pdf (accessed June 16, 2016).

U.S. Department of Agriculture. 2016. *Quick Stats 2.0*. U.S. Department of Agriculture, National Agricultural Statistics Service, Washington, DC. https://quickstats.nass.usda.gov/ (accessed September 25, 2016).

Weatherhead, P. J. 1989. Sex ratios, host-specific reproductive success, and impact of brown-headed cowbirds. *Auk* 106:358–366.

Weatherhead, P. J. 1990. Nest defense as sharable parental care in red-winged blackbird. *Animal Behaviour* 39:1173–1178.

Weatherhead, P. J. 2005. Effects of climate variation on timing of nesting, reproductive success, and offspring sex ratios of red-winged blackbirds. *Oecologia* 144:168–175.

Weatherhead, P. J., S. Tinker, and H. Greenwood. 1982. Indirect assessment of avian damage to agriculture. *Journal of Applied Ecology* 19:773–782.

White, S. B., R. A. Dolbeer, and T. A. Bookhout. 1985. Ecology, bioenergetics, and agricultural impacts of a winter-roosting population of blackbirds and starlings. *Wildlife Monographs* 93:1–42.

Wilson, E. A. 1985. Blackbird depredation on rice in southwestern Louisiana. MS Thesis. Louisiana State University, Baton Rouge, LA.

Wilson, E. A., E. A. LeBoeuf, K. M. Weaver, and D. J. LeBlanc. 1989. Delayed seeding for reducing blackbird damage to sprouting rice in southwestern Louisiana. *Wildlife Society Bulletin* 17:165–171.

Wywialowski, A. P. 1996. Field corn lost to wildlife in 1993. *Wildlife Society Bulletin* 24:264–271.

Yasukawa, K., R. L. Knight, and S. K. Skagen. 1987. Is courtship intensity a signal of male parental care in red-winged blackbird (*Agelaius phoeniceus*)? *Auk* 104:628–634.

Yasukawa, K., J. L. McClure, R. A. Boley, and J. Zanocco. 1990. Provisioning of nestlings by male and female red-winged blackbirds, *Agelaius phoeniceus*. *Animal Behavior* 40:153–166.

Yasukawa, K., and W. A. Searcy. 1995. Red-winged blackbird (*Agelaius phoeniceus*). No. 184. In *The Birds of North America*, ed. P. G. Rodewald. Cornell Lab of Ornithology, Ithaca, NY. https://birdsna.org/Species-Account/bna/species/rewbla (accessed September 25, 2016).

Ziolkowski, D., Jr., K. Pardieck, and J. R. Sauer. 2010. On the road again for a bird survey that counts. *Birding* 42:32–40.

CHAPTER 3

Ecology of Yellow-Headed Blackbirds

Daniel J. Twedt
Patuxent Wildlife Research Center
Memphis, Tennessee

CONTENTS

3.1	Description	44
	3.1.1 Definitive Basic Plumage	44
	3.1.2 Formative Plumage	46
	3.1.3 Juvenile Plumage	46
	3.1.4 Size	47
3.2	Distribution	47
	3.2.1 Breeding	47
	3.2.2 Habitat	48
	3.2.3 Winter	48
3.3	Life History	48
	3.3.1 Nests	49
	3.3.2 Eggs	50
	3.3.3 Incubation	51
	3.3.4 Nestlings and Fledglings	51
	3.3.5 Reproductive Success	52
	3.3.6 Survival	52
	3.3.7 Second-Year Birds	53
	3.3.8 Diet	53
	3.3.9 Feather Molt	54
3.4	Behavior	54
	3.4.1 Vocalizations	54
	3.4.2 Movements	55
	3.4.3 Displays	55
	3.4.4 Socialization	56
	3.4.5 Migration	56
3.5	Populations	57
3.6	Agricultural Damages	58
	References	59

Figure 3.1 Adult male yellow-headed blackbird. (Courtesy of Polly Wren Neldner.)

The yellow-headed blackbird (*Xanthocephalus xanthocephalus*) is a passerine species in the family Icteridae. The species names, both common and scientific, are derived from the bright saffron–yellow head of the male: *xantho* from the Greek word *xanthos*, meaning "yellow," and *cephalus* from the Greek *kephalos*, referring to the head. The species, however, is sexually dimorphic, males being larger and possessing more vibrant plumage than the drabber, more diminutive females. The brilliant yellow head, neck, and breast of the adult male shown in Figure 3.1 is indicative of the species, which otherwise is predominately black except for white wing patches. Females and young, hatching-year males have more subdued dull black and brown body feathers with a yellowish breast but lack the prominent yellow head.

Found throughout nonforested regions of western North America, this migratory species nests in emergent vegetation of deep-water wetlands. These social blackbirds are polygynous and typically nest within grouped territories. The predominately insectivorous diet of breeding birds and their nestlings shifts postbreeding to a diet of mostly seeds. Postbreeding flocks of yellow-headed blackbirds forage in uplands (e.g., grain fields) but return to roost within emergent wetlands. Flocking is sustained during migration to the southwestern and south-central United States and Mexico and continues through winter.

Thomas Say (Swenk 1933) and John Richardson (1831) collected the species in 1820, but Charles Lucien Bonaparte (1825) provided the first detailed description. The first reference to *X. xanthocephalus* was by David Starr Jordan (1884, page 92). George Ammann (1938) and R.W. Fautin (1940, 1941a, 1941b) provided foundational accounts describing yellow-headed blackbirds during the breeding season.

3.1 DESCRIPTION

3.1.1 Definitive Basic Plumage

The definitive basic plumage (Howell and Pyle 2015) of male yellow-headed blackbirds, as shown in Figure 3.1, embodies the species' name. The male's forehead, occiput, nape, sides of neck, auricular, posterior malar, throat, and upper breast are yellow to orange-yellow. Their lores, eye rings, chin, and anterior malar region are dark black. Primary coverts and outer greater wing coverts are white, often tipped with dark brown or black (Oberholser 1974), and flash distinctively during flight. The remaining wing, body, and tail feathers are black, but their feather tips may grade to brown and raw sienna,

Figure 3.2 Female yellow-headed blackbird perched on bulrush (*Scirpus* spp.) stems. (Courtesy of Mike Wisnicki.)

Figure 3.3 Flock of male yellow-headed blackbirds in harvested grain field during fall/winter (note the dark feather tips of fresh basic plumage on the yellow feathers of the head and neck—these tips will wear off by spring, thereby unveiling their striking yellow heads). (Courtesy of Barry Zimmer.)

especially when fresh. Contrasting with their black body plumage is an anal circlet of orange-yellow feathers with whitish margins. The bill, gape, and legs are dark black in males and dull black in females. The iris is dark olive-brown to black. Female definitive basic plumage, as shown in Figure 3.2, is less evocative of the species name. Although similar to the formative plumage of young males and females, the definitive basic plumage of females is more brightly colored (Crawford and Hohman 1978). The throat, upper portion of breast, and anal circlet are orange-yellow. The supercilium and malar region are dull orange-yellow. There is extensive brown flecking in the auricular and submalar regions and a distinctive brownish submalar stripe. The chin and throat are cream without brown flecking. The crown, nape, back, and sides of neck are raw sienna to dark brown throughout the remaining body and flight feathers. Dark brown feathers on the anterior belly are streaked with white.

Definitive basic plumage is worn from acquisition in late summer to the following summer. Yellow-headed blackbirds do not have an alternate plumage (Ammann 1938). Even though Roberts (1936, 687) indicated that the winter dress displayed in Figure 3.3 is "brightened by a partial molt," it is likely this brightening is due to feather wear (Ammann 1938). The golden-brown feather tips on the head and nape of males, which are visible in Figure 3.3 (Jaramillo and Burke 1999, Plate 24), wear off during the winter to reveal a uniformly yellow head. In females, feather wear changes the auricular region from brown to yellow, flecked with brown.

3.1.2 Formative Plumage

The formative plumage (Howell et al. 2003), acquired when birds are fledglings, is similar in both sexes but more extensively yellow in males. The supercilium, forehead, chin, throat, upper portion of breast, anal circlet, auricular, and malar regions are orange-yellow with brown flecking in the supermalar and auricular regions. The crown, nape, back, and sides of neck are raw sienna grading to dark brown in the remaining body feathers. In most males, the feathers of the lower malar stripe, anterior neck, and posterior border of breast have orange-yellow bases beneath raw sienna or dark brown tips (Oberholser 1974). Dark clove-brown juvenile remiges and rectrices are retained in the formative plumage of both males and females.

The formative plumage of females consists of a cream to white chin and prominent brownish submalar stripe that may be flecked with white. The auricular regions of females are light buff-yellow to light brown. Their upper breast is dark buff-yellow. Formative plumage is worn throughout winter through the following breeding season, at which time the head, neck, and breast regions of females in formative plumage are paler than those of females in definitive basic plumage (Crawford and Hohman 1978).

3.1.3 Juvenile Plumage

The juvenile plumage of yellow-headed blackbirds shown in Figure 3.4 is initiated within the nest and completed soon after fledging. Males and females have similar juvenile plumage, with forehead, crown, nape, most of the malar region, and breast being buff flecked with brown. Head, chin, throat, belly, under-tail coverts, and legs are pale buff to cream, whereas the middle of the back is broadly pink-cinnamon or buff (Oberholser 1974). Remiges and rectrices are dark clove-brown with

Figure 3.4 Fledgling yellow-headed blackbird with buffy juvenile plumage on head and belly—darker clove-brown wings and tail will be retained through formative plumage until molted after their first breeding season. (Courtesy of Gary Kurtz.)

tertial feathers tipped dull cinnamon. The greater and median wing coverts are tipped dull white-cream to cinnamon, which results in two conspicuous wing bars (Ammann 1938; Oberholser 1974). In hatchlings, the bill is brown with the gape, legs, and feet reddish pink, whereas the legs and feet of fledglings are buff-tan.

3.1.4 Size

Male yellow-headed blackbirds are significantly larger than females, with mass dependent on age, sex, and breeding locale. Adult, after-second-year males typically weigh between 90 and 100 g, whereas adult females are between 50 and 60 g (Twedt 1990). Second-year birds have about 8% less mass, males being approximately 80 g and females 44 g (Twedt and Crawford 1995). The body mass of yellow-headed blackbirds decreases during breeding (0.20 g/d, $r = -0.44$; Searcy 1979) and increases postbreeding with the deposition of premigratory fat. A subtle geographic cline in yellow-headed blackbird size exists within the Great Plains (Twedt et al. 1994).

3.2 DISTRIBUTION

3.2.1 Breeding

The breeding range of yellow-headed blackbirds (Figure 3.5) is estimated at 4.7 million km^2, with 70% of this area within the United States (Partners in Flight Science Committee 2013). Within this range, their distribution is limited by the availability of suitable, deep-water, emergent wetland habitat. Breeding sites span from central British Columbia (east of coastal range), northern Alberta (including the Peace–Athabasca Delta), central Saskatchewan, southern Manitoba, and extreme southwestern Ontario through Minnesota and Wisconsin to northwestern Indiana, northern Illinois (Bohlen 1989; Ward 2005b), Iowa, northwestern Missouri (Robbins and Easterla 1992), Kansas (Thompson and Ely 1992), western Oklahoma (Shackford and Tyler 1987), the panhandle of Texas, central New Mexico, northern and western Arizona, California, and predominately east of the Cascade Mountains in Oregon and Washington.

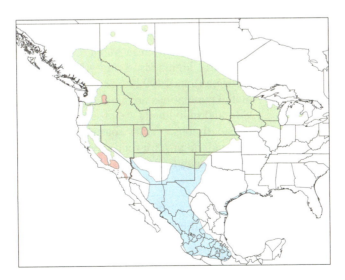

Figure 3.5 Range of the yellow-headed blackbird; green = breeding, blue = winter, and red = resident. (Courtesy of Daniel Twedt.)

At the periphery of their breeding range, small populations of yellow-headed blackbirds are often isolated. In Arizona, they breed in the Chino Valley, Mogollon Plateau, along the Gila River and Picacho Reservoir (Monson and Phillips 1981), and along the Colorado River, including extreme northeastern Baja California (Howell and Webb 1995). In California, they nest along the Salton Sea, Central Valley, Klamath Basin, Modoc Plateau, Mono Basin, and Owens Valley (Small 1994; Jaramillo 2008). Isolated breeding locations occur near Vancouver, British Columbia, in coastal Washington, and in Fern Ridge Reservoir, Oregon (Gilligan et al. 1994). Isolated eastern breeding populations are in northwestern Ohio (Peterjohn 1989), southeastern Ontario (Cadman et al. 1987), northwestern Indiana (Keller et al. 1986), and eastern and central Michigan (Brewer et al. 1991). Formerly interconnected populations in Illinois, Wisconsin, and Iowa are increasingly isolated and threatened with extirpation (Ward 2005a).

3.2.2 Habitat

Yellow-headed blackbirds nest predominately in prairie wetlands but breeding colonies are also common in wetlands associated with quaking aspen (*Populus tremuloides*) parklands and mountain meadows. Isolated colonies are associated with wetlands and lakes in boreal forests and arid regions. Nests are attached to emergent vegetation within deep-water areas (typically 30–60 cm deep) of palustrine wetlands. Breeding yellow-headed blackbirds forage both within wetlands and the surrounding grasslands, croplands, or savanna.

Postbreeding, yellow-headed blackbirds generally forage in highly disturbed sites such as ripening, harvested, or plowed agricultural fields, but they can also be found in meadows, pastures, and farmyards. Foraging birds may occur in large flocks of up to several thousand birds. Flocks may be species and sex specific, or they may be mixed flocks comprised of several icterid species, typically red-winged blackbirds (*Agelaius phoeniceus*) and common grackles (*Quiscalus quiscala*). During the day, yellow-headed blackbirds loaf in wetland vegetation, shrubby vegetation, and in woodlots. Foraging birds may range over several kilometers before returning at dusk to roost in emergent wetland vegetation.

3.2.3 Winter

Yellow-headed blackbirds spend the winter months primarily in southern Arizona, New Mexico, and southwestern Texas through Mexico to Veracruz, Oaxaca, Guerrero, and Nayarit (Howell and Webb 1995). Their distribution in Mexico is primarily within disturbed fields of the central plateau's agricultural highlands. Some yellow-headed blackbirds winter within their breeding range in California (Small 1994). Generally, this species only winters in areas that lack severe cold, but isolated populations are regularly recorded during the winter in Utah and Washington (Orians 1980), and individuals are observed at other northern locations (eBird 2016). Small flocks are consistently observed in coastal regions of the Gulf of Mexico in Texas and Louisiana, and isolated individuals are routinely observed in other coastal areas of the Gulf of Mexico and the Atlantic Ocean (eBird 2016). Vagrants have been reported from Caribbean islands and Europe.

3.3 LIFE HISTORY

Yellow-headed blackbirds are polygynous, with up to eight females within each male territory, although not all territorial males attract females. Male–female pair bonds appear to be established upon arrival at breeding sites and are not maintained after the breeding season. The breeding site fidelity of males is moderate: 51% of adult males and 19% of second-year males

(Searcy 1979), although males changed territories in 43% of between-year opportunities (Beletsky and Orians 1994). Males exhibit greater site fidelity than females (Ward and Weatherhead 2005).

Males arrive before females at breeding locations and establish territories within emergent vegetation and on clumps of floating vegetation over deep-water areas of wetlands (Fautin 1940). Territories are less likely to be established on wetlands if trees or cliffs project >30° above the horizon (Orians 1980). Nests are attached to robust emergent vegetation over relatively deep water near-open areas or at vegetation edges (Orians and Willson 1964). Where red-winged blackbirds have established territories within the deep-water areas of wetlands, later-arriving yellow-headed blackbirds may displace them and usurp their territories (Orians and Willson 1964; Willson 1966). Territory size is variable and dictated primarily by habitat quality as determined by foraging opportunity. Territory area is commonly 120–900 m^2, but larger territories of >3000 m^2 occur. Conversely, in areas where adults forage primarily in the uplands, territory size may be <100 m^2 (Willson 1966; Orians 1980).

More aggressive males obtain larger territories (Lightbody and Weatherhead 1987b). After-second-year males are dominant over smaller second-year males, such that attempts by second-year males to hold territories are often thwarted. Dominance, however, is not related to the degree or hue of yellow coloration of males, as territorial males whose yellow heads were experimentally blackened were able to attract females to their territories, defend their territories, and in some cases usurp the territory of another "unblackened" male (Rohwer and Røskaft 1989). Once established, territories are vigorously defended against conspecific males. Males that lack territories will challenge settled males for several weeks after territory establishment.

Early in territory establishment, males may remain on territory only during the morning and evening, foraging in surrounding uplands during midday. Time spent on territory increases, and by the time females arrive males spend nearly all day on their territories (Rohwer and Røskaft 1989).

Upon arrival at the breeding locations, females join males in foraging flocks and move among established territories, with some females continuing to flock after others have initiated egg-laying (Fautin 1941a). During movements among territories, nest sites are selected by females. Females may settle randomly within suitable habitat on a territory (Lightbody and Weatherhead 1987a) but tend to select nest sites within marshes of intermediate width, with moderately dense vegetation and extensive channeling. Thus female density is likely related to the density of live and senescent vegetation, as sparse vegetation provides inadequate support for nests, whereas dense vegetation without open-water channels affords increased predator access (Orians and Wittenberger 1991).

On more productive wetlands, as denoted by higher rates of dragonfly emergence, males tend to gather more females into their harems, and females settle in higher densities. The number of females within a territory is correlated with amount of edge (Willson 1966). Territories on which females settle early in the breeding season tend to attract more females than do late-settled territories (Orians 1980). Mean distance among nests within territories ranged from 4 to 14 m (Ammann 1938). Females defend a small area surrounding their nest site, but they do not respect the territorial boundaries established by males. Within a harem, females do not appear to be cooperative nor are they competitive (Lightbody and Weatherhead 1987a).

3.3.1 Nests

Selection of a nesting site and nest construction, exclusively the bailiwick of the female, may begin as early as 3 days after arrival on at the breeding site. Nests are attached to robust dead, or less commonly live, emergent vegetation (Orians and Willson 1964), typically cattails (*Typha* spp.) or bulrush-tule (*Scirpus* spp.) but also spikerush (*Eleocharis* spp.), and reeds (*Phragmites* spp.). Nest sites in Manitoba had mean stem densities of 80–104 stems per m^2 and were located <6.5 m from open water (Lightbody and Weatherhead 1987b).

Nests are over relatively deep water near open areas or at vegetation edges: 50–110 cm in British Columbia (Willson 1966), 16–76 cm in Wisconsin (Minock and Watson 1983), and 22–61 cm in

Figure 3.6 Nest of yellow-headed blackbird affixed to cattails (*Typha* spp.) containing an egg and newly hatched chicks. (Courtesy of Andrew Sabai.)

Saskatchewan (Miller 1968). Nests are placed 15–53 cm above the water surface (Miller 1968). Shallow water level under nests may increase vulnerability to predators (Kapilow et al. 1980) or result in abandonment of nests (Bent 1958) or territories (Lederer et al. 1975). The probability of nest abandonment increases later in the nesting season (Ortega and Cruz 1991).

Nests, as depicted in Figure 3.6, are compact, rigid, open cups constructed solely by females from long strands of wet vegetation. Generally open from above, exposure from the sides and below is dependent upon the vegetation at the nest site. Nests anchored to bulrush may be very exposed, whereas nests constructed within dense cattails may be indiscernible. Regardless of exposure, at least some of the rim is free from vegetation, presumably to allow access for entry and maintenance of young. Nest construction typically spans 4 days but may be completed within 2 days or extended for 10 days (Ortega and Cruz 1991).

Constructed of woven and plaited strips, the nest is often of the same vegetation as the supporting emergent vegetation to which it is attached. The rim of the nest is smooth and horizontal. Ammann (1938) provides a detailed description of nest construction wherein the female initially weaves wet stands of vegetation around a few (four to five) upright stalks of dead vegetation using her bill. Vegetation strands are held in her bill and head movements used to weave strands around support stems. Loose ends are pulled over the rim and anchored into the interior of the nest. At the end of each building bout, females stamp their feet in rapid succession on the bottom and sides of nest. After the outer wall is completed, an inner cup of grasses or other plumose vegetation is added in arcs parallel to the circumference. Upon completion the nest is about the size of a soup bowl: outside diameter 13–14 cm; outside height 13–15 cm (maximum = 55 cm); inside diameter 6.5–7.5 cm; inside depth 6.4 cm (Ammann 1938).

3.3.2 Eggs

Eggs are sub-elliptical, grayish-white to pale greenish-white, profusely and evenly blotched and speckled with shades of brown, rufous, and pearl gray. Eggs have a length of about 26 mm, a breadth of 18 mm (Ortega and Cruz 1991), and a mass of 4.5 g. The eggs of second-year birds are smaller than those of after-second-year birds (Crawford 1977).

Egg-laying typically commences within 2 days after the nest is completed but may be delayed for up to 7 days. In Utah, 62% of females laid their first egg the day after their nest was complete, 24% on the second day after completion, 10% on the third, and 4% on the fourth (Fautin 1941a). Normally one egg is laid per day, often between dawn and 6:30 a.m., on successive days until a clutch of four is attained. Yellow-headed blackbirds are a determinant laying species, such that eggs removed from the nest are not replaced. Clutches exceeding four eggs may occur, but clutch size typically declines with nest initiation date (Ammann 1938; Willson 1966; Arnold 1992). Reduced clutch size may be due to nesting of younger birds (Crawford 1977) or re-nesting by females with failed nests (Fautin 1941a). Females typically do not raise second broods but may re-nest after nest failure or rarely a successful, early breeding female may re-nest (M. Ward, personal communication, April 1, 2016). Usually re-nest attempts have a two-egg clutch in a newly constructed nest (Orians 1980; Harms et al. 1991).

The date of egg-laying is related to arrival at breeding locations. Egg-laying in Utah ranged from May 7 through June 22 (Fautin 1941a). The mean date of first egg-laying in Washington was May 18 (SD = 7.1 d, n = 274; Orians 1980) and May 26 in Manitoba (SD = 5.6 d, n = 351; Arnold 1992). Although new clutches may commence through mid-June (Willson 1966), all clutches in Iowa initiated after June 5 belonged to second-year females (Crawford 1977).

The nests of yellow-headed blackbirds are rarely parasitized by brown-headed cowbirds (*Molothrus ater*; Harms et al. 1991), possibly due to their agonistic behavior toward cowbirds and because nest initiation by yellow-headed blackbirds occurs 1–2 weeks before peak egg-laying by brown-headed cowbirds in red-winged blackbird nests (Ortega and Cruz 1991). Nevertheless, artificially added egg-shaped and egg-sized objects are accepted (Lyon et al. 1992), and cross-fostered cowbird eggs were incubated and young reared by yellow-headed blackbirds (Ortega and Cruz 1988, 1991). Yet yellow-headed blackbirds rejected 33% of experimentally added red-winged blackbird eggs (Dufty 1994).

3.3.3 Incubation

Only females develop incubation patches and incubate eggs. Incubation generally begins when the second egg is laid, but individuals vary the onset of incubation. In Utah, 32% began incubating after the first egg was laid, 58% after the second egg, and 10% after the third egg. Incubation lasts 12–14 days, with longer incubation periods more common later in the breeding period (Fautin 1941a). During incubation, females spent 64% of time on nests—averaging 9 minutes (range 1–41) on and 5 minutes (range 1–18) away (Fautin 1941a).

Although males do not incubate, they are vigilant against territorial intruders. Circulating testosterone levels of adult males are elevated upon arrival on breeding grounds and remain high during the active breeding season. Males nesting at high densities have higher testosterone levels than males nesting at low densities, but no association was found between testosterone level and breeding success (Beletsky et al. 1990).Testosterone levels decline toward the end of the breeding season when males may also begin to feed young.

Most eggs hatch in the morning (Ammann 1938). Typically a crack develops near the egg's greatest diameter and is pushed outward and elongated until, over a period of approximately 30 minutes, the shell is broken into two pieces around its circumference. Repeated body movements and convulsions, separated by rest periods of circa 1 minute, are used to push the shell halves apart and liberate the young. Eggs deposited before the onset of incubation hatch synchronously, but eggs deposited after the onset of incubation hatch asynchronously.

3.3.4 Nestlings and Fledglings

The sex ratio at hatching is 1:1 (Willson 1966; Patterson and Emlen 1980; Richter 1983). After hatching, the altricial nestlings are brooded exclusively by females for up to 3 days. Chicks gain control over their body temperature at about 6 days old. Typically emerging at just over 3 grams, nestlings gain 3–5 g of mass per day (Fautin 1941b), such that at 10 days of age females are 30–40 g and males are 45–55 g (Willson 1966; Richter 1984; Ortega and Cruz 1992).

Upon hatching, their bodies are yellowish-pink to salmon-pink, with red to pink gapes. Their eyes are covered with skin until opening—3 days after hatching. Their bills are brownish, tarsus and toes are vinaceous, and claws are buffy. Patches of sparse, pinkish-buff neossoptiles are present at hatching on the head and along the spinal tract, with smaller patches on the wings and along the femoral and ventral tracts (Fautin 1941b). The color of the natal down is vinaceous-buff, but it darkens to brown when wet.

The prejuvenile molt begins about 2 days after hatching, as primaries, secondaries, tertiaries, and wing coverts emerge. Rectrices erupt at approximately 4 days old (Ammann 1938). At the time of fledging, most neossoptiles have broken off, although some may remain for several days after fledging. Plumage differences can be discerned between sexes, females having darker backs and buff wing bars, whereas males have lighter, tawny backs and whiter wing bars (Ortega and Cruz 1992). Prejuvenile molt is complete in 27 days, except for the primaries and rectrices, which are not fully grown in until the young are about 40 days old (Ammann 1938; Twedt 1990).

The young remain in nest for 9–14 days but leave the nest before they are fully able to fly. These "fledglings" continue to receive parental care for several days after departure from the nest. After leaving, the young do not return to the nest but hide among the dead vegetation near the surface of the water. They are adroit at hopping and moving about the wetland vegetation. Flight skills, however, are quickly developed such that within 4–5 days after departing the nest the young can make short flights.

The young beg for food soon after hatching. They are fed from the time they hatch until they develop flight skills. Food items are brought to the nest by adults in their bill (Orians 1980). Initially, only females feed the young in the nest, typically making 11–16 feeding trips to a nest each hour (Willson 1966) and delivering between 300 and 2,000 calories per hour from dawn to late afternoon (Orians 1980). During the first few days after hatching, food delivered to nests may be divided equally among nestlings but feeding of older nestlings is less egalitarian, as the oldest and heaviest young are fed first or exclusively (Fautin 1941b). Males may assist in feeding after young are about 4 days old but almost exclusively at the primary (i.e., first initiated) nest in their territory (Gori 1988b, 1990; Cash and Johnson 1990). Feeding by males does not affect the number of young fledged per nest (Lightbody and Weatherhead 1987b). Once capable of flight, young birds flock with adults to forage in uplands but return to the wetlands to roost.

3.3.5 Reproductive Success

Between 45% and 62% of nests successfully fledge young, with earlier initiated nests more successful than later initiated nests (Lightbody and Weatherhead 1987b). Nests constructed over deeper water, at depths >52 cm, are more successful than nests built over shallower water depths (Ortega 1991). Successful nests fledge an average of 1.5–2.2 young (Lightbody and Weatherhead 1987b). Nests with clutches of four eggs fledge more young than do nests with other clutch sizes.

3.3.6 Survival

Starvation associated with low delivery rates of prey results in nestling mortality (Willson 1966; Orians 1980). Nest mortality also results from flooded nests or increased predation correlated with low water levels (Kapilow et al. 1980). High winds or heavy rainfall may result in nestling mortality (Ammann 1938; Fautin 1941a; Twedt and Crawford 1995). Mortality among nestlings is usually in inverse order of hatching, with the youngest perishing first (Willson 1966; Richter 1982). Brooding of newly hatched nestlings and improved thermoregulation of older nestlings explicates increased mortality noted between 3 and 6 days of age (Fautin 1941b).

Eggs and young are preyed upon in the nest. Marsh wrens (*Cistothorus palustris*) puncture unguarded eggs (Picman 1988). Other avian predators of eggs and young include gulls (*Larus* spp.), magpies (*Pica* spp.), common grackle, American bittern (*Botaurus lentiginosus*), American coot (*Fulica americana*), and other Rallidae. Reptilian and mammalian predators of eggs and young include the following: bull snake (*Pituophis melanoleucus*), garter snakes (*Thamnophis* spp.), blue racer (*Coluber constrictor*), mink (*Mustela vison*), red fox (*Vulpes vulpes*), raccoon (*Procyon lotor*), deer mouse (*Peromyscus maniculatus*), and striped skunk (*Mephitis mephitis*).

In Washington, 6.6% (range = 3.5%–11.3%) of nests were destroyed during laying (Harms et al. 1991). Nest predation rate in Colorado, from laying to fledging, was 48.5% (Ortega and Cruz 1991). Fledglings and adults are preyed upon opportunistically, notably by great horned owl (*Bubo virginianus*) and barn owl (*Tyto alba*) while in roosts. Northern harriers (*Circus cyaneus*) harass breeding colonies and foraging flocks. Remains of yellow-headed blackbird were found in 26 of 346 food samples from nestling northern harriers and in 27 of 228 pellets from great horned owls.

The maximum age of a male, banded, yellow-headed blackbird was 9 years when encountered, but a captive male lived 16 years (Ammann 1938; Kennard 1975). Annual survival rates estimated from band encounter data were 58% and 75% for adult males and females and 45% and 41% for immature males and females, respectively (Bray et al. 1979). Survival rate–based annual return to the same wetland, assuming no dispersal, was 51% for adult males (Searcy 1979).

3.3.7 Second-Year Birds

During their first spring after hatching, upon return from their winter range, females build nests, breed, and rear young (Bent 1958). Second-year females, however, tend to nest later than do after-second-year females, by an average of 16 days (Crawford 1977). Conversely, second-year males are typically unable to establish territories due to harrying from older males (Ammann 1938).

3.3.8 Diet

Foods consumed by yellow-headed blackbirds vary with age, season, and sex. Nestlings and dependent young are fed invertebrates, particularly emergent aquatic insects. Emergent odonates, including damselflies (Zygoptera), predominate in the diet of nestlings, with the remainder composed of Coleoptera (Hydrophilidae, Dytiscidae, and Carabidae), Ephemeroptera, Trichoptera, Gastropoda, Hemiptera, Homoptera, Hydrachnidia, Diptera, Orthoptera, and Arachnida (Fautin 1941b; Willson 1966; Orians 1980; Fischer and Bolen 1981). Nestling starvation may occur when odonate emergence is low.

After fledging, invertebrate consumption declines and varies between sexes and among seasons. Females consume more insects during spring than do males, whereas males consume more cultivated grains. By fall, weed seeds dominate the female diet, but the bulk of the male diet is cultivated crops. Juvenile birds consume more insects and smaller weed seeds than do adults. The mean dry weight of esophageal contents increased from about 0.2 g in early July to >0.6 g by mid-August in males and from <0.2 g to about 0.4 g in females (Twedt et al. 1991). The increase in consumption coincides with the energetic demands of molt and premigratory fat deposition. Diet varies geographically with exploitation of the most available resources (Twedt et al. 1991). Dietary overlap exists with sympatric blackbird species (Homan et al. 1994).

The overlap of foraging area and breeding territory is determined by availability of resources. Where aquatic insects are abundant, foraging may be exclusively within territorial boundaries, mostly at the water surface during periods of odonate emergence (Orians 1980). Where aquatic insects are less abundant, territory size is reduced and foraging shifts to upland habitats.

Postbreeding birds forage predominately in agricultural fields, including wheat, oat, barley, milo, millet, sorghum, sunflower, and corn. Flocks also forage in plowed, bare, fallow, or grassy fields.

Unharvested grains are consumed by perching birds. Sunflower is hulled before consumption. Once a foraging site is established, flocks may return to the same location over several days. Large foraging flocks, often mixed with other icterid species, may exhibit "rolling" movements within agricultural fields while gleaning weed seeds and grains. This rolling appearance occurs when birds foraging at the rear of the flock fly over foraging birds and land at the vanguard of the flock.

3.3.9 Feather Molt

Yellow-headed blackbirds, like most North American blackbird species, exhibit a complex basic strategy of molt cycles (Howell et al. 2003). In this molt strategy, birds initiate their first complete molt, the prejuvenile molt, as nestlings. Juvenile plumage is acquired by a complete prejuvenile molt (Humphrey and Parkes 1959), which begins when the nestlings are ~2 days old and is completed in about 40 days when fledged birds are independent.

Molt complexity arises after fledging, from an inserted first-cycle molt (i.e., preformative) molt, which is not repeated in subsequent molt cycles. Yellow-headed blackbirds exhibit a partial or incomplete preformative molt that is characteristic of most North American birds (Pyle 1997), during which juvenile flight feathers (remiges and rectrices) are retained while molting head and body feathers to attain formative plumage (Howell et al. 2003). Preformative molt may begin as soon as 20 days after hatching, with nearly all individuals having completed preformative molt by early September (Ammann 1938; Twedt and Linz 2015).

Juvenile primaries, secondaries, rectrices, primary coverts, tertials, underwing coverts, and alulae are retained for the first year (Ammann 1938; Twedt and Crawford 1995) then replaced during prebasic molt (Howell et al. 2003). Definitive basic plumage is acquired by the prebasic molt, a complete molt in which all feathers are replaced, which commences at about 1 year of age and is repeated annually thereafter. Yellow-headed blackbirds molt primaries sequentially from P1 through P9, whereas secondaries are molted from the extremities, S1 and S8, toward the center (Twedt 1990). Similarly, specific replacement sequences have been noted for other feather tracts (Ammann 1938; Oberholser 1974; Twedt and Linz 2015). No alternate plumage occurs, so prealternate molt is absent (Ammann 1938).

3.4 BEHAVIOR

3.4.1 Vocalizations

Male yellow-headed blackbirds produce two distinct songs and six separate calls, whereas females sing only one song and emit four different calls. The male accenting song is musical, directed toward birds at long distances, and usually accompanied by a symmetrical song spread display (see Section 3.4.3). This song lasts approximately 1.5 seconds and consists of several fluid introductory notes that may or may not be followed by a highly variable trill. The male buzzing song is longer (circa 4 seconds), more nasal, directed toward birds near the singer, and usually accompanied by asymmetrical song spread (see Section 3.4.3). All males sing a similar buzzing song of introductory notes that are distinctly separated from the subsequent, very prolonged, trill, represented phonetically as *kuk—koh-koh-koh—waaaaaaaa*. Both the accenting song and the buzzing song are nearly always delivered from a perch, with all males singing both song types. Males sing most often on the breeding grounds during morning and evening with delivery rates of approximately one per minute, peaking during egg-laying and incubation. Buzzing songs were delivered 0.54 times per minute, whereas accenting songs were delivered 0.35 times per minute (Rohwer and Røskaft 1989).

Female chatter, although not musical, appears to be functionally equivalent to the song of males. Its delivery may or may not be accompanied by song spread display. This female "song," consisting of rapidly repeated, harsh, nasal, and raspy *cheee-cheee-cheee* notes, is used during aggressive encounters and by females departing nests (Orians and Christman 1968).

Newly hatched birds emit a weak, high-pitched food call, which becomes louder and deeper as nestlings age. As flight develops, food calls become more broken with distinct notes and gradually develop into check and chuck calls. The check (*tsheck*) call is a loud single note without defined harmonics that is most frequently used during the breeding season, during feeding, and during flight. Check calls are given several times per hour during the breeding season but may be given as often as 30 times per minute when threatened. During autumn, a softer chuck (*clerrk*) call is often emitted, probably functioning as a flock communication. The gradual development of flocking calls and the age-related improvement in song quality suggest that these are learned vocalizations.

A variety of other calls are emitted by yellow-headed blackbirds, depending on circumstances: Males emit a two-noted *chuck-uck* call, a growl, and a hawk alarm call, whereas the female repertoire includes a scream in response to predators or nestling disturbance (Nero 1963). Postbreeding, roosting congregations are highly vocal during evening arrival and before departure at dawn, although singing at this time is uncommon.

3.4.2 Movements

While foraging on ground or moving about wetland vegetation, yellow-headed blackbirds walk or occasionally hop short distances. Individuals will climb up or slide down vegetation to attain a perch. Their flight is slightly undulating, with feathers pressed against the body and tail held out behind. During the breeding season, adults forage in nearby uplands (<1 km), whereas postbreeding birds may range >10 km before returning to a wetland roost.

Territorial fights during which males peck and grasp each other's plumage and roll about on the ground may ensue from boundary displays, song spreads, or supplanting flights. During aerial fights at territorial boundaries, combatants fly up face to face and strike at each other with bill and feet, but these encounters seldom result in injury.

During the breeding season, females are weakly aggressive toward other females, displaying agitation and vocalization toward 8% of introduced mounts (Lightbody and Weatherhead 1987a). Even so, a female emitting an alarm call rouses the assistance of numerous other females; they fly to her aid and mobbing behavior ensues.

3.4.3 Displays

Males engage in 13 characterized displays, whereas the female repertoire has only seven displays (Orians and Christman 1968). Most commonly, perched birds have their feathers relaxed, tail tilted slightly below the body axis with wings folded above the tail, neck retracted, and bill horizontal. In a sleeked posture, males have their feathers pressed against their body, neck extended, and legs crouched. In an alert posture, feathers are less strongly sleeked with neck extended and tail flicked upward. The check call often accompanies an alert posture. A head forward display with head bent forward is common during the nonbreeding season and territorial disputes.

Males display two primary types of song spreads, each accompanied by a distinct song type. In the symmetrical song spread, accompanied by the accenting song (see above), the wings are spread to expose the white patches, the tail is spread and lowered, and the head is directed upward at 30°–45°. Their wings may be arched in a V-shape over their back. The symmetrical song spread is typically produced in response to other males flying over their territory or the arrival of a new female, but it also precedes display flights and nest-site displays.

Asymmetrical song spread, displayed as the buzzing song is delivered, has the head held up and turned sharply to the left, with the bill pointing higher than 45° and wings slightly spread

(Twedt and Crawford 1995). Asymmetrical song spread is used primarily in territorial disputes. Female song spread is similar to the male's asymmetrical song spread but more subdued.

Territorial males conduct nearly 12 display flights per hour but expel less than one intruder per hour (Rohwer and Røskaft 1989). Males have two flight displays: a flight-stall display (Orians and Christman 1968) and a bill-up flight (Nero 1963). Other postures include a bill-down posture, a bill-up posture, and a crouch display (Nero 1963). Females use a flight display that is reminiscent of the bill-up flight of males, wherein they point their bill upwards, beat their wings rapidly but shallowly, and dangle their feet.

Sexual chasing is common during the breeding season. Males overtake rapidly flying females in midflight and use their bills to grasp the female's rump. Following a female's flight display, females assume a precopulatory posture in which their legs are flexed with body held horizontal or slightly tipped forward, with raised bill and closed tail. During copulation males balance on the female's back using rapid wing movements. Males may mount repeatedly before copulation is successful, and multiple successive copulations may occur. Extra-pair copulations, generally with intruding males from nearby territories, accounted for half of observed copulations in Manitoba (Lightbody and Weatherhead 1987b).

3.4.4 Socialization

Breeding is facilitated by social interactions, with isolated breeding territories being rare. Breeding colonies act as information exchange centers to enhance foraging; members of the colony locate productive foraging areas by following successful foragers, especially neighbors (Gori 1988a). Postbreeding birds are highly social, with birds foraging in close-knit flocks and roosting proximate to each other.

3.4.5 Migration

Annual migration is between their breeding range in the northern Great Plains and western United States, as identified in Figure 3.5, and their winter range in the southwestern United States and Mexico. As diurnal migrants, they move in long, irregular, and loose flocks that may be composed nearly exclusively of adult males or a mix of females and immature males (Crase and DeHaven 1972). They also congregate in mixed-species flocks with red-winged blackbirds, common grackles, Brewer's blackbirds (*Euphagus cyanocephalus*), and brown-headed cowbirds. Flocks of 1,000–2,000 birds have been observed during spring, whereas in fall flocks of >5,000 migrants roost in wetlands (Crase and DeHaven 1972).

Fall migrants arrive in Arizona as early as July (Phillips et al. 1964), but most yellow-headed blackbirds appear to depart their breeding grounds from late August through mid-September (Fautin 1941b; Twedt and Linz 2015). Encounters with banded birds indicate that yellow-headed blackbirds breeding on the Great Plains migrate from northwest to southeast across the prairie provinces of Canada, then essentially north–south across the central United States (Royall et al. 1971).

In spring, adult (after-second-year) males are the first to arrive on breeding grounds, with adult females arriving 7–14 days after these males. Second-year birds arrive after adults, with males again arriving about 7 days before females. Arrival dates vary regionally, with adult males arriving on breeding grounds from mid-March through early May (Miller 1968; Crawford 1978).

Throughout eastern North America, yellow-headed blackbirds are rare but regular fall migrants and increasingly they are also detected during spring (eBird 2016). Most encounters during fall are from coastal locations but encounters during spring have no coastal affinity.

Yellow-headed blackbirds depart breeding grounds earlier and return to breeding grounds later than sympatric red-winged blackbirds or common grackles. Earlier completion of molt may account for their early fall departure (Twedt and Linz 2015). However, a lower metabolic rate and associated cold intolerance has been hypothesized to account for their later arrival on breeding grounds (Twedt 1990). Their winter distribution lends credence to thermodynamic sensitivity, as males comprise most of the population on the northern wintering range in Arizona, whereas smaller bodied females compose about 80% of populations on their southern wintering range in Jalisco and Nayarit, Mexico (Phillips et al. 1964).

Males have elevated testosterone levels upon arrival on the breeding grounds (Beletsky et al. 1990). If increased testosterone levels stimulate spring migration, differences in endocrine production, relative to other blackbird species, may account for temporal differences in migration.

3.5 POPULATIONS

Yellow-headed blackbirds are monotypic but exhibit a clinal difference in body size within the Great Plains breeding range (Twedt et al. 1994). However, genetic variation within this same population failed to differentiate subpopulations (Twedt et al. 1992).

Their population exhibited no significant continental trend (−0.6%) based on North American Breeding Bird Surveys from 1966 to 2015 (Sauer et al. 2017), as the 90% credible interval ($CI_{90\%}$) associated with this trend included zero (−1.0%, 0.7%). However, these same data suggest an ongoing range contraction along the eastern and northern edges of their range. A negative trend was detected within the Eastern Breeding Bird Survey Region (−5.2%; $CI_{90\%}$ = −7.3%, −3.2%), with declines also noted within several Bird Conservation Regions including Boreal Hardwoods Transition (−8.3%; $CI_{90\%}$ = −15.2%, −2.5%), Boreal Taiga Plains (−5.3%; $CI_{90\%}$ = −7.3%, −2.8%), Eastern Tallgrass Prairie (−5.2%; $CI_{90\%}$ = −8.9%, −1.1%), and Prairie Hardwood Transition (−3.7%; $CI_{90\%}$ = −6.2%, −1.2%). There is also evidence of diminishing populations along the western edge of their range, as populations have significantly declined (−7.1%; $CI_{90\%}$ = −11.6%, −3.3%) in Coastal California (Sauer et al. 2017).

Based on data from Breeding Bird Surveys, wherein yellow-headed blackbirds were detected on 682 of 4,003 survey routes at a rate of 2.9 ± 0.4 birds per route, their continental population was estimated at 11 million birds, with 9 million of those in the U.S. population (Partners in Flight Science Committee 2013). A revised continental population estimate of 15 million birds (Rosenberg et al. 2016) more closely aligns with an estimated 11.6 million (±3.4 million) yellow-headed blackbirds breeding in Minnesota, North and South Dakota, Manitoba, and Saskatchewan, a figure that was based on area searches of General Land Office quarter sections (Nelms et al. 1994; Linz et al. 2000). The offspring from this breeding population swell the premigratory fall population to 16.8–18 million (±5.0 million) birds (Peer et al. 2003; Homan et al. 2004).

Because the distribution of yellow-headed blackbirds within their historical range is dictated by the availability of suitable habitat, drainage and conversion of wetlands to agriculture and encroachment of forests have reduced the availability of suitable breeding habitat, which in turn has diminished the species' range (Ward 2005a). Of 27 breeding colonies in Iowa that were active during 1960–1962, three locations were no longer active during 1983–1984, with habitat conditions no longer suitable at two of these abandoned sites (Brown 1988). Similarly, marsh drainage in California has reduced or eliminated some breeding populations (Small 1994).

Because populations tend to fluctuate naturally with wetland conditions on the breeding grounds, population recovery after drought is probably rapid. However, in isolated populations, failure to attract immigrants may result in population decline (Ward 2005a). Moreover, colonization of unoccupied sites may be slow, because potential colonizers may evaluate the suitability of a site using information on numbers of young fledged per nest (Ward 2005b). Thus, young inexperienced birds that are unlikely to return to their natal wetland (McCabe and Hale 1960; Searing and Schieck 1993) or newly arriving immigrants are more likely to colonize unoccupied sites than are older experienced birds (Ward et al. 2010).

Further northward and westward range contraction is anticipated along the eastern portion of the species range as global climate changes (Matthews et al. 2004). Local populations fluctuate with wetland conditions and their range may be contracting due to habitat loss, lack of migratory connectivity, and climate change, but the continental population of yellow-headed blackbirds is not in jeopardy.

Because they typically breed in large, deep-water wetlands, future habitat loss due to drainage and conversion to agriculture is less likely than for those species breeding in shallow-water wetlands. Moreover, conversion of upland habitats that surround wetlands from grasslands to cultivation of small grains, corn, and sunflower has provided an abundant food supply for postbreeding individuals and has probably increased the survival of fledged young.

Regional populations of yellow-headed blackbirds seem correlated with water levels within wetlands with emergent vegetation: increasing with more water and decreasing as water levels decline (Lederer et al. 1975; Nelms et al. 1994). This correlation is due to both increased quantity and quality of suitable breeding habitat. Specifically, as dense stands of vegetation (e.g., cattails) are broken into smaller fragments by increased water level, smaller vegetation clumps with greater edge habitat are created. These smaller vegetation patches tend to provide more desirable nest sites (Ellarson 1950). Indiscriminate use of herbicides, such as glyphosate, may remove nesting habitat and render wetlands unsuitable for breeding, but judicial application may enhance breeding opportunities by increasing channeling and suitable nests sites (Linz et al. 1996).

Colonization of new breeding sites may be dependent on factors other than solely habitat. Attempts to induce colonization of marshes where yellow-headed blackbirds were not present have not been successful. In Wisconsin, red-winged blackbirds fledged 100 cross-fostered yellow-headed blackbirds over 3 years; males that were assumed to be young fledged at the marsh returned to establish territories but failed to attract females (McCabe and Hale 1960). Similarly, in British Columbia, cross-fostered yellow-headed blackbird young fledged, but these birds did not return to their natal marsh (Searing and Schieck 1993). The presence of conspecific birds, via decoys and audio recordings, was not sufficient to attract colonizing yellow-headed blackbirds to seemingly suitable wetlands.

Breeding yellow-headed blackbirds tend to be faithful to their breeding location, often returning to the same marsh and territory in subsequent years. Indeed, 51% of adult males and 19% of second-year males return the next year to breed in the same wetland (Searcy 1979). However, over 5 years, 60% of males changed breeding marshes at least once and changed territories during 43% of between-year opportunities. Many males (30%) skip territory ownership in ≥1 year during this period, either due to dispersal from the study area or becoming nonbreeding "floaters" (Beletsky and Orians 1994). The decision to disperse is likely related to low reproductive success, as males with few mates were more likely to disperse than were males with high reproductive output. Even so, compared to site-faithful males, males that disperse have reduced reproductive success for the first year— likely due to relegation to marginal territories within more productive wetlands. Acquisition of more central territories and increased harem sizes in subsequent years, however, ultimately improves the long-term reproductive performance of dispersed males (Ward and Weatherhead 2005). Females also disperse in response to poor reproductive success but appear to be motivated by poor success within the breeding site rather than their individual reproductive performance. Unlike males, females suffer no reduction in reproductive success following dispersal (Ward and Weatherhead 2005).

3.6 AGRICULTURAL DAMAGES

Where available, commercial crop grains comprise a major portion of the postbreeding diet of yellow-headed blackbirds, with larger bodied birds tending to consume more crops than smaller birds. That is, males consume more grains than females, and within each sex adults consume more crops than do immature birds (Twedt et al. 1991). Crops depredated include corn, sunflower, oats, wheat, milo, sorghum, and durum (Twedt et al. 1991).

Depredation to sunflower crops by blackbirds throughout the northern Great Plains was estimated at over $5.4 million loss, but less than 20% ($1.1 million) of this loss was attributed to yellow-headed blackbirds. Because males consumed an average of 248 g of sunflower compared to 139 g consumed by females, most (64%) of the economic loss caused by yellow-headed blackbirds was attributed to males (Peer et al. 2003).

Foraging in agricultural fields may be a two-edged sword for yellow-headed blackbirds. Although grain crops contribute mightily to their diet, they may concomitantly be exposed to agricultural pesticides. Granular carbamates may be gleaned by foraging birds, especially during spring, and result in mortality (Mineau 2005). Elsewhere, pesticides being aerially applied to agricultural fields may drift into adjacent wetland breeding colonies, resulting in mortality of nestlings, either directly or indirectly by reduction in aquatic invertebrate food sources.

The early departure of yellow-headed blackbirds from their breeding grounds, compared to the later departure of sympatric red-winged blackbirds and common grackles, likely reduces depredation near breeding sites. Although little information exists on the diet of yellow-headed blackbirds during migration and winter, it is likely that commercial grains, including those pilfered from animal feedlots, remain an important food resource throughout this period. However, during migration and winter, most crop fields foraged by yellow-headed blackbirds are likely to have been harvested. Grain gleaned from harvested fields is not begrudged by farmers, as this consumption results in no financial loss.

Because of their penchant for eating crops, yellow-headed blackbirds (along with other icterids) are often persecuted (Twedt et al. 1991). While marauding agricultural fields, mixed-species flocks of blackbirds have been harried or harassed by various methods, including hazing with aircraft, frightening devices (including firearms), or frisson-inducing agents (Woronecki et al. 1967). In areas where depredations are particularly severe, yellow-headed blackbirds may be killed alongside other blackbird species damaging crops via shooting, toxic bait (e.g., DRC-1339), chemical sprays, traps, or nets (see Chapter 10). Alternatively, habitats may be altered to reduce their occupancy (Linz and Homan 2011).

Localized use of lethal methods to control crop depredation probably has little long-term impact on continental populations of yellow-headed blackbirds. On the other hand, if lethal measures are broadly directed at breeding or roosting populations, these actions could devastate local breeding populations. Moreover, any increase in their mortality within isolated breeding populations at the edges of their range could extirpate those populations. Indeed, the loosely "colonial" breeding habits of yellow-headed blackbirds make them vulnerable to local extirpation (McCabe 1985). Isolated populations on the periphery of the breeding range are particularly at risk due to lack of habitat connectivity to support immigration (Ward 2005a).

REFERENCES

Ammann, G. A. 1938. The life history and distribution of the yellow-headed blackbird. PhD Dissertation. University of Michigan, Ann Arbor, MI.

Arnold, T. W. 1992. Variation in laying date, clutch size, egg size, and egg composition of yellow-headed blackbirds (*Xanthocephalus*): A supplemental feeding experiment. *Canadian Journal Zoology* 70:1904–1911.

Beletsky, L. D., and G. H. Orians. 1994. Site fidelity and territorial movements of males in a rapidly declining population of yellow-headed blackbirds. *Behavioral Ecology and Sociobiology* 34:257–265.

Beletsky, L. D., G. H. Orians, and J. C. Wingfield. 1990. Steroid hormones in relation to territoriality, breeding density, and parental behavior in male yellow-headed blackbirds. *Auk* 107:60–68.

Bent, A. C. 1958. *Life histories of North American blackbirds, orioles, tanagers, and allies.* U.S. National Museum Bulletin 211. Smithsonian Institution, Washington, DC.

Bohlen, H. D. 1989. *The birds of Illinois.* Indiana University Press, Bloomington, IL.

Bonaparte, C. L. 1825. *American ornithology, or the natural history of birds inhabiting the United States, not given by Wilson*. Vol. 1. Carey, Lea & Carey, Philadelphia, PA.

Bray, O. E., A. M. Gammell, and D. R. Anderson. 1979. Survival of yellow-headed blackbirds banded in North Dakota. *Bird-Banding* 50:252–255.

Brewer, R., G. A. McPeek, and R. J. Adams, Jr. 1991. *The atlas of breeding birds of Michigan*. Michigan State University Press, East Lansing, MI.

Brown, M. 1988. Yellow-headed blackbird nesting in Iowa: A twenty year follow up. *Iowa Bird Life* 58:38–39.

Cadman, M. D., P. F. J. Eagles, and F. M. Helleiner. 1987. *Atlas of the breeding birds of Ontario*. University of Waterloo Press, Waterloo, Ontario, Canada.

Cash, K. J., and L. S. Johnson. 1990. Male parental investment and female competence in yellow-headed blackbirds. *Auk* 107:205–212.

Crase, F. T., and R. W. DeHaven. 1972. Current breeding status of the yellow-headed blackbird in California. *California Birds* 3:39–42.

Crawford, R. D. 1977. Breeding biology of year-old and older female red-winged and yellow-headed blackbirds. *Wilson Bulletin* 89:73–80.

Crawford, R. D. 1978. Temporal patterns of spring migration of yellow-headed blackbirds in North Dakota. *Prairie Naturalist* 10:120–122.

Crawford, R. D., and W. L. Hohman. 1978. A method for aging female yellow-headed blackbirds. *Bird-Banding* 49:201–207.

Dufty, A. M., Jr. 1994. Rejection of foreign eggs by yellow-headed blackbirds. *Condor* 96:799–801.

eBird. 2016. *eBird: An online database of bird distribution and abundance*. eBird, Ithaca, NY. http://www.ebird.org (accessed March 10, 2016).

Ellarson, R. S. 1950. The yellow-headed blackbird in Wisconsin. *Passenger Pigeon* 12:99–109.

Fautin, R. W. 1940. The establishment and maintenance of territories by the yellow-headed blackbird in Utah. *Great Basin Naturalist* 1:75–91.

Fautin, R. W. 1941a. Incubation studies of the yellow-headed blackbird. *Wilson Bulletin* 53:107–122.

Fautin, R. W. 1941b. Development of nestling yellow-headed blackbirds. *Auk* 58:215–232.

Fischer, D. H., and E. G. Bolen. 1981. Nestling diets of red-winged and yellow-headed blackbirds on playa lakes of west Texas. *Prairie Naturalist* 13:81–84.

Gilligan, J., M. Smith, D. Rogers, and A. Contreras. 1994. *Birds of Oregon: Status and distribution*. Cinclus Publications, McMinnville, OR.

Gori, D. F. 1988a. Colony facilitated foraging in yellow-headed blackbirds: Experimental evidence for information transfer. *Ornis Scandinavica* 19:224–230.

Gori, D. F. 1988b. Adjustment of parental investment with mate quality by male yellow-headed blackbirds (*Xanthocephalus*). *Auk* 105:672–680.

Gori, D. F. 1990. Response to cash and Johnson. *Auk* 107:206–208.

Harms, K. E., L. D. Beletsky, and G. H. Orians. 1991. Conspecific nest parasitism in three species of new world blackbirds. *Condor* 93:967–974.

Homan, H. J., G. M. Linz, W. J. Bleier, and R. B. Carlson. 1994. Dietary comparisons of adult male common grackles, red-winged blackbirds, and yellow-headed blackbirds in north central North Dakota. *Prairie Naturalist* 26:273–281.

Homan, H. J., L. B. Penry, and G. M. Linz. 2004. *Linear modeling of blackbird populations breeding in central North America*. National Sunflower Association, Mandan, ND. http://www.sunflowernsa.com/uploads/research/133/133.pdf (accessed September 12, 2016).

Howell, S. N. G., C. Corben, P. Pyle, and D. I. Rogers. 2003. The first basic problem: A review of molt and plumage homologies. *Condor* 105:635–653.

Howell, S. N. G., and P. Pyle. 2015. Use of "definitive" and other terms in molt nomenclature: A response to Wolfe et al. (2014). *Auk: Ornithological Advances* 132:365–369.

Howell, S. N. G., and S. Webb. 1995. *A guide to the birds of Mexico and northern Central America*. Oxford University Press, Oxford, UK.

Humphrey, P. S., and K. C. Parkes. 1959. An approach to the study of molts and plumages. *Auk* 76:1–31.

Jaramillo, A. 2008. *Yellow-headed blackbird account*. California bird species of special concern: A ranked assessment of species, subspecies, and distinct populations of birds of immediate conservation concern in California, Studies of Western Birds 1, Western Field Ornithologists, Camarillo, CA and California Department of Fish and Game, Sacramento, CA, vol. 1, pp. 444–450.

Jaramillo, A., and P. Burke. 1999. *New world blackbirds: The icterids*. Christopher Helm, London, UK.

Jordan, D. S. 1884. *Manual of the vertebrates of the northern United States, including the district east of the Mississippi River, and north of North Carolina and Tennessee, exclusive of marine species*. 4th ed. Jansen, McClurg, Chicago, IL.

Kapilow, L., J. Maffi, and R. J. Lederer. 1980. Yellow-headed and red-winged blackbird nesting studies: A case for long-term research. *Bioscience* 51:209–214.

Keller, C. E., S. A. Keller, and T. C. Keller. 1986. *Indiana birds and their haunts*. 2nd ed. Indiana University Press, Bloomington, IN.

Kennard, J. H. 1975. Longevity records of North American birds. *Bird-Banding* 46:55–73.

Lederer, R. J., W. S. Mazen, and P. J. Metropulos. 1975. Population fluctuation in a yellow-headed blackbird marsh. *Western Birds* 6:1–6.

Lightbody, J. P., and P. J. Weatherhead. 1987a. Interactions among females in polygynous yellow-headed blackbirds. *Behavioral Ecology and Sociobiology* 21:23–30.

Lightbody, J. P., and P. J. Weatherhead. 1987b. Polygyny in the yellow-headed blackbird: Female choice versus male competition. *Animal Behavior* 35:1670–1684.

Linz, G. M., D. C. Blixt, D. L. Bergman, and W. J. Bleier. 1996. Responses of red-winged blackbirds, yellow-headed blackbirds, and marsh wrens to glyphosate-induced alterations in cattail density. *Journal of Field Ornithology* 67:167–176.

Linz, G. M., and H. J. Homan. 2011. Use of glyphosate for managing invasive cattail (*Typha* spp.) to protect crops near blackbird (Icteridae) roosts. *Crop Protection* 30:98–104.

Linz, G. M., B. D. Peer, H. J. Homan, R. L. Wimberly, D. L. Bergman, and W. J. Bleier. 2000. Has an integrated pest management approach reduced blackbird damage to sunflower? *Human conflicts with wildlife: Economic considerations*. NWRC Special Symposium. National Wildlife Research Center, Fort Collins, CO, vol. 3, pp. 132–137.

Lyon, B. E., L. D. Hamilton, and M. Magrath. 1992. The frequency of conspecific brood-parasitism and the pattern of laying determinacy in yellow-headed blackbirds. *Condor* 94:590–597.

Matthews, S., R. O'Connor, L. R. Iverson, and A. M. Prasad. 2004. *Atlas of climate change effects in 150 bird species of the Eastern United States*. General Technical Report NE-318. U.S. Department of Agriculture, Forest Service, Northeastern Research Station, Newtown Square, PA.

McCabe, R. A. 1985. The loss of a large colony of yellow-headed blackbirds (*Xanthocephalus*) from southern Wisconsin. *International Ornithological Congress* 18:1034.

McCabe, R. A. and J. B. Hale. 1960. An attempt to establish a colony of yellow-headed blackbirds. *Auk* 77:425–432.

Miller, R. S. 1968. Conditions of competition between redwings and yellow-headed blackbirds. *Journal Animal Ecology* 37:43–62.

Mineau, P. 2005. *Direct losses of birds to pesticides—Beginnings of a quantification*. General Technical Report PSW-GTR-191. U.S. Department of Agriculture, Forest Service, Pacific Southwest Research Station, Albany, CA, Vol. 2, pp. 1065–1070.

Minock, M. E., and J. R. Watson. 1983. Red-winged and yellow-headed blackbird nesting habitat in a Wisconsin marsh. *Journal Field Ornithology* 54:324–326.

Monson, G., and A. R. Phillips. 1981. *Annotated checklist of the birds of Arizona*. 2nd ed. University of Arizona Press, Tucson, AZ.

Nelms, C. O., W. J. Bleier, D. L. Otis, and G. M. Linz. 1994. Population estimates of breeding blackbirds in North Dakota, 1967, 1981–82, and 1990. *American Midland Naturalist* 132:256–263.

Nero, R. W. 1963. Comparative behavior of the yellow-headed blackbird, red-winged blackbird, and other icterids. *Wilson Bulletin* 75:376–413.

Oberholser, H. C. 1974. *The bird life of Texas*. University of Texas Press, Austin, TX.

Orians, G. H. 1980. *Some adaptations of marsh nesting blackbirds*. Princeton University Press, Princeton, NJ.

Orians, G. H., and G. M. Christman. 1968. *A comparative study of the behavior of red-winged, tricolored, and yellow-headed blackbirds*. University of California Press, Los Angeles, CA.

Orians, G. H., and M. F. Willson. 1964. Interspecific territories of birds. *Ecology* 45:736–744.

Orians, G. H., and J. F. Wittenberger. 1991. Spatial and temporal scales in habitat selection. *American Naturalist* 137(Suppl):S29–S49.

Ortega, C. P. 1991. The ecology of blackbird/cowbird interactions in Boulder County, Colorado. PhD Dissertation, University of Colorado, Boulder, CO.

Ortega, C. P., and A. Cruz. 1988. Mechanisms of egg acceptance by marsh dwelling blackbirds. *Condor* 90:349–358.

Ortega, C. P., and A. Cruz. 1991. A comparative study of cowbird parasitism in yellow-headed blackbirds and red-winged blackbirds. *Auk* 108:16–24.

Ortega, C. P., and A. Cruz. 1992. Differential growth patterns of nestling brown-headed cowbirds and yellow-headed blackbirds. *Auk* 109:368–376.

Partners in Flight Science Committee. 2013. *Population Estimates Database, version 2013.* http://rmbo.org/pifpopestimates (accessed February 9, 2016).

Patterson, C. B., and J. M. Emlen. 1980. Variation in nestling sex ratios in the yellow-headed blackbird. *American Naturalist* 115:743–747.

Peer, B. D., H. J. Homan, G. M. Linz, and W. J. Bleier. 2003. Impact of blackbird damage to sunflower: Bioenergetic and economic models. *Ecological Applications* 13:248–256.

Peterjohn, B. G. 1989. *The Birds of Ohio.* Indiana University Press, Bloomington, IN.

Phillips, A., J. Marshall, and G. Monson. 1964. *The birds of Arizona.* University of Arizona Press, Tucson, AZ.

Picman, J. 1988. Behavioral interactions between North American marsh-nesting blackbirds and marsh wrens and their influence on reproductive strategies of these passerines. *International Ornithological Congress* 19:2624–2634.

Pyle, P. 1997. Molt limits in North American passerines. *North American Bird Bander* 22:49–90.

Richardson, J. 1831. *Fauna Boreali-Americana. Part 2: The birds.* John Murray, London, UK.

Richter, W. 1982. Hatching asynchrony: The nest failure hypothesis and brood reduction. *American Naturalist* 120:828–832.

Richter, W. 1983. Balanced sex ratios in dimorphic altricial birds: The contribution of sex-specific growth dynamics. *American Naturalist* 121:158–171.

Richter, W. 1984. Nestling survival and growth in the yellow-headed blackbird, *Xanthocephalus xanthocephalus. Ecology* 65:597–608.

Robbins, M. B. and D. A. Easterla. 1992. *Birds of Missouri: Their distribution and abundance.* University of Missouri Press, Columbia, MO.

Roberts, T. S. 1936. *Manual for the identification of the birds of Minnesota and neighboring states.* University of Minnesota Press, Minneapolis, MN.

Rohwer, S., and E. Røskaft. 1989. Results of dyeing male yellow-headed blackbirds solid black: Implications for the arbitrary identity badge hypothesis. *Behavioral Ecology and Sociobiology* 25:39–48.

Rosenberg, K. V., J. A. Kennedy, R. Dettmers, R.P. Ford, D. Reynolds, C.J. Beardmore, P.J. Blancher, et al. 2016. *Partners in Flight Landbird Conservation Plan: 2016 Revision for Canada and Continental United States.* Partners in Flight Science Committee. http://www.partnersinflight.org/ (accessed September 25, 2016).

Royall, W. C., J. L. Guarino, J. W. DeGrazio, and A. Gammell. 1971. Migration of banded yellow-headed blackbirds. *Condor* 73:100–106.

Sauer, J. R., D. K. Niven, J. E. Hines, D. J. Ziolkowski, Jr, K. L. Pardieck, J. E. Fallon, and W. A. Link. 2017. The North American Breeding Bird Survey, results and analysis 1966–2015. Version 2.07.2017 USGS Patuxent Wildlife Research Center, Laurel, MD. https://www.mbr-pwrc.usgs.gov/bbs/bbs.html (accessed April 30, 2017).

Searcy, W. A. 1979. Size and mortality in male yellow-headed blackbirds. *Condor* 81:304–305.

Searing, G. F., and J. Schieck. 1993. *Yellow-headed blackbird transplant program, Vancouver, British Columbia.* LGL, Sidney, British Columbia, Canada.

Shackford, J. S., and J. D. Tyler. 1987. A nesting yellow-headed blackbird colony in Texas County, Oklahoma. *Bulletin Oklahoma Ornithological Society* 20:9–12.

Small, A. 1994. *California birds: Their status and distribution.* Ibis, Vista, CA.

Swenk, M. H. 1933. The exact type localities of the birds discovered in Nebraska by Thomas Say on the Long expedition. *Nebraska Bird Review* 1:33–35.

Thompson, M. C., and C. Ely. 1992. *Birds in Kansas.* University Kansas Museum Natural History Public Education Series No. 12, Lawrence, KS.

Twedt, D. J. 1990. Diet, molt, and geographic variation of yellow-headed blackbirds, *Xanthocephalus xanthocephalus.* PhD Dissertation. North Dakota State University, Fargo, ND.

Twedt, D. J., W. J. Bleier, and G. M. Linz. 1991. Geographic and temporal variation in the diet of yellow-headed blackbirds. *Condor* 93:975–986.

Twedt, D. J., W. J. Bleier, and G. M. Linz. 1992. Genetic variation in male yellow-headed blackbirds from the northern Great Plains. *Canadian Journal Zoology* 70:2280–2282.

Twedt, D. J., W. J. Bleier, and G. M. Linz. 1994. Geographic variation in yellow-headed blackbirds. *Condor* 96:1030–1036.

Twedt, D. J., and R. D. Crawford. 1995. Yellow-headed blackbird (*Xanthocephalus xanthocephalus*). In *The Birds of North America Online*, ed. A. Poole. Cornell Lab of Ornithology, Ithaca, NY.

Twedt, D. J., and G. M. Linz. 2015. Flight feather molt in yellow-headed blackbirds (*Xanthocephalus xanthocephalus*) in North Dakota. *Wilson Journal of Ornithology* 127:622–629.

Ward, M. P. 2005a. The role of immigration in the decline of an isolated migratory bird population. *Conservation Biology* 19:1528–1536.

Ward, M. P. 2005b. Habitat selection by dispersing yellow-headed blackbirds: Evidence of prospecting and the use of public information. *Oecologia* 145:1432–1939.

Ward, M. P., T. J. Benson, B. Semel, and J. Herkert. 2010. The use of social cues in habitat selection by wetland birds. *Condor* 112:245–251.

Ward, M. P., and P. J. Weatherhead. 2005. Sex-specific differences in site fidelity and the cost of dispersal in yellow-headed blackbirds. *Behavioral Ecology and Sociobiology* 59:108–114.

Woronecki, P. P., J. L. Guarino, and J. W. De Grazio. 1967. Blackbird damage control with chemical frightening agents. *Vertebrate Pest Conference* 3:54–56.

Willson, M. F. 1966. Breeding ecology of the yellow-headed blackbird. *Ecological Monograph* 36:51–77.

CHAPTER 4

Ecology and Management of the Common Grackle

Brian D. Peer
Western Illinois University
Moline, Illinois

Eric K. Bollinger
Eastern Illinois University
Charleston, Illinois

CONTENTS

4.1	Taxonomy	66
4.2	Distribution	67
	4.2.1 Breeding	67
	4.2.2 Nonbreeding	67
	4.2.3 Historical Changes	68
4.3	Life History	68
	4.3.1 Timing	68
	4.3.2 Nests	68
	4.3.3 Eggs	68
	4.3.4 Incubation	69
	4.3.5 Parental Care	69
	4.3.6 Brood Parasitism	69
4.4	Diet	70
4.5	Feather Molt and Plumages	70
4.6	Behavior	71
	4.6.1 Territories	71
	4.6.2 Mating System	71
	4.6.3 Predatory Behavior	71
4.7	Migration	71
4.8	Populations	72
4.9	Agricultural Damages	72
4.10	Summary and Future Research	73
References		73

Figure 4.1 Male common grackle. (Courtesy of Larry Slomski.)

The common grackle (*Quiscalus quiscula*; hereafter "grackle"; Figure 4.1) is a large charismatic blackbird species (Icteridae) with striking yellow eyes, iridescent bronze or purple plumage, and a long tail. It is found from the eastern half of North America westward to the Rocky Mountains. It is at home in both suburban and rural habitats. Grackles are one of the earliest nesting passerines in the spring in the Midwest. Preferred nesting sites are generally windbreaks and tree plantings consisting of conifers, which provide early access to nesting cover.

The grackle is an opportunistic forager found on lawns, agricultural fields, fast food parking lots, and wading in shallow water. While primarily insectivorous during the breeding season, it also feeds on seeds, including sprouting corn and rice, as well as ripening sunflower crops. In late summer and fall, grackles may forage in mixed flocks with red-winged blackbirds (*Agelaius phoeniceus*), yellow-headed blackbirds (*Xanthocephalus xanthocephalus*), and, to a lesser extent, brown-headed cowbirds (*Molothrus ater*). Flocks can number in the tens of thousands, causing significant agricultural losses. It has a reputation for eating other birds' eggs and nestlings, a reputation that may be overstated.

The grackle population has spread from eastern North America to the north and west, however, this expansion has appeared to slow. It is declining throughout much of its range in eastern North America (Sauer et al. 2017). Recent reports indicate that the grackle is among the top 50 "common" North American bird species that have experienced dramatic population declines over the last 40 years, having lost more than 50% of their population (Butcher 2007; Rosenberg et al. 2016). What makes the grackle's decline even more puzzling and disconcerting is that it is a habitat generalist, unlike many of the other species of global conservation concern, which are habitat specialists (Rosenberg et al. 2016). The reason for the grackle's decline is unclear.

Unlike other widespread blackbird species that are among the most studied birds in North America (e.g., brown-headed cowbird, red-winged blackbird; Peer et al. 2013), the grackle is an "understudied" bird. According to the Web of Science (2016), a mere 44 papers have been published on it since 2002. In this paper, we review its ecology and management in relation to agricultural crop damage and advocate that lethal controls measures be reexamined in the wake of this species' precipitous decline.

4.1 TAXONOMY

Grackles are in one of the four clades of the Icteridae family—grackles and allies, orioles, caciques and oropendolas, and meadowlarks and allies. Grackles are monophyletic and *Q. quiscula* is the most basal member of *Quiscalus* (Powell 2014). There are three subspecies. *Quiscalus quiscula quiscula* (Florida grackle) is restricted to the south below a line from Louisiana to North Carolina. *Quiscalus quiscula stonei* (purple grackle) occurs immediately north from Louisiana to southeastern New York. *Quiscalus quiscula versicolor* (bronzed grackle) is the most widely distributed and occurs to the north and west of the other two subspecies (American Ornithologists' Union [AOU] 1957). *Quiscalus quiscula stonei* has a bronze body with purplish tail; individuals south and east have purple bodies with bluish tails (Peer and Bollinger 1997a).

4.2 DISTRIBUTION

4.2.1 Breeding

The grackle occurs from the Florida Keys north to Nova Scotia, Prince Edward Island, and Newfoundland (Figure 4.2) (Peer and Bollinger 1997a). It extends to the northwest into Alberta, the Northwest Territories (ebird 2016), and eastern British Columbia and to the southwest to New Mexico and west Texas (Peer and Bollinger 1997a).

4.2.2 Nonbreeding

The grackle winters primarily from the northern United States (Minnesota east to Maine) and occasionally further north and west (Peer and Bollinger 1997a). To the southwest, it ranges to New Mexico and Texas and southeast to Florida, but it mostly withdraws north from the Florida Keys in winter (Robertson and Woolfenden 1992; Peer and Bollinger 1997a). It does not appear to occur in Mexico (Peer and Bollinger 1997a; ebird 2016).

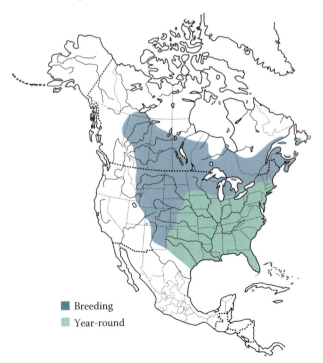

Figure 4.2 Range and trend map of the common grackle. (Birds of North America, Cornell Laboratory of Ornithology.)

4.2.3 Historical Changes

Similar to great-tailed grackles (*Q. mexicanus*) and brown-headed cowbirds, common grackles were rare in western North America within the past century, but at least through the early 1990s these three have demonstrated the greatest range expansion of any native pest bird species in the West (Marzluff et al. 1994). More recently, however, this increase has slowed as grackles have declined throughout their range including the West (Sauer et al. 2017).

4.3 LIFE HISTORY

4.3.1 Timing

In the Midwest, birds begin arriving at breeding sites in early March with breeding beginning in late March and early April. In the central portion of its range, the first eggs are laid from late March through late April and the latest eggs are laid at the end of June (Peterson and Young 1950; Howe 1978; Peer and Bollinger 1997b). The birds typically have one brood but renest in response to predation (Peer and Bollinger 1997a).

4.3.2 Nests

The nests are large bulky cups of woven grasses, stems, and some leaves that are lined with mud and then fine grasses (Peer and Bollinger 1997a). The choice of nest substrate and location is fairly plastic. Conifers are highly preferred, including northern white cedar (*Thuja occidentalis*), eastern red cedar (*Juniperus virginiana*), scotch pine (*Pinus sylvestris*), blue spruce (*Picea pungens*), and loblolly pine (*Pinus taeda*) (Peer and Bollinger 1997a). Common deciduous trees include Siberian elm (*Ulmus pumila*), Siberian peashrub (*Caragana arborescens*), box elder (*Acer negundo*), and hawthorn (*Crataegus rotundifolia*) (Homan et al. 1996). They occasionally nest in shrubs, human-made structures, cattails, cliffs, birdhouses, tree cavities, on the ground, and in the occupied nests of osprey (*Pandion haliaetus*) and great blue heron (*Ardea herodias*) (Peer and Bollinger 1997a). Grackles nest in loose colonies averaging 10–18 nests/colony or alone in coniferous plantings, shelterbelts, in suburbia, near water, and near agricultural fields (Peer and Bollinger 1997a). Nest height ranges from 0.2 to 22 m with means from 1.2 to 6.1 m (Peer and Bollinger 1997a).

4.3.3 Eggs

The background color of grackle eggs is typically light blue to pearl gray with dark brown or black scrawls. However, there is a high degree of inter- and intraclutch egg variation and background color can range from immaculate to dark brown (Figure 4.3) (Peer and Bollinger 1997b; Peer and Rothstein 2010). In 28% of clutches, the last-laid egg has less maculation and in some cases is immaculate (Peer and Rothstein 2010). There is no clear adaptive explanation for this variation. Possible explanations include that females typically begin incubation once the fourth egg has been laid, thus the last-laid egg is covered the majority of the time and does not need to be camouflaged (e.g., Ruxton et al. 2001); the female could be pigment-depleted by the time the last egg is laid (Lowther 1988); or a lighter colored egg could allow more light to penetrate, hastening the development of the embryo (e.g., Maurer et al. 2011, 2015; Birkhead 2016).

Clutch size ranges from one to seven eggs and the modal clutch size is five (Peer and Bollinger 1997a). Egg mass increases with laying sequence in clutches of more than four, possibly to increase the chances of survival for the last-laid egg (Howe 1976).

Figure 4.3 Example of intraclutch egg variation in the common grackle. (From Peer and Rothstein 2010. With permission.)

4.3.4 Incubation

The female incubates the eggs (Peterson and Young 1950), and nearly half of males desert their mate at this time (Wiley 1976). The mean incubation period is 13.5 d (range from 11.5 to 15 d; Peer and Bollinger 1997a). Incubation usually begins after the last egg has been laid in clutches of four or less, and the young hatch synchronously. In clutches greater than four and after the penultimate egg has been laid, the young hatch asynchronously (Howe 1978).

4.3.5 Parental Care

The female broods the young, but males have occasionally been observed brooding (Maxwell and Putnam 1972). Both sexes feed the young, with females providing more food. The primary food items delivered range from 70% to 88% animal material and 6% to 26% vegetable matter (Peer and Bollinger 1997a). Males and females remove fecal sacs from the nest (Maxwell and Putnam 1972).

4.3.6 Brood Parasitism

There are only 30 recorded instances of parasitism by the brown-headed cowbird, despite thousands of grackle nests having been observed (Peer et al. 2001). Grackles are at the upper size limit for suitable hosts of the cowbird (Peer and Bollinger 1997b). Grackles recognize cowbirds as a threat and respond aggressively to them at the nest; however, they only reject 13% of experimentally introduced cowbird eggs (Peer and Bollinger 1997b; Peer et al. 2010). Parasitism may be infrequent because grackles nest earlier in the season than most songbirds, but it is possible they were parasitized more often in the past but have subsequently lost most of their egg rejection behavior in the absence of parasitism (Peer and Bollinger 1997b; Peer and Sealy 2004; Peer and Rothstein 2010; Peer et al. 2010). The most likely reason for the loss of egg rejection, in the absence of its utility, is that grackles may have rejected their own oddly colored eggs because of the extreme intraclutch egg variability (Peer and Bollinger 1997b; Peer and Sealy 2004; Peer and Rothstein 2010; Peer et al. 2010).

4.4 DIET

The diet is varied and shifts significantly and predictably across seasons. During the breeding season, grackles eat primarily invertebrates, which compose roughly 20%–30% of their year-round diet (Beal 1900; Meanley 1971). Insects predominate in this invertebrate component, with larval beetles, grasshoppers, and caterpillars being among the most common taxa (Beal 1900; Meanley 1971; Homan et al. 1994). During the remainder of the year, agricultural grains and other seeds, including acorns, are consumed, making up 70%–75% of the yearly diet (Peer and Bollinger 1997a). Corn is probably the most commonly eaten grain or seed, but this varies regionally and seasonally, presumably in response to availability (Beal 1900; Robertson et al. 1978; White et al. 1985; Homan et al. 1994). For example, sunflower seeds are the dominant food eaten in the fall in North Dakota, making up over 40% of the diet from August 15 to October 21, whereas small grains (wheat, barley, and oats), and not corn, were the primary food eaten from July 1 to August 14 (Homan et al. 1994). In Arkansas, rice is the major component of the fall and winter diet (Meanley 1971). Acorns are commonly eaten throughout the grackle's range in the fall (Beal 1900; Dolbeer et al. 1978; White et al. 1985). Finally, both wild and cultivated fruits can be regionally important foods, as they ripen in the late summer and early fall (Beal 1900; Malmborg and Willson 1988).

Grackles forage primarily on the ground but will opportunistically feed in trees and even in water (Beeton and Wells 1957; Whoriskey and FitzGerald 1985). They generally forage in large flocks that frequently number in the thousands and often include other blackbird species and European starlings (*Sturnus vulgaris*) (Peer and Bollinger 1997a). Grackles typically use their bills to uncover food on the ground, frequently using it to toss aside leaves to find small acorns. The upper mandible has a downward projecting keel that extends below the tomia and appears to be an adaptation to "saw" open acorns by scoring the nut and cracking by adduction (Wetmore 1919; Schorger 1941; Beecher 1951). Grackles have been observed eating eggs and nestlings from other birds' nests (Bent 1958; Sealy 1994). However, when large samples of grackle intestinal contents were examined, vertebrate prey (including eggs and nestlings) was <1% of their diets (Beal 1900; Meanley 1971; White et al. 1985). In addition, they have been known on rare occasions to kill and eat adult songbirds (Middleton 1977; Davidson 1994), as well as frogs (Ernst 1944), salamanders (Hamilton 1951), and fish, especially for individuals nesting near large bodies of water (Hamilton 1951; Whoriskey and FitzGerald 1985). Grackles will follow farmers plowing fields to eat exposed grubs (Beal 1900; Bent 1958) and even mice (Bent 1958). Overall, the grackle is aptly described as a very opportunistic forager.

4.5 FEATHER MOLT AND PLUMAGES

Grackles go through a typical sequence of molts for a passerine. Hatchlings have a sparse coat of natal down described by Dwight (1900) as "pale sepia-brown in color." They next acquire their juvenal plumage via a partial or complete prejuvenal molt. This plumage is generally dull brown; the tail feathers have some purple and are darker (Dwight 1900). In addition, the juvenal body feathers are edged in paler brown. There is very obscure streaking and barring on the underparts that is somewhat more obvious in females.

Grackles have no Basic I plumage and hatching-year birds molt directly into a definitive basic plumage (Wood 1945; Oberholser 1974). This is a complete molt, although some hatching-year birds retain juvenal underwing coverts and tertials (Selander and Giller 1960). The typical prebasic molt occurs from July to October (Stone 1937) and begins with the lesser wing coverts, then moves to the greater coverts, secondaries, forehead, crown, nape, rump, primary coverts, uppertail coverts, scapulars, proximal primaries, breast, chin, distal remiges, and median rectrices (Wood 1945). Axillars are sometimes retained until only the primaries and rectrices remain to be molted. Proximal remiges are lost and replaced quickly relative to the distal four. Auriculars are sometimes lost after the rest of the head feathers. Finally, feathers along the belly and back are typically lost before those along the sides.

The definitive basic plumage is dominated by iridescent black feathers with either a metallic greenish blue or dark metallic purple head, neck, and upper breast depending on whether the bird belongs to the bronzed or purple races (Dwight 1900; Wood 1945; Oberholser 1974). The neck and upper breast can also be a brassy green, and the lores and malar are usually velvety black. Most of the remaining plumage is various shades of black and metallic blacks, purples, bronzes, and greens, again varying somewhat predictably between races. Females are similar to males but duller, especially on their breasts and bellies. The tail is approximately 40% of the total body length and is keel-shaped in both sexes but somewhat shorter in females.

4.6 BEHAVIOR

4.6.1 Territories

The female chooses the nest site and the pair defends the small area near it (Maxwell 1970), although conspecifics have been observed visiting other nests without consequence (Howe 1976; B. Peer unpublished data). Loose aggregations often occur, and these groups display some behaviors typical of colonial birds, such as common defense (Peer and Bollinger 1997b; B. Peer unpublished data).

4.6.2 Mating System

The mating system is monogamous, but polygyny may occur (Wiley 1976; Howe 1978). This aspect needs further study using molecular techniques.

4.6.3 Predatory Behavior

Grackles have a reputation for depredating other birds' eggs, nestlings, and even other adult birds (Peer and Bollinger 1997a). These observations are largely anecdotal. No studies have quantified how frequently this behavior occurs. On the other hand, studies that have quantified their diets have found that vertebrate prey items are uncommon (see Section 4.5, "Diet").

4.7 MIGRATION

Grackles are generally described as partial, short-distance migrants, with seasonal movements between breeding and wintering sites for most populations (Peer and Bollinger 1997a). However, populations breeding in states along the Gulf of Mexico are largely nonmigratory (Bent 1958). Like many short-distance migrants, populations breeding in the northern parts of the grackle's range travel greater distances during migration than those breeding farther south. For example, grackles breeding at 44°–45° N latitude move about 1,000 km to their wintering grounds, whereas those breeding at 34°–35° N only move about 200 km (Dolbeer 1982). Banding data suggest that females migrate, on average, roughly 100 km farther than males and that young (e.g., hatch-year) birds migrate 100–300 km farther than older males (Dolbeer 1982).

Fall migratory pathways are oriented primarily toward the Gulf of Mexico (Burtt and Giltz 1977), and fall migration typically peaks in late October through early November (Dolbeer 1982; Peterjohn 1989). Spring migration, on the other hand, peaks in late February and March (Meanley and Dolbeer 1978; Dolbeer 1982). Males typically arrive at breeding sites about 1 week before females (Peterson and Young 1950; Wiens 1965). There is some evidence of natal philopatry (Bergman and Homan 1994).

Grackles, like other icterids, migrate during the day, often with red-winged blackbirds, brown-headed cowbirds, and European starlings. They often move in large flocks and roost at night in large, communal aggregations (Peer and Bollinger 1997a). Orientation and navigation mechanisms have not been studied in grackles specifically, but Edwards et al. (1992) found magnetic material in the head and neck, suggesting that grackles may use the earth's magnetic field.

4.8 POPULATIONS

The grackle has declined throughout its range (Figure 4.4). According to the Breeding Bird Survey (BBS) there has been a −1.75% annual decline survey-wide from 1966 to 2015 (Sauer et al. 2017): −2.18% in the eastern BBS Region, −1.11% in the central region, and −1.01% in the West. Similarly, in Louisiana where grackles depredate rice, there has been a 3.65% annual decline from 1966 to 2015. Peer et al. (2003) estimated the breeding population of grackles in the sunflower-growing region of the northern Great Plains at approximately 13 million and the fall population at approximately 19 million. In North Dakota and South Dakota, which is in the core of this region, the trend has been a 1.88% and 0.44% increase, respectively, from 1966–2015 (Sauer et al. 2017). The current North American population is estimated at 69 million (Rosenberg et al. 2016), and in 2013 it was ranked as the 27th most common bird (PIF 2013), down from 11th based on BBS data in 1992–1993 (Peterjohn et al. 1994).

4.9 AGRICULTURAL DAMAGES

Grackles are one of the most economically significant avian pest species in North America (DeGrazio 1978; Dolbeer 1980; Mott 1984; USDA 2015). Most agricultural damage is spatially concentrated near communal roosts that are used by grackles and other blackbirds in the fall through spring. These roosts can number in the tens of millions. For example, grackles, along with

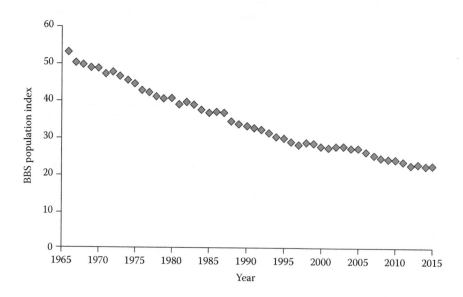

Figure 4.4 Population decline of the common grackle in North America, 1966 to 2015, based on breeding bird survey data. (Sauer et al. 2017.)

red-winged blackbirds, are the major avian taxa responsible for bird damage to ripening corn in the United States (Besser and Brady 1986), with annual losses in the tens of millions of dollars. In addition, the common grackle is one of the most important species damaging sprouting corn and rice that also results in monetary losses in the tens of millions of dollars per year (Stone and Mott 1973; USDA 2015). Ripening rice is also damaged (Meanley 1971), as are other small grains (Pierce 1970; Homan et al. 1994), sunflowers (Stone and Mott 1973; Homan et al. 1994), peanuts (Mott et al. 1972), blueberries (Mott and Stone 1973), and sweet cherries (Virgo 1971). Finally, grackles can also become pests in livestock feedlots, especially near roost locations (White et al. 1985). The USDA-APHIS has estimated that it killed 489,773 grackles in the past 5 years, primarily using the avicide DRC-1339 (USDA online). In 2015, 134,741 common grackles were killed by the USDA-APHIS, which likely had a negligible effect on the overall population when considering that annual grackle adult survivorship is 51.6% (Peer and Bollinger 1997a).

4.10 SUMMARY AND FUTURE RESEARCH

The grackle is a widespread blackbird species that is responsible for a portion of the avian agricultural damage in North America. However, it has undergone a dramatic population decline in the past 40 years and the reason for this decline is unclear. It is a generalist in both its nesting and foraging behavior, making these declines all the more alarming. Alteration of habitat may be responsible, especially the removal of shelterbelts and windbreaks, but this requires additional study. Despite the population decline, control measures are still employed to reduce crop losses. The arbitrary nature of lethal control is inefficient and probably does not reduce crop losses, and thus the use of these control measures must be reexamined. Similar to our appeal 20 years ago, albeit a much more urgent one now, population monitoring and monitoring of movements of birds to and from the breeding grounds would be helpful to determine the cause of this decline. In general, the grackle is an understudied species, especially when considering its widespread range. Studies of its mating system and colony dynamics using molecular methods is warranted, as are studies on vocalizations, colonialism, and molt (Peer and Bollinger 1997a).

REFERENCES

American Ornithologists' Union. 1957. *Checklist of North American birds*. 5th ed. American Ornithologists' Union, Baltimore, MD.
Beal, F. E. L. 1900. *Food of the bobolink, blackbirds, and grackles*. Bulletin No. 13. U.S. Department of Agriculture, Division Biological Survey, Washington, DC.
Beecher, W. J. 1951. Adaptations for food-getting in the American blackbirds. *Auk* 68:411–440.
Beeton, A. M., and L. Wells. 1957. A bronzed grackle (*Quiscalus quiscula*) feeding on live minnows. *Auk* 74:263–264.
Bent, A.C. 1958. Life histories of North American blackbirds, orioles, tanagers, and allies. *U.S. Natl. Mus. Bull.* No. 211.
Bergman, D. L., and H. J. Homan. 1994. Common grackle philopatry in North Dakota. *South Dakota Bird Notes* 46:10.
Besser, J. F., and D. J. Brady. 1986. *Bird damage to ripening field corn increases in the United States from 1971 to 1981*. Fish and Wildlife Leaflet 7. U.S. Fish and Wildlife Service, Washington, DC.
Birkhead, T. 2016. *The Most Perfect Thing: Inside (and Outside) a Bird's Egg*. Bloomsbury, USA.
Burtt, H. E., and M. L. Giltz. 1977. Seasonal directional patterns of movements and migrations of starlings and blackbirds in North America. *Bird-Banding* 48:259–271.
Butcher, G. S. 2007. Common birds in decline. A state of the birds report. *Audubon* 109:58–62.
Davidson, A. H. 1994. Common grackle predation on adult passerines. *Wilson Bulletin* 106:174–175.

DeGrazio, J. W. 1978. World bird damage problems. *Vertebrate Pest Conference* 8:9–24.

Dolbeer, R. A. 1980. *Blackbirds and corn in Ohio.* Resource Publication 136. U.S. Department of the Interior Fish and Wildlife Service, Washington, DC.

Dolbeer, R. A. 1982. Migration patterns for age and sex classes of blackbirds and starlings. *Journal of Field Ornithology* 53:28–46.

Dolbeer, R. A., P. P. Woronecki, A. R. Stickley, Jr., and S. B. White. 1978. Agricultural impact of a winter population of blackbirds and starlings. *Wilson Bulletin* 90:31–44.

Dwight, J., Jr. 1900. The sequence of plumages and moults of passerine birds of New York. *Annual New York Academy Science* 13:73–360.

eBird. 2016. eBird: An online database of bird distribution and abundance [web application]. eBird, Ithaca, NY. http://www.ebird.org (accessed September 24, 2016).

Edwards, H. H., G. D. Schnell, R. L. Dubuois, and V. H. Hutchison. 1992. Natural and induced remnant magnetism in birds. *Auk* 109:43–56.

Ernst, S. G. 1944. Observation on the food of the bronzed grackle. *Auk* 61:644–645.

Hamilton, W. J., Jr. 1951. The food of nestling bronzed grackles, *Quiscalus quiscula versicolor*, in central New York. *Auk* 68:213–217.

Homan, H. J., G. M. Linz, and W. J. Bleier. 1994. Effect of crop phenology and habitat on the diet of common grackles (*Quiscalus quiscula*). *American Midland Naturalist* 131:381–385.

Homan, H. J., G. M. Linz, W. J. Bleier, and R. B Carlson. 1996. Colony-site and nest-site use by common grackles in North Dakota. *Wilson Bulletin* 108:104–114.

Howe, H. F. 1976. Egg size, hatching asynchrony, sex and brood reduction in the common grackle. *Ecology* 57:1195–1207.

Howe, H. F. 1978. Initial investment, clutch size, and brood reduction in the common grackle (*Quiscalus quiscula* L.). *Ecology* 59:1109–1122.

Lowther, P. E. 1988. Spotting pattern of the last laid egg of the house sparrow. *Journal of Field Ornithology* 59:51–54.

Malmborg, P. K., and M. F. Willson. 1988. Foraging ecology of avian frugivores and some consequences for seed dispersal in an Illinois woodlot. *Condor* 90:173–186.

Marzluff, J. M., R. B. Boone, and G. W. Cox. 1994. Historical changes in populations and perceptions of native pest bird species in the west. *Studies in Avian Biology* 15:202–220.

Maurer, G., S. J. Portugal, and P. Cassey. 2011. Review: An embryo's eye view of avian eggshell pigmentation. *Journal of Avian Biology* 42:494–504.

Maurer, G., S. J. Portugal, M. E. Hauber, I. Miksik, D. G. Russell, and P. Cassey. 2015. First light for avian embryos: Eggshell thickness and pigmentation mediate variation in development and UV exposure in wild bird eggs. *Functional Ecology* 29:209–218.

Maxwell, G. R., II. 1970. Pair formation, nest building, and egg laying of the common grackle in northern Ohio. *Ohio Journal of Science* 70:284–291.

Maxwell, G. R., II., and L. S. Putnam. 1972. Incubation, care of young, and nest success of the common grackle (*Quiscalus quiscula*) in northern Ohio. *Auk* 89:349–359.

Meanley, B. 1971. *Blackbirds and the southern rice crop.* Resource Publication 100. U.S. Department of Interior, U.S. Fish and Wildlife Service, Washington, DC.

Meanley, B., and R. A. Dolbeer. 1978. Source of common grackles and red-winged blackbirds wintering in Tennessee. *Migrant* 49:25–28.

Middleton, A. L. A. 1977. Predatory behavior by common grackles. *Canadian Field-Naturalist* 91:187.

Mott, D. F., J. F. Besser, R. R. West, and J. W. DeGrazio. 1972. Bird damage to peanuts and methods for alleviating the problem. *Vertebrate Pest Conference* 5:118–120.

Mott, D. F., and C. P. Stone. 1973. *Bird damage to blueberries in the United States.* Special Scientific Report – Wildlife 172. U.S. Department of Interior, U.S. Fish and Wildlife Service, Washington, DC.

Oberholser, H. C. 1974. *The birdlife of Texas.* Vol. 2. University Texas Press, Austin, TX.

Peer, B. D., and E. K. Bollinger. 1997a. Common grackle (*Quiscalus quiscula*). In *The birds of North America*, Vol. 271, eds. A. Poole and F. Gill. Academy of Natural Sciences, Philadelphia, PA.

Peer, B. D., and E. K. Bollinger. 1997b. Explanations for the infrequent cowbird parasitism on common grackles. *Condor* 99:151–161.

Peer, B. D., H. J. Homan, G. M. Linz, and W. J. Bleier. 2003. Impact of blackbird damage to sunflower: bioenergetic and economic models. *Ecological Applications* 13:248–256.

Peer, B. D., H. J. Homan, and S. G. Sealy. 2001. Infrequent cowbird parasitism on common grackles revisited: New records from the northern Great Plains. *Wilson Bulletin* 113:90–93.

Peer, B. D., J. W. Rivers, and S. I. Rothstein. 2013. Cowbirds, conservation, and coevolution: Potential misconceptions and directions for future research. *Chinese Birds* 4:15–30.

Peer, B. D., and S. I. Rothstein. 2010. Phenotypic plasticity in common grackles in response to repeated brood parasitism. *Auk* 127:293–299.

Peer, B. D., S. I. Rothstein, and R. A. McCleery. 2010. Intraclutch variation in egg appearance constrains rejection of brown-headed cowbird eggs in common grackles. *Auk* 127:759–764.

Peer, B. D., and S. G. Sealy. 2004. Fate of grackle defenses in the absence of brood parasitism: Implications for long-term brood parasite-host coevolution. *Auk* 121:1172–1186.

Peterjohn, B. G. 1989. *The birds of Ohio.* Indiana University Press, Bloomington, IN.

Peterjohn, B. G., J. R. Sauer, and W. A. Link. 1994. The 1992 and 1993 summary of the North American Breeding Bird Survey. *Bird Populations* 2:46–61.

Peterson, A., and H. Young. 1950. A nesting study of the bronzed grackle. *Auk* 67:466–476.

Pierce, R. A. 1970. Bird depredations on rice and other grains in Arkansas. *Bird Control Seminar* 5:101–109.

Powell, A. F. L. A., F. K. Barker, S. M. Lanyon, K. J. Burns, J. Klicka, and I. J. Lovette. 2014. A comprehensive species-level molecular phylogeny of the new world blackbirds (Icteridae). *Molecular Phylogenetics and Evolution* 71:94–112.

Robertson, R. J., P. J. Weatherhead, F. J. S. Phelan, G. L. Holroyd, and N. Lester. 1978. On assessing the economic and ecological impact of winter blackbird flocks. *Journal of Wildlife Management* 42:53–60.

Robertson, W. B., Jr., and G. E. Woolfenden. 1992. *Florida Bird Species: An Annotated List.* FOS Special Publication 6. Florida Ornithological Society, Gainesville, FL.

Rosenberg, K. V., J. A. Kennedy, R. Dettmers, et al. 2016. *Partners in Flight Landbird Conservation Plan: 2016 Revision for Canada and Continental United States.* Partners in Flight Science Committee. http://www.partnersinflight.org/ (accessed September 25, 2016).

Ruxton, G. D., M. Broom, and N. Colegrave. 2001. Are unusually colored eggs a signal to potential conspecific brood parasites? *American Naturalist* 157:451–458.

Sauer, J. R., D. K. Niven, J. E. Hines, D. J. Ziolkowski, Jr, K. L. Pardieck, J. E. Fallon, and W. A. Link. 2017. The North American Breeding Bird Survey, results and analysis 1966–2015. Version 2.07.2017 USGS Patuxent Wildlife Research Center, Laurel, MD. https://www.mbr-pwrc.usgs.gov/bbs/bbs.html (accessed April 30, 2017).

Schorger, A. W. 1941. The bronzed grackle's method of opening acorns. *Wilson Bulletin* 53:238–240.

Selander, R. K. and D. R. Giller. 1960. First year plumages of the brown-headed cowbird and redwinged (sic) blackbird. *Condor* 62:202–214.

Sealy, S. G. 1994. Observed acts of egg destruction, egg removal, and predation on nests of passerine birds at Delta Marsh, Manitoba. *Canadian Field-Naturalist* 108:41–51.

Stone, C. P., and D. F. Mott. 1973. *Bird damage to sprouting corn in the United States.* Special Scientific Report Wildlife 173. U.S. Department of Interior, U.S. Fish and Wildlife Service, Washington, DC.

Stone, W. 1937. *Bird studies at old Cape May, II.* Delaware Valley Ornithology Club, Philadelphia, PA.

USDA. 2015. *Program Data Report G – 2015 Animals Dispersed/Killed or Euthanized/Freed.* https://www.aphis.usda.gov/wildlife_damage/pdr/PDR-G_Report.php (accessed October 23, 2016).

U.S. Department of Agriculture. 2015. *Environmental Assessment: Managing blackbird damage to sprouting rice in southwestern Louisiana.* U.S. Department of Agriculture Wildlife Services, Washington, DC.

Virgo, B. D. 1971. Bird damage to sweet cherries in the Niagara Peninsula, Ontario. *Canadian Journal of Plant Science* 51:415–423.

Web of Science. 2016. *Thomson Reuters.* https://www.webofknowledge.com/ (accessed September 7, 2016).

Wetmore, A. 1919. Notes on the structure of the palate in the Icteridae. *Auk* 36:190–197.

White, S. B., R. A. Dolbeer, and T. A. Bookhout. 1985. Ecology, bioenergetics, and agricultural impacts of a winter-roosting population of blackbirds and starlings. *Wildlife Monographs* 93:1–42.

Whoriskey, F. G., and G. J. FitzGerald. 1985. The effects of bird predation on an estuarine stickleback (Pisces: Gasterosteidae) community. *Canadian Journal of Zoology* 63:301–307.

Wiens, J. A. 1965. Behavioral interactions of red-winged blackbirds and common grackles on a common breeding ground. *Auk* 82:356–374.

Wiley, R. H. 1976. Affiliation between the sexes in common grackles I: Specificity and seasonal progression. *Zeitschrift für Tierpsychology* 40:59–79.

Wood, H. B. 1945. The sequence of molt in purple grackles. *Auk* 62:455–456.

CHAPTER 5

The Brown-Headed Cowbird: Ecology and Management of an Avian Brood Parasite

Brian D. Peer
Western Illinois University
Macomb, Illinois

Virginia E. Abernathy
Australian National University
ACT, Australia

CONTENTS

5.1	Taxonomy	79
	5.1.1 *Molothrus* Clade	79
	5.1.2 Subspecies	79
5.2.	Distribution	79
	5.2.1 The Cowbird's Dynamic Distribution	79
	5.2.2 Range Expansion and Conservation Concerns	80
5.3	Life History	80
	5.3.1 Brood Parasitic Lifestyle	81
	5.3.2 Factors Affecting Host Choice	82
5.4	Diet	83
5.5	Feather Molt	84
5.6	Behavior	84
	5.6.1 Mating System	84
	5.6.2 Female Breeding Ranges and Territoriality	84
	5.6.3 Depredation of Host Nests	85
	5.6.4 Cowbird Nestling Behavior	85
5.7	Migration	86
5.8	Populations	86
5.9	Agricultural Damages	88
5.10	Management	88
	5.10.1 Management for Endangered Host Species	88
	5.10.2 Agricultural Management	92
5.11	Coda and Future Research	92
	References	92

The brown-headed cowbird (*Molothrus ater*) is one member of the blackbird family (Icteridae) subject to population management due to the damage it inflicts on agricultural crops (Figure 5.1) (U.S. Department of Agriculture 2015) and, more controversially, because it is an obligate avian brood parasite (Peer et al. 2013a). Brown-headed cowbirds (henceforth "cowbirds") lay their eggs in the nests of other birds, which often raise fewer and in some cases none of their own young (Peer et al. 2013b). There are approximately 100 species of avian brood parasites, however, the cowbird is arguably the most vilified and one of the most maligned birds in North America by laypersons and scientists alike (Peer et al. 2013a, 2013b). This is in stark contrast to the most famous brood parasite, the common cuckoo (*Cuculus canorus*), which is revered and whose precipitous decline in Great Britain has generated concern (e.g., Douglas et al. 2010). Much of the vilification of the cowbird is based on misinformation about this species (Peer et al. 2013b). Cowbirds are no different than other "exploiters" in ecological communities such as hawks or snakes. Instead of appreciating the unique adaptations cowbirds have evolved for parasitism, most humans have a negative view of them because they do not care for their young. Much of the harm caused by cowbirds is actually a result of anthropogenic habitat degradation. Cowbirds have been in North America for at least 1 million years, thus any recent negative impact on hosts by cowbirds is due to human anthropogenic change to their habitats (Rothstein and Peer 2005).

Cowbirds are a greater management concern due to their effects on endangered songbirds rather than the damage they cause to agricultural crops. Cowbirds may have the potential to cause agricultural damage over a wider area compared to other blackbirds because of their ability to commute long distances between nest searching and feeding areas (e.g., Rothstein et al. 1984). The most significant issue for agriculture is the depredation of rice in southwestern Louisiana, where it is estimated that blackbirds in general cause $4 million worth of damage annually (Wilson et al. 1990). The vast majority of this damage, however, is caused by red-winged blackbirds (*Agelaius phoeniceus*) (Cummings et al. 2005). Relatively little research has been done to determine precisely how much of this damage is caused by cowbirds, yet the USDA killed >3.4 million cowbirds from 2009 to 2015 in Louisiana (U.S. Department of Agriculture 2015). Management decisions have apparently been based on a few, dated studies and models based on erroneous information, which we detail in this chapter.

Figure 5.1 Female brown-headed cowbird (left) and male (right). (Courtesy of Jim Rivers.)

5.1 TAXONOMY

5.1.1 *Molothrus* Clade

The cowbird is in the order Passeriformes and family Icteridae (Lowther 1993). There are four major subclades within the Icteridae: the meadowlarks and allies, caciques and oropendolas, orioles, and grackles and allies, of which the genus *Molothrus* is a member. Within the *Molothrus* clade, the screaming cowbird (*Molothrus rufoaxillaris*) is the most basal species, and the brown-headed cowbird is the most derived, with the bronzed cowbird (*Molothrus aeneus*) as its sister species (Powell et al. 2014). The *Molothrus* clade is about 2.8–3.8 million years old (Rothstein et al. 2002) and comprises the only obligate brood parasites within Icteridae (Lanyon 1992).

5.1.2 Subspecies

There are three subspecies of *M. ater*: *M. a. ater*, which occurs in eastern North America, *M. a. obscurus*, in western North America, and *M. a. artemisiae*, in the Great Basin and western Great Plains (Lowther 1993). *M. a. obscurus* is the only subspecies that is sympatric with the bronzed cowbird (Rothstein 1978). These subspecies may be maintained through assortative mating (Eastzer et al. 1985).

Several differences exist between these subspecies. *M. a. obscurus* is smaller than *M. a. artemisiae* (Rothstein et al. 1986) and has one type of flight whistle, whereas *M. a. artemisiae* has multiple whistle dialects (Rothstein and Fleischer 1987; Fleischer and Rothstein 1988). Additionally, *M. a. obscurus* nestlings have yellow rictal flanges, whereas *M. a. ater* and *M. a. artemisiae* have white flanges (Rothstein 1978). Using these flange differences, it has been found that *M. a. obscurus* occurs on the western slope of the Sierra Nevada mountain range, and *M. a. artemisiae* occurs on the eastern slope. Morphological evidence (nestling flange color and body size) and mito-chondrial DNA (mtDNA) indicate that these subspecies interbreed in parts of the Sierra Nevada range (Fleischer and Rothstein 1988; Fleischer et al. 1991). There may be an additional hybrid zone between *M. a. obscurus* and *M. a. artemisiae* in Colorado (Ortega and Cruz 1992).

5.2 DISTRIBUTION

5.2.1 The Cowbird's Dynamic Distribution

The cowbird's distribution is dynamic and reliant on landscape features (Morrison and Hahn 2002). Currently, cowbirds are widespread in North America, breeding from southeast Alaska and much of western Canada to southeast Canada and throughout most of the United States (Lowther 1993). They are year-round residents in the eastern half of the United States and in parts of the southwest, including north and central Mexico, and along the west coast, wintering in the southernmost extremes of their range (southern Baja California, southern Mexico, and south Florida; Lowther 1993).

It is widely reported that this current distribution is the result of European settlers clearing large tracts of forest, allowing the cowbird to expand its range over the past 300–400 years (Lowther 1993; Rothstein and Peer 2005). Cowbirds require open areas in which to feed and often feed in association with large ungulates (see Section 5.4). Immediately before the arrival of Europeans, cowbirds were mainly restricted to the grasslands of the Great Plains in central North America (Lowther 1993). Thus, the deforestation initiated by the Europeans allowed cowbirds to expand into most of the United States and into Canada (Rothstein and Peer 2005; Peer et al. 2013b). This, however, was not the first time the cowbird distribution had changed. Prior to the arrival of Europeans, Native Americans altered the landscape in the cowbird's favor, by burning forests for farming and to increase game populations (Pyne 1977). After contact with Europeans, the Native American populations collapsed, allowing forests to return before the early 1800s (Rothstein and Peer 2005; Peer et al. 2013b). Indeed, the dense forests Europeans found when they arrived in North America may have been a recent phenomenon that resulted from the rapidly dwindling Native American population (Mann 2005).

However, cowbirds were likely more widespread, especially during the Pleistocene. Fossils of cowbirds from the Late Pleistocene and Holocene (10,000–500,000 years ago) have been found in California, Oregon, New Mexico, Texas, Kansas, Florida, and Virginia (Lowther 1993), indicating that cowbirds were much more pervasive. The landscape and flora and fauna during this time were radically different from now (Pielou 1991). Megafauna such as mammoths, mastodons, bison, horses, and llamas were abundant, likely creating a virtual cowbird paradise, providing evidence that cowbirds were widespread prior to significant man-made changes to landscapes (Rothstein and Peer 2005; Peer et al. 2013b).

5.2.2 Range Expansion and Conservation Concerns

Since European settlement, the cowbird's distribution has continued to expand, first into the eastern third of the United States by the early 1800s (Rothstein 1994). By the 1900s cowbirds were present in the Canadian Maritime Provinces and the Pacific Slope (Rothstein 1994). Cowbirds had arrived in coastal southern California by 1920 and in Oregon by the 1940s (Rothstein 1994). In the Sierra Nevada range, cowbirds were absent before the 1930s but became very common by the 1970s (Rothstein et al. 1980). By the 1950s and 1960s, cowbirds had expanded to the southeastern United States and northwest Canada (Rothstein 1994) and only began breeding in southern Florida in the late 1980s and early 1990s (Cruz et al. 2000). They are still an uncommon breeding species in Florida (Post and Sykes 2011; Sauer et al. 2017). The recent expansion of cowbirds has caused concern, because it was assumed that they would begin exploiting new, naïve hosts that were defenseless against parasitism (Cruz et al. 2000). However, those fears have not materialized (see Section 5.10.1).

5.3 LIFE HISTORY

Scott and Ankney (1983) found that cowbirds lay eggs in "clutches" of one to seven eggs, with an average of 4–4.6 eggs, similar to nonparasitic icterids. Each clutch was separated by 1–2 days of nonlaying (Scott and Ankney 1983; Fleischer et al. 1987), though females can lay almost daily throughout the breeding season if provided with supplemental nutrition (Jackson and Roby 1992). It was previously believed that cowbirds had a higher annual fecundity than nonparasitic passerines, with the potential to lay 26–40 eggs per season (Scott and Ankney 1980; Fleischer et al. 1987; Jackson and Roby 1992; Holford and Roby 1993). However, more recent studies using molecular techniques to monitor the laying of individual females in the wild have demonstrated that the mean realized fecundity (i.e., the number of eggs laid in nests that could potentially fledge a cowbird nestling) is only between one and nine eggs per season (Hahn et al. 1999; Woolfenden et al. 2003). Eggs are laid before sunrise and a female cowbird only spends an average of 20–40 s on a host nest in order to lay her egg, which could aid a cowbird in avoiding detection by the host (Scott 1991; Sealy et al. 1995; Peer and Sealy 1999). Eggs are incubated from 10 to 14 days, and the average is 11 days (Lowther 1993; Peer and Bollinger 2000).

Nestlings can be found from mid-April to late August, with the majority from early May to early July (Lowther 1993). Nestlings leave the nest between 8 and 13 days, and fledglings become independent at 25–39 days (Woodward and Woodward 1979; Woodward 1983). Males become sexually mature after 1 year, but typically do not mate until year two due to constraints in song learning by this brood parasite (O'Loghlen and Rothstein 2003). In contrast, females typically breed after 1 year (Lowther 1993; Alderson et al. 1999).

5.3.1 Brood Parasitic Lifestyle

Cowbirds are generalist brood parasites and have been known to parasitize 248 host species, though only 172 have successfully reared a cowbird nestling (Lowther 2015). Common hosts include warblers, vireos, sparrows, flycatchers, thrushes, the northern cardinal (*Cardinalis cardinalis*), and

Figure 5.2 Parasitized red-winged blackbird nest with three blackbird eggs (bluish background with black scrawls) and two cowbird eggs (white with brown/gray spots). (Courtesy of Brian Peer.)

Figure 5.3 Parasitized nest of a northern cardinal with five host eggs and a single cowbird egg. The cowbird egg (upper left) is often very similar in appearance to cardinal eggs. (Courtesy of Brian Peer.)

other blackbirds (Lowther 1993; Figures 5.2 and 5.3). Many traits possessed by cowbirds that are beneficial for brood parasitism also occur in their nonparasitic blackbird relatives (Mermoz and Ornelas 2004). For example, compared to nonparasitic icterids, cowbirds do not have an accelerated rate of embryonic development (Kattan 1995; Strausberger 1998) or an accelerated nestling growth rate (Mermoz and Ornelas 2004). They do not have mimetic eggs for their hosts (Rothstein 1990; cf. Peer et al. 2000) and their incubation period is not significantly shorter than nonparasitic icterids (Mermoz and Ornelas 2004). However, the cowbird incubation period is shorter than many of its hosts, and cowbird nestling growth rate tends to be faster, increasing the chances that cowbird young will hatch first and grow more quickly, providing them a competitive advantage in the nest (Rothstein and Robinson 1998). The lower energy content and smaller size of their eggs, though no different from other icterids (Mermoz and Ornelas 2004), also allow cowbirds to hatch more quickly than many of their hosts (Kattan 1995; Strausberger 1998). The increased thickness of cowbird eggshells is an adaptation to brood parasitism (Mermoz and Ornelas 2004; Jaeckle et al. 2012). Thicker eggshells are more difficult for small-billed hosts and other female cowbirds to puncture and remove from a nest and protect the cowbird egg from being damaged during rapid laying events (Spaw and Rohwer 1987; B. Peer unpublished data). Cowbird eggshells also exhibit greater porosity than some hosts, including icterids, allowing increased gas flux and faster development (Jaeckle et al. 2012). Cowbirds may have higher fecundity (Scott and Ankney 1980; but see Section 5.3), faster and earlier egg laying than other passerines (Scott 1991; Sealy et al. 1995), and the females possess larger hippocampi for remembering the locations of host nests (Sherry et al. 1993). Together, these traits have allowed the cowbird to become a successful obligate brood parasite.

5.3.2 Factors Affecting Host Choice

Parasitism frequencies vary regionally (e.g., red-winged blackbirds and dickcissels [*Spiza americana*]; Linz and Bolin 1982; Searcy and Yasukawa 1995; Jensen and Cully 2005), which may be in part due to variation in local host communities (Barber and Martin 1997). Individual female cowbirds behave as generalists, using multiple host species throughout a breeding season, or as specialists, using one or a few hosts (Alderson et al. 1999; Woolfenden et al. 2003; Strausberger and Ashley 2005). Cowbirds often parasitize hosts in a nonrandom fashion, using only a fraction of the available species in an area (Strausberger and Ashley 1997; Chace et al. 2005; Patten et al. 2006; but see Rivers et al. 2010).

The suitability of a host is determined by its size, incubation period, diet, breeding season, and whether it has evolved defenses against brood parasitism. Larger hosts are more capable of raising cowbirds with their own young (Rothstein 1975), but if the cowbird nestling does not hatch earlier than the host young, it may not survive well with a larger host (Peer and Bollinger 1997). Likewise, hosts with shorter incubation periods may not be as suitable, because cowbirds would have more difficulty hatching before the host young and may not be able to compete as well for food. Diet is also important for host choice, because females must select hosts that feed their young a primarily insectivorous diet (Peer and Bollinger 1997). Nestling cowbirds cannot survive on the diets supplied by strictly granivorous or frugivorous species, although hosts such as the cedar waxwing (*Bombycilla cedrorum*), American goldfinch (*Spinus tristis*), and house finch (*Haemorhous mexicanus*) are often parasitized (Rothstein 1976; Middleton 1991; Kozlovich et al. 1996). A host can avoid parasitism if its peak breeding season is not aligned with that of the cowbird. The American goldfinch breeds later than many other passerines (Middleton 1991), whereas the common grackle (*Quiscalus quiscula*) is one of the earliest breeders (Peer and Bollinger 1997). Therefore, cowbird parasitism is reduced in these host species in part due to nonoverlapping breeding seasons.

Though many cowbird hosts accept parasitism (Peer and Sealy 2004a), some defend themselves and thus are poor choices as hosts. Host defenses include egg ejection (Rothstein 1975; Peer and Sealy 2004a), burial of the cowbird egg in the nest lining (Sealy 1995), nest desertion (Hosoi and Rothstein 2000), and mobbing of the parasite to prevent it from laying although this is not always an effective measure (Ellison and Sealy 2007). Hosts that eject parasitic eggs (rejecters) tend to do this 75%–100% of the time (Peer and Sealy 2004a). Therefore, a cowbird egg laid in a rejecter's nest is likely to fail, yet cowbird eggs have been occasionally found in such host's nests and in other unsuitable host nests (e.g., waterfowl, gulls; Lowther 2015). Additionally, hosts that appear to be suitable due to lack of defenses, similar nestling diet, and breeding season are sometimes not parasitized by cowbirds (Ortega and Cruz 1991).

Habitat and landscape features also affect host choice, both at the species and individual levels. Cowbird abundance decreases with increasing distance from forest edges and foraging areas; thus not surprisingly parasitism frequencies are typically higher on nests at the edges of forests and near cowbird foraging areas (Robinson et al. 1995a; Chace et al. 2005; Patten et al. 2006). The preferred breeding habitat appears to be at the edge, and cowbirds are not as common in heavily wooded areas or widespread grasslands (Lowther 1993; but see Robinson et al. 1995a). For example, in the Midwest cowbirds prefer hosts nesting in woodlands, followed by shrublands, and lastly grasslands (Peer et al. 2000). In the heart of the Great Plains, where there may not be as much alternative habitat available, grassland hosts are parasitized at much higher frequencies (Rivers et al. 2010). Within habitats, hosts nesting closer to perches are parasitized more often than hosts farther from perches (Patten et al. 2011).

Female cowbirds may preferentially choose individual hosts based on host age (Smith 1981), and more vocal and aggressive hosts may have higher rates of parasitism than quieter hosts (Smith and Arcese 1994; Clotfelter 1998; but see Gill et al. 1997). Better concealed nests are less likely to be parasitized (Briskie et al. 1990; Saunders et al. 2003; cf. Smith 1981), and in some cases nests higher in the forest canopy are parasitized less frequently (Briskie et al. 1990; Hahn and Hatfield 2000).

5.4 DIET

Primary food items include arthropods and seeds, with agricultural grains comprising a relatively small portion of the cowbird diet (Lowther 1993). The majority of grains consumed are seeds from grasses and weeds, although it varies seasonally and geographically (Lowther 1993). Insects comprise a larger percentage of the diet during the breeding season and females consume more calcium for egg laying (Ankney and Scott 1980). Female cowbirds also remove host eggs (see Section 5.6.3), which provides additional calcium. However, unlike the brood parasitic female common cuckoos, which always consume the eggs they remove, cowbirds often puzzlingly drop host eggs to the ground without consuming them (Sealy 1992; B. Peer unpublished data).

Of the agricultural crops consumed, corn and rice are the most frequent, although corn is typically eaten at feedlots and waste from harvested fields. In Ohio, 26% of the diet consisted of corn from August to October (Williams and Jackson 1981); 22%–46% of the diet was corn in Tennessee during the winter months (Dolbeer et al. 1978; White et al. 1985); and 65% of the stomach contents in Ohio during the winter consisted of corn (Dolbeer and Smith 1985). Cowbird damage to rice is more problematic because it occurs to ripening crops. Meanley (1971) found that 46% of the annual diet in Arkansas consisted of rice, and it increased to 68% from August to October (Meanley 1971). Cowbirds also consume rice pests, including the rice stinkbug (*Oebalus pugnax*) and rice water weevil (*Lissorhoptrus oryzophilus*; Meanley 1971). One method of reducing damage to rice is to treat seeds with caffeine (Avery et al. 2005).

Cowbirds forage on the ground in open areas and in association with large ungulates (bison, cattle, horses, to name a few) if they are present—hence the name *cowbird* (Friedmann 1929; Goguen et al. 2005). It is a commensal relationship in which the cowbird benefits when the quadrupeds stir up insects, making them more readily available for consumption; the quadrupeds are apparently unaffected (Friedmann 1929). Not surprisingly, given the cowbirds' predilection for eating corn and feeding in association with cattle, they are often found foraging in feedlots (Ortega 1998). Cowbirds are also unusual in that their foraging sites are located up to 7–16 km from the areas in which they search for host nests (Rothstein et al. 1984; Curson and Mathews 2003). They typically spend the mornings laying, searching for host nests, and maintaining pair bonds and forage later in the day in open areas (Ortega 1998).

5.5 FEATHER MOLT

Molothrus ater ater has no alternate plumage; *M. a. obscurus* has a partial pre-alternate molt of the head feathers (Lowther 1993; Ortega 1998). The prebasic molt occurs from July to October (Lowther 1993; Ellis et al. 2012) and results in an increase in the basal metabolic rate by 13% (Lustick 1970). It also suppresses constitutive innate immune function, especially in females (Ellis et al. 2012). Because female cowbirds are host generalists, they are exposed to a wide variety of pathogens when they visit many different types of host nests, and their immune system is predictably more robust compared to cowbird species that parasitize fewer hosts and other icterids (Hahn and Reisen 2011). This reduction in immunity during molt may translate into higher mortality for female cowbirds, and it has even been suggested to account for the male-biased secondary sex ratio (Ellis et al. 2012) that can be observed after their first breeding season (Darley 1971).

5.6 BEHAVIOR

5.6.1 Mating System

The cowbird mating system takes almost every form from monogamy (Robinson et al. 1995b), to promiscuity (Woolfenden et al. 2002; Strausberger and Ashley 2003), to polygyny (Teather and Robertson 1986), and polyandry (Friedmann 1929). This variation may be driven by a male-biased sex ratio (see Section 5.10.2) or cowbird density (Yokel 1989; Strausberger and Ashley 2003). Cowbirds appear to be socially monogamous, having one social partner, but also engage in extra-pair copulations (Robinson et al. 1995a; Strausberger and Ashley 2003). Cowbird pair bonds can endure for several seasons (Dufty 1982; Woolfenden 2000). The degree of promiscuity varies among populations, however, and could be due to higher cowbird density (Elliott 1980; Strausberger and Ashley 2003).

5.6.2 Female Breeding Ranges and Territoriality

Cowbirds are highly social, often seen foraging in large, mixed flocks with other blackbirds (Lowther 1993). Territorial and aggressive behavior has rarely been observed in cowbirds at feeding areas (Rothstein et al. 1984; Teather and Robertson 1985) and males do not defend territories, although they appear to have breeding ranges (Rothstein et al. 1984; Yokel 1986). Females maintain distinct areas for breeding, but these areas often overlap (Rothstein et al. 1984; Fleischer 1985). The size of female breeding ranges also varies considerably within and among populations (Table 5.1). Breeding ranges in eastern North America are similar in size, whereas those in western North America are much larger. This may be explained by the fact that both cowbirds and hosts are less common in the west and are spread over larger areas compared to areas in the east (Dufty 1982; Rothstein et al. 1984; Hahn et al. 1999; Strausberger and Ashley 2003).

The large breeding and feeding ranges and the distances cowbirds must travel between them lead to cowbirds having large home ranges of 405–3,198 ha (Rothstein et al. 1984; Goguen and Mathews 2001). Understanding this aspect of cowbird behavior is important from a management

Table 5.1 Variation in Female Cowbird Breeding Range Sizes (ha)

Average Breeding Range Size (ha)	Total Range of Breeding Range Sizes (ha)	Study Site	Source
78 ± 43.67	40–150	Sierra Nevada	Rothstein et al. 1984
64	20.7–188.9	New Mexico	Goguen and Mathews 2001
20.4 ± 6.68	9.9–33.2	New York	Dufty 1982a
9.38 ± 7.9	2.6–32.2	New York	Hahn et al. 1999
9.10 ± 8.31	0.43–27.04	Illinois	Strausberger and Ashley 2003
8.58 ± 3.25	5.45–15.51	Ontario	Teather and Robertson 1985
4.5 ± 0.4	0.9–13.4	Ontario	Darley 1983

perspective, because cowbird abundance and parasitism frequencies are directly related to distance from feeding sites (Morrison and Hahn 2002; Chace et al. 2005; Howell et al. 2007). Increasing this distance and forcing cowbirds to make longer commutes could be more effective in reducing parasitism frequencies of endangered species (see Section 5.10.1; Chace et al. 2005).

5.6.3 Depredation of Host Nests

Female cowbirds often remove a host egg the day before or after parasitizing a nest (Sealy 1992) and have been observed eating host eggs (Scott et al. 1992). Eating eggs has obvious nutritional benefits, and egg removal may also increase incubation efficiency of the parasitic egg and reduce competition for the cowbird nestling (Peer 2006). Female cowbirds, and less frequently males, also peck host eggs or remove host nestlings when they discover nests too far into incubation for successful parasitism (Arcese et al. 1996; Granfors et al. 2001; Peer 2006; Dubina and Peer 2013). This behavior forces the host to renest and provides the cowbird with another opportunity to parasitize it (Arcese et al. 1996). Females destroy more eggs in nests that are later in development than nests early in incubation and destroy more eggs in nests where laying appears to be complete, suggesting that cowbirds track host reproductive cycles (Swan et al. 2015).

Brood parasites may engage in "mafia" tactics whereby they force hosts to accept their offspring; otherwise the parasite will destroy the host's eggs or nestlings (Zahavi 1979). One study found that cowbirds may utilize this behavior when parasitizing prothonotary warblers (*Protonotaria citrea*). Nests in which cowbird eggs were experimentally removed experienced a significantly higher rate of predation (Hoover and Robinson 2007). However, questions remain concerning this behavior, in part due to the fact that it was unknown whether the same females that laid the eggs returned to depredate the nests (Peer et al. 2013b). Additional study of this fascinating behavior is warranted.

5.6.4 Cowbird Nestling Behavior

Nestling size difference, rather than begging calls, is more important in affecting provisioning rates by parents, with parents feeding larger nestlings who reach higher in the nest more than smaller nestlings (Lichtenstein and Sealy 1998; Glassey and Forbes 2003; Rivers 2007). This behavior would explain why cowbirds fare better with smaller or intermediate-sized nestlings (Kilner 2003). Lengthy and intense cowbird begging may stimulate host nestlings to beg more intensely, thereby increasing feeding rates of the parents, but this effect has only been observed in smaller hosts (Dearborn 1998; Glassey and Forbes 2003). Interestingly, cowbirds are less virulent in populations in which they are more likely to share a nest with a sibling, indicating that kin selection constrains their virulence and aggressive begging behavior (Rivers and Peer 2016).

5.7 MIGRATION

M. a. ater is a short distant migrant, whereas the movements of *M. a. obscurus* and *M. a. artemesiae* are less clear because they winter within much of their breeding ranges (Ortega 1998). They migrate during the day in mixed flocks with European starlings (*Sturnus vulgaris*) and other blackbirds, including common grackles, red-winged blackbirds, and less often Brewer's (*Euphagus cyanocephalus*) and rusty blackbirds (*Euphagus carolinus*; Lowther 1993), and can be found foraging in open habitats (B. Peer, per obs). Roosts in the fall include early successional forests in New Jersey (Lyon and Caccamise 1981).

Cowbirds move ~40 km per day in the spring (Lowther 1993) and leave Mexico and the southern United States around March 1. They typically arrive in the central United States in early to mid-March (B. Peer, unpubl. data) and reach the northernmost portions of their range by mid-April (Friedmann 1929). Similar to other songbirds, males arrive on the breeding grounds before females (Friedmann 1929), and older birds arrive before younger individuals (Darley 1983). Males also depart first in the fall (Rothstein et al. 1980; Ortega 1998). *M. a. ater* arrives relatively late during the breeding season and departs sooner than many of its hosts at the end of the season in late July and early August (Rothstein et al. 1980; Ortega 1998).

Cowbirds gather with other blackbirds and starlings in winter roost sites in the southern states. They winter in Kentucky in deciduous trees around the margin of conifer stands (Robertson et al. 1978); in Oklahoma they roost in cattails (Lowther 1993). Roost sizes and composition vary. In Louisiana, which according to Christmas Bird Count (CBC) data has more cowbirds than any other state, has roosts that averaged 8.7 million cowbirds from 1974 to 1993 (Ortego 2000). These numbers fluctuated dramatically, with numbers as low as 28,000 and highs of 38,000,000 (Ortego 2000).

Cowbirds are more likely to return to the same breeding site than wintering location (Ortega 1998), but they are less likely to return to a previous breeding site compared to other blackbirds and starlings (Dolbeer 1982). The average annual distance between breeding sites in one study was 44 km for females and 95 km for males; the annual wintering distance was 244 km for females and 288 km for males (Dolbeer 1982). Young cowbirds in eastern North America tend to return to their natal area to breed (Payne 1976), whereas those in the west have higher dispersal rates than adults, regardless of sex (Anderson et al. 2005, 2012). This variation could be due to differences in habitat type and host density in the east versus west. Such high dispersal rates could reduce the effectiveness of control programs, because if new cowbirds are constantly immigrating into an area large numbers of cowbirds might need to be killed each year to keep parasitism frequencies low (Rothstein and Cook 2000; Rothstein et al. 2003).

5.8 POPULATIONS

The current size of the cowbird population in North America is estimated to be 120 million (Rosenberg et al. 2016). One of numerous myths (see Section 5.10.1) concerning cowbirds is that their populations are rapidly increasing, and as a result their hosts are in serious jeopardy (Rothstein and Peer 2005; Peer et al. 2013b). The reality is that cowbirds have steadily *declined*. Partners in Flight reports a 23% population decline from 1970 to 2014 (Rosenberg et al. 2016), and Breeding Bird Survey (BBS) data indicate a –0.66% annual decline from 1966 to 2015 and a –0.07% annual decline from 2005 to 2015 (Sauer et al. 2017; Figure 5.4). In 1966, the relative abundance of the cowbird survey-wide was 17.6, and in 2015 that value had dropped to 12.8 (Sauer et al. 2017). The decline is also significant in the majority of states that permit cowbird trapping programs (Peer et al. 2013b). Likewise, parasitism rates have declined substantially. In Missouri, Cox et al. (2012) found that parasitism of three common hosts—Acadian flycatchers (*Empidonax virescens*), indigo buntings (*Passerina cyanea*), and northern cardinals (*C. cardinalis*)—declined significantly over a 20-year period. Similarly, Rivers et al. (2010) found that parasitism on dickcissels (*S. americana*), their most important host in that region, declined significantly over several decades in Kansas.

The decline of cowbirds in Louisiana has been even greater according to both Christmas Bird Count Data (National Audubon Society 2010) and BBS data (1967–2015: –1.08% annual decline; Sauer et al. 2017, Figure 5.5). The relative abundance of cowbirds in 1967 was 20.4, and in 2015 it had dropped precipitously to 12.1 (Sauer et al. 2017). Roadside surveys conducted in the rice growing district of southwestern Louisiana from March to May 1979 revealed that the cowbird was the second most common blackbird encountered to the red-winged blackbird, but they were

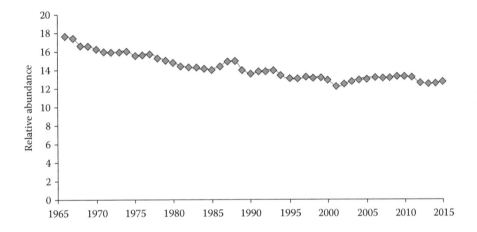

Figure 5.4 Population trend of the brown-headed cowbird surveywide from the Breeding Bird Survey, 1966–2013. (Sauer et al. 2017.)

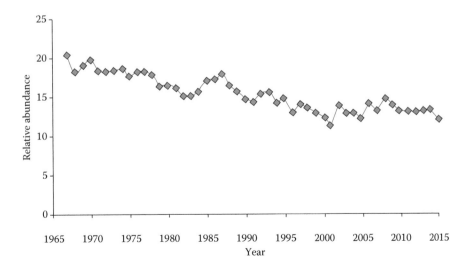

Figure 5.5 Population trend of the brown-headed cowbird in Louisiana from the Breeding Bird Survey, 1967–2013. (Sauer et al. 2017.)

a distant second (Wilson 1985). Notably, these data are also >35 years old. Cowbirds represented only 20%–21% of blackbirds observed during the spring and 9%–20% of blackbirds in the fall, whereas red-winged blackbirds were observed 77%–80% and 77%–91%, respectively (Wilson 1985). Undoubtedly, control programs implemented for endangered songbirds and rice depredation have contributed to the decline of cowbirds (see Section 5.10).

5.9 AGRICULTURAL DAMAGES

In other blackbirds, such as the red-winged blackbird, yellow-headed blackbird (*Xanthocephalus xanthocephalus*), and common grackle, molt and migration coincide, which leads to greater damage to maturing corn and sunflower crops (see Chapters 2 through 4 of this volume). In contrast, cowbirds begin migrating southward sooner (Lowther 1993), and most of the agricultural damage they cause is to sprouting rice in the southern states (Meanley 1971). Cowbirds also have the potential to commute long distances between areas in which their hosts nest and feeding locations (Rothstein et al. 1984), which could cause damage over a greater area when compared to other blackbird species.

In addition to cowbirds, red-winged blackbirds, common grackles, and boat-tailed grackles (*Quiscalus major*) cause the most damage to rice (U.S. Department of Agriculture 2015). Late-winter and early spring planting from February to March is concurrent with the large number of resident and wintering blackbirds in the rice growing region of Louisiana. Not surprisingly, crops near roost sites suffer the greatest harm (Neff and Meanley 1957; Pierce 1970). Farmers noted losses in the 1920s (Kalmbach 1937), and in 1980 Congress advised Wildlife Services to begin working with farmers to help alleviate bird-related damages (U.S. Department of Agriculture 2015).

Louisiana ranked third in acreage planted for rice with 160,000 ha in 2012, and the estimated value of the rice crop in 2014 was $454 million (U.S. Department of Agriculture 2015). From 1973–1982 estimated losses from blackbird damage was about $US11.3 million nationally (Besser 1985), whereas damage in 2001 was reported to be valued at $US21.5 million nationwide (Cummings et al. 2005). However, the amount of damage caused by cowbirds is poorly documented. The primary justification for cowbird control is a single study published almost five decades ago (Meanley 1971). A more recent and thorough review indicated that the amount of rice and agricultural grains consumed by cowbirds is relatively minor (Lowther 1993). Cowbirds are also cited as the second most common bird found in a roadside survey in the rice growing region of Louisiana (Wilson 1985) and the second most abundant bird observed in the Louisiana CBC (see Section 5.7). However, as noted above, they are a very distant second to red-winged blackbirds, and cowbird populations have been declining in Louisiana and throughout their range. What is lacking, and desperately needed, are field studies of food choice by cowbirds, updated population surveys, and bioenergetics and economic models similar to what have been constructed for blackbirds depredating sunflower crops (e.g., Peer et al. 2003).

5.10 MANAGEMENT

5.10.1 Management for Endangered Host Species

Like all birds native to North America, the cowbird is protected by the Migratory Bird Treaty Act (see Chapter 1 of this volume). However, it is unique in that it is the target of lethal control measures because it depredates agricultural crops and because it is a brood parasite. The dogma and misinformation associated with lethal trapping programs to benefit endangered warblers and vireos has been described in detail elsewhere (Rothstein and Peer 2005; Peer et al. 2013b), and therefore we will briefly summarize it here. The cowbird is the most frequently studied avian brood parasite in the world, but it is also one of the most hated birds in the world (Peer et al. 2013a, b). Until the 1980s the majority of publications on cowbirds concentrated on their basic biology and interactions with host species. Three publications that implicated cowbirds in the decline of songbirds changed the research focus to cowbird management and the "cowbird problem." The first was by Brittingham and Temple (1983), who reported that cowbird populations were increasing. However, their data were based on Christmas Bird Counts, which were misleading due to a number of confounds described in Rothstein and Peer (2005). In contrast, BBS data indicate that cowbirds have declined since 1966 (see Section 5.8). Terborgh (1989) then suggested that four factors were the cause of declining bird populations in North America, including habitat loss in the breeding and wintering grounds, increased nest predation, and cowbird parasitism. Finally, Robinson (1992; see also Robinson et al. 1995a) reported extremely high rates of cowbird parasitism in forest fragments in Illinois, and most of these nests were also depredated (cf. Bollinger et al.1997). Research on management of cowbirds and their hosts increased through the mid-2000s, likely due to these publications, but the trend reversed after 2005, when most publications began focusing on basic biology once again (Peer et al. 2013b). This change was perhaps in part due to the paper by Rothstein and Peer (2005), who pointed out that much of what was believed about cowbirds was actually false and outlined the following myths associated with cowbird ecology and evolution.

Myth 1: Cowbird populations are increasing.

As addressed above, cowbird populations have steadily declined since 1966. This myth is particularly relevant to managers focused on agricultural damage because we are in need of more up-to-date population modeling of this species in regions where crop losses are of concern.

Myth 2: Cowbirds are new to North America.

While cowbirds are of a recent origin when compared to other brood parasitic lineages (Rothstein et al. 2002; Spottiswoode et al. 2011), they have been in North America for at least 1 million years (Rothstein et al. 2002). Contrary to suggestions that cowbirds increased their range only after the arrival of Europeans and their subsequent modification of the habitat, fossil evidence and a Pleistocene landscape filled with megafauna indicates that cowbirds have been in North America and parasitizing hosts for a very long period of time (see Section 5.2.1). Therefore, any host populations that could not withstand parasitism went extinct, and those that are now at risk must be due to human activities (Rothstein and Peer 2005; Peer et al. 2013b).

Myth 3: Cowbird parasitism limits or reduces host populations.

Parasitism is one of many ecological factors (e.g., predation, competition, resource availability, disease, weather, etc.) that limit the population sizes of birds (Newton 1998). Despite these limitations, bird populations are still sustained, in some cases through source–sink population dynamics, in which small populations in fragmented habitats are supplemented by recruitment from larger populations that produce a surplus of young (Donovan et al. 1995). Some populations may be limited by both predation and cowbird parasitism. Edge habitat that is preferred by cowbirds is also preferred by nest predators. However, when a nest is parasitized by a cowbird some host young often survive, whereas a depredated nest produces no young (Rothstein and Peer 2005; Peer et al. 2013b). All host populations that are at risk from cowbird parasitism have also suffered significant habitat losses, and parasitism aggravates the effects of habitat loss for most species of conservation concern (Rothstein and Peer 2005; Peer et al. 2013b).

Myth 4: New hosts are defenseless against parasitism and are at risk when coming into contact with cowbirds.

Numerous studies have demonstrated that antiparasite behaviors such as egg ejection often evolve within a lineage and are retained for long periods of time even through speciation events (Peer and Sealy 2004b; Peer et al. 2007, 2011a, 2011b). Hosts that nest beyond the range of cowbirds or other brood parasites still respond to experimental parasitism by ejecting the foreign eggs. For example, gray catbirds (*Dumetella carolinensis*) on Bermuda, where there are no cowbirds, reject foreign eggs at a frequency similar to those nesting in mainland North America (Rothstein 2001). Likewise, island scrub jays (*Aphelocoma insularis*) reject eggs despite the fact that there are no breeding cowbirds on Santa Cruz Island and they have maintained this behavior as long as 150,000 years (Peer et al. 2007). Lastly, bohemian waxwings (*Bombycilla garrulus*) nesting beyond the range of the cowbird in Alaska reject 100% of experimental cowbird parasitism. They evolved this trait approximately 3 million years ago and retain it despite its lack of current utility (Peer et al. 2011a). There are only a few instances in which hosts' defenses have appeared to decline and in neither has the trait disappeared completely, suggesting that if brood parasitism were renewed the trait would increase rapidly to benefit the host (Peer and Sealy 2004b; Kuehn et al. 2014).

Myth 5. Control programs always result in increased host population sizes.

Indeed, if an individual removes a cowbird from the nest, the reproductive success of that individual typically increases, but control programs do not necessarily translate into larger host population sizes (Peer et al. 2013b). Cowbird culling has been implemented as a part of the management plans for the endangered southwestern willow flycatcher (*Empidonax traillii extimus*) and least Bell's vireo (*Vireo bellii pusillus*) in California, and the black-capped vireo (*Vireo atricapilla*) and to a lesser extent the golden-cheeked warbler (*Setophaga chrysoparia*) in Texas, and one of the best-known examples is the Kirtland's warbler (*Setophaga kirtlandii*) in Michigan and Wisconsin (Morrison et al. 1999; Smith et al. 2000; Ortega et al. 2005). There is no compelling evidence that cowbird trapping prevented the Kirtland's Warbler from extinction (Rothstein and Peer 2005; cf. USDA APHIS 2015). That is not to say that implementing the control program was not the correct response; it clearly was at the time (Rothstein and Peer 2005; Peer et al. 2013b). This warbler is an extreme habitat specialist requiring jack pine (*Pinus banksiana*) trees between 6 and 24 years old to nest (DeCapita 2000). Cowbird control was initiated in 1972 following the 1971 census, which reported only 201 singing males (Mayfield 1972). More than 150,000 cowbirds have been killed in an effort to benefit the warbler populations (U.S. Fish and Wildlife Service 2011), yet the number of males recorded remained constant at ~200 until the early 1990s, when nesting increased (DeCapita 2000). Thus, despite 20 years of cowbird removal and parasitism frequencies near nil, the population did not increase, indicating that cowbirds were not responsible for the lack of increase in warblers. Currently, the warbler population is at an all-time high of 2,365 singing males and now it also breeds in Wisconsin and Ontario (Richard 2008; U.S. Fish and Wildlife Service 2016). Cowbird control has remained constant, and it was only the increase in habitat that led to the increased population size of this rare warbler species (Rothstein and Peer 2005; Peer et al. 2013b). Least Bell's vireo and black-capped vireo populations increased after cowbird control programs were implemented, but there was a concomitant increase in habitat for both species (Rothstein and Peer 2005). Southwestern willow flycatcher numbers have not increased in response to cowbird culling, however, the riparian habitat on which they are dependent has been decimated (Rothstein and Peer 2005).

Because cowbird control has benefited some populations, it is believed that it should be implemented any time a species is affected by parasitism (Rothstein and Peer 2005). There are also immediate demonstrable effects when female cowbirds are killed rather than waiting on habitat restoration. However, this viewpoint is simplistic because it ignores the fact that cowbirds are native and parasitism is a natural process. If someone were to advocate the killing of raptors that feed on other birds the outcry would be predictable (Rothstein and Peer 2005). There are also numerous adverse effects associated with cowbird control (reviewed in Rothstein and Peer 2005), and these include the killing of nontarget species, control decisions based on reports instead of peer-reviewed science, special interest groups that encourage random killing of cowbirds without scientific justification, businesses with profit motives, and the prospect that cowbirds may be keystone species that benefit entire communities.

This myth is also relevant to crop damage management. Do we know how much damage is reduced by killing a specific number of cowbirds? What is the minimum number of birds that need to be killed to achieve acceptable losses, and must they be killed in a small area or over a wider range to reduce damage to these acceptable levels?

Rothstein and Peer (2005) recommended that prior to beginning any additional cowbird control programs, wildlife managers should consider seven questions based on the United States Fish and Wildlife Service [USFWS] (2002) recovery plan for the southwestern willow flycatcher. These questions are especially important considering the current funding climate and whether these funds would be better spent restoring habitat. First, is there a legal obligation to control cowbird populations? The Endangered Species Act states that any harmful effects on endangered species should be ameliorated, and this includes cowbird parasitism. However, cowbird control is instituted so routinely that there is almost no attempt to evaluate whether it is actually necessary (Rothstein and Peer 2005; cf. Wilsey et al. 2013). Second, are cowbirds the immediate factor limiting host populations? In some cases, cowbird parasitism exacerbates the effects of habitat loss on endangered species. However, in other situations, like the southwestern willow flycatcher, parasitism is a problem in some populations but not others (Rothstein and Peer 2005). Third, under what demographic criteria should control be started? In other words, is there a specific critical parasitism frequency at which a control program should be implemented? This question obviously requires extensive knowledge of the species, its population dynamics, and the effects of parasitism on individual fitness. Fourth, what are the goals of the cowbird control program? If a program is initiated, specific goals should be included along with regular external peer review, so that it does not continue indefinitely when its efficacy may be limited. A better approach may be to conduct control programs on an experimental basis to ensure that they are effectively addressing the primary problem limiting a particular population (USFWS 2002). Fifth, can landscape modification alleviate the negative effects of cowbird parasitism? Cowbird abundance has declined over the past 100 years in reforested areas (Chace et al. 2005). Therefore, a long-term solution to benefit endangered hosts may be habitat restoration or increasing the distance cowbirds must travel between forested breeding habitat and open landscapes for foraging, as opposed to cowbird control programs (Robinson et al. 1995b; Chace et al. 2005; Ortega et al. 2005; Peer et al. 2013b). Sixth, can we determine when cowbird control can be ended by using models? Rothstein and Peer (2005) describe the least Bell's vireo cowbird control program at Camp Pendleton, which began in the 1980s, when there were only around 50 nesting pairs of vireos that also experienced a 50% parasitism rate. There has been a 20-fold increase in the number of vireos, and even if cowbirds also showed a similar increase, the parasitism rate would be a mere 1/20th of what it was prior to cowbird control. Seventh, if cowbird control is required, can it be done more effectively? We described the problems associated with cowbird control programs and these must be addressed, in addition to doing a better job of educating the public about the need for such programs from both an ethical and financial viewpoint.

This set of guidelines can be modified for the management of cowbirds and indeed all other birds responsible for agriculture damage described in this book. For example:

1. Is there a legal obligation to lethally control cowbirds to alleviate agricultural damage?
2. At what level of crop loss should lethal control be initiated?
3. Are cowbirds responsible for the majority of damage or are other species to blame? If they are not, then effort should be invested in controlling the other species.
4. What are the stipulated goals of the control program? In other words, is there a specific level of damage that can be tolerated and is there a scenario where lethal control can be ended?
5. If lethal measures are still required, can they be done more effectively? Habitat restoration has led to the decrease of cowbird parasitism and cowbird populations, so can the same approach reduce crop losses? Modifying habitat by reducing cattail in wetlands has benefited sunflower growers in the northern Great Plains (Leitch et al. 1997). Thus, perhaps a similar long-term approach can be adopted to mitigate agricultural damage.

5.10.2 Agricultural Management

More cowbirds are killed annually by the USDA than any other native species ($n = 542,231$ in 2014; USDA 2015) and cowbirds represented 20% of all animals killed by the USDA in 2015. The number of cowbirds taken is second only to the number of nonnative European starlings killed ($n = 1,140,309$ in 2014). Louisiana Wildlife Services killed an estimated 3.4 million cowbirds from 2009 to 2015 using DRC-1339 and an additional 1,340 from 2009 to 2014 using firearms and traps (USDA 2015).

Louisiana Wildlife Services Environmental Assessment (USDA 2015) suggests that they could kill 1 million cowbirds annually based on their "take models." However, these models are seriously flawed. First, they estimate that the fall population of cowbirds could peak at 51.7 million, based in part on a 1:1 secondary sex ratio and female cowbirds laying 40 eggs annually. Both of these values are incorrect. It is well known that cowbirds have a decidedly male-biased secondary sex ratio, owing to the fact that the female parasitic lifestyle apparently leads to increased mortality (see Section 5.5). All 21 studies reviewed by Ortega (1998) reported a male-biased sex ratio, and in some cases it was as high as 6.3 male : 1 female. Second, genetic studies have revealed that previous estimates that female cowbirds lay 40 eggs per season (e.g., Scott and Ankney 1980) are likely incorrect. The maximum number of eggs laid by a female cowbird in a season was 13 and the mean was only three to four eggs per season (Alderson et al. 1999; Strausberger and Ashley 2005; see Section 5.3).

These issues, combined with the fact that cowbird control to alleviate damage to the sprouting rice crop has apparently been justified on the basis of a single dietary study nearly five decades old (Meanley 1971), raises doubt as to whether such drastic control measures are warranted. At a minimum, additional research is required to determine the dietary preferences of cowbirds. Obviously the take models must be updated to consider that cowbird populations are declining and to use accurate demographic information, similar to what has been done with the red-winged blackbird (Peer et al. 2003).

5.11 CODA AND FUTURE RESEARCH

Cowbirds are unique in that they are culled to benefit endangered songbird populations and to decrease agricultural losses. Cowbird control for songbird management has been carefully scrutinized, but the continuation of these programs requires additional study and justification. The same needs to be done for management to alleviate crop losses. It is especially warranted given that cowbird populations are declining and have been for >40 years. Indeed, lethal control of cowbirds has undoubtedly contributed to their decline, in addition to increased forested habitat. Studies on diet must be conducted considering that management plans are based on a single study conducted 50 years ago. In view of the number of cowbirds killed, this is urgently needed. Once this basic information is obtained, then more realistic models of economic damage can be constructed and a more responsible management plan can be implemented (e.g., Peer et al. 2003).

REFERENCES

Alderson, G. W., H. L. Gibbs, and S. G. Sealy. 1999. Determining the reproductive behaviour of individual brown-headed cowbirds using microsatellite DNA markers. *Animal Behaviour* 58:895–905.

Anderson, K. E., M. Fujiwara, and S. I. Rothstein. 2012. Demography and dispersal of juvenile and adult brown-headed cowbirds (*Molothrus ater*) in the eastern Sierra Nevada, California, estimated using multistate models. *Auk* 129:307–318.

Anderson, K. E., S. I. Rothstein, R. C. Fleischer, and A. L. O'Loghlen. 2005. Large-scale movement patterns between song dialects in brown-headed cowbirds (*Molothrus ater*). *Auk* 122:803–818.

Ankney, C. D., and D. M. Scott. 1980. Changes in nutrient reserves and diet of breeding brown-headed cowbirds. *Auk* 97:684–696.

Arcese, P., J. N. M. Smith, and M. I. Hatch. 1996. Nest predation by cowbirds and its consequences for passerine demography. *Proceedings of the National Academy of Sciences of the United States of America* 93:4608–4611.

Avery, M. L., S. J. Werner, J. L. Cummings, J. S. Humphrey, M. P. Milleson, J. C. Carlson, T. M. Primus, and M. J. Goodall. 2005. Caffeine for reducing bird damage to newly seeded rice. *Crop Protection* 24:651–657.

Barber, D. R., and T. E. Martin. 1997. Influence of alternate host densities on brown-headed cowbird parasitism rates in black-capped vireos. *Condor* 99:595–604.

Besser, J. F. 1985. *A grower's guide to reducing bird damage to US agriculture crops*. Bird Damage Research Report 340. U.S. Fish and Wildlife Service Denver Wildlife Research Center, Denver, CO.

Bollinger, E. K., B. D. Peer, and R. W. Jansen. 1997. Status of Neotropical migrants in three forest fragments in Illinois. *Wilson Bulletin* 109:521–526.

Briskie, J. V., S. G. Sealy, and K. A. Hobson. 1990. Differential parasitism of least flycatchers and yellow warblers by the brown-headed cowbird. *Behavioral Ecology and Sociobiology* 27:403–410.

Brittingham, M. C., and S. A. Temple. 1983. Have cowbirds caused forest songbirds to decline? *BioScience* 33:31–35.

Chace, J. F., C. Farmer, R. Winfree, D. R. Curson, W. E. Jensen, C. B. Goguen, and S. K. Robinson. 2005. Cowbird (*Molothrus* spp.) ecology: A review of factors influencing distribution and abundance of cowbirds across spatial scales. *Ornithological Monographs* 57:45–70.

Clotfelter, E. D. 1998. What cues do brown-headed cowbirds use to locate red-winged blackbird host nests? *Animal Behaviour* 55:1181–1189.

Cox, W. A., F. R. Thompson, III, B. Root, and J. Faaborg. 2012. Declining brown-headed cowbird (*Molothrus ater*) populations are associated with landscape-specific reductions in brood parasitism and increases in songbird productivity. *PLoS One* 7:e47591.

Cruz, A., J. W. Prather, W. Post, and J. W. Wiley. 2000. The spread of shiny and brown-headed cowbirds into the Florida region. In *Ecology and management of cowbirds and their hosts: Studies in the conservation of North American passerine birds*, eds. J. N. M. Smith, T. L. Cook, S. I. Rothstein, S. K. Robinson, and S. G. Sealy, 47–57. University of Texas Press, Austin, TX.

Cummings, J., S. Shwiff, and S. Tupper. 2005. Economic impacts of blackbird damage to the rice industry. In *Proceedings of the 11th Wildlife Damage Management Conference*, eds. D. L. Nolte, K. A. Fagerstone, 317–322. University of Nebraska–Lincoln, Lincoln, NE.

Curson D. R., and N. E. Mathews. 2003. Reproductive costs of commuting flights in Brown-headed Cowbirds. *J. Wildl. Manage.* 67:520–529.

Darley, J. A. 1971. Sex ratio and mortality in the brown-headed cowbird. *Auk* 88:560–566.

Dearborn, D. C. 1998. Begging behavior and food acquisition by brown-headed cowbird nestlings. *Behavioral Ecology and Sociobiology* 43:259–270.

DeCapita, M. E. 2000. Brown-headed cowbird control on Kirtland's warbler nesting areas in Michigan, 1972–1995. In *Ecology and management of cowbirds and their hosts: Studies in the Conservation of North American passerine birds*, eds. J. N. M. Smith, T. L. Cook, S. I. Rothstein, S. K. Robinson, and S. G. Sealy, 333–341. University of Texas Press, Austin, TX.

Dolbeer, R. A. 1982. Migration patterns for age and sex classes of blackbirds and starlings. *Journal of Field Ornithology* 53:28–46.

Dolbeer, R. A., and C. R. Smith. 1985. Sex-specific feeding habits of brown-headed cowbirds in northern Ohio in January. *Ohio Journal of Science* 85:104–107.

Dolbeer, R. A., P. P. Woronecki, A. R. Stickley, Jr., and S. B. White. 1978. Agricultural impacts of a winter population of blackbirds and starlings. *Wilson Bulletin* 90:31–44.

Donovan, T. M., R. H. Lamberson, A. Kimber, F. R. Thompson, III, and J. Faaborg. 1995. Modeling the effects of habitat fragmentation on source and sink demography of Neotropical migrant birds. *Conservation Biology* 9:1396–1407.

Douglas, D. J. T., S. E. Newson, D. I. Leech, D. G. Noble, and R. A. Robinson. 2010. How important are climate-induced changes in host availability for population processes in an obligate brood parasite, the European cuckoo? *Oikos* 119:1834–1840.

Dubina, K. M., and B. D. Peer. 2013. Egg pecking and discrimination by female and male brown-headed cowbirds. *Journal of Ornithology* 154:553–557.

Dufty, A. M. 1982. Movements and activities of radio-tracked brown-headed cowbirds. *Auk* 99:316–327.

Eastzer, D. H., A. P. King, and M. J. West. 1985. Patterns of courtship between cowbird subspecies: Evidence for positive assortment. *Animal Behaviour* 33:30–39.

Elliott, P. F. 1980. Evolution of promiscuity in the brown-headed cowbird. *Condor* 82:138–141.

Ellis, V., L. Merrill, S. I. Rothstein, A. L. O'Loghlen, and J. C. Wingfield. 2012. Physiological consequences of molt in brown-headed cowbirds (Molothrus ater). *Auk* 129:231–238.

Ellison, K., and S. G. Sealy. 2007. Small hosts infrequently disrupt laying by brown-headed and bronzed cowbirds. *Journal of Field Ornithology* 78:379–389.

Fleischer, R. C. 1985. A new technique to identify and assess the dispersion of eggs of individual brood parasites. *Behavioral Ecology and Sociobiology* 17:91–99.

Fleischer, R. C., and S. I. Rothstein. 1988. Known secondary contact and rapid gene flow among subspecies and dialects in the brown-headed cowbird. *Evolution* 42:1146–1158.

Fleischer, R. C., S. I. Rothstein, and L. S. Miller. 1991. Mitochondrial DNA variation indicates gene flow across a zone of known secondary contact between two subspecies of the brown-headed cowbird. *Condor* 93:185–189.

Fleischer, R. C., A. P. Smyth, and S. I. Rothstein. 1987. Temporal and age-related variation in the laying rate of the parasitic brown-headed cowbird in the eastern Sierra Nevada, California. *Canadian Journal of Zoology* 65:2724–2730.

Friedmann, H. 1929. *The cowbirds: A study in the biology of social parasitism*. C. C. Thomas, Springfield, IL.

Gill, S. A., P. M. Grieef, L. M. Staib, and S. G. Sealy. 1997. Does nest defence deter or facilitate cowbird parasitism? A test of the nesting-cue hypothesis. *Ethology* 103:56–71.

Glassey, B., and S. Forbes. 2003. Why brown-headed cowbirds do not influence red-winged blackbird parent behaviour. *Animal Behaviour* 65:1235–1246.

Goguen, C. B., D. R. Curson, and N. E. Mathews. 2005. Behavioral ecology of the brown-headed cowbird in a bison-grazed landscape. *Ornithological Monographs* 57:71–83.

Goguen, C. B., and N. E. Mathews. 2001. Brown-headed cowbird behavior and movements in relation to livestock grazing. *Ecological Applications* 11:1533–1544.

Granfors, D. A., P. J. Pietz, and L. A. Joyal. 2001. Frequency of eggs and nestling destruction by female brown-headed cowbirds at grassland and nests. *Auk* 118:765–769.

Hahn, D. C., and J. S. Hatfield. 2000. Host selection in the forest interior: Cowbirds target ground-nesting species. In *Ecology and management of cowbirds and their hosts: Studies in the conservation of North American passerine birds*, eds. J. N. M. Smith, T. L. Cook, S. I. Rothstein, S. K. Robinson, and S. G. Sealy, 120–127. University of Texas Press, Austin, TX.

Hahn, D. C., and W. K. Reisen. 2011. Heightened exposure to parasites favors the evolution of immunity in brood parasitic cowbirds. *Evolutionary Biology* 38:214–224.

Hahn, D. C., J. A. Sedgwick, I. S. Painter, and N. J. Casna. 1999. A spatial and genetic analysis of cowbird host selection. *Studies in Avian Biology* 18:204–217.

Holford, K. C., and D. D. Roby. 1993. Factors limiting fecundity of captive brown-headed cowbirds. *Condor* 95:536–545.

Hoover, J. P., and S. K. Robinson. 2007. Retaliatory mafia behavior by a parasitic cowbird favors host acceptance of parasitic eggs. *Proceedings of the National Academy of Sciences of the United States of America* 104:4479–4483.

Hosoi, S. A., and S. I. Rothstein. 2000. Nest desertion and cowbird parasitism: Evidence for evolved responses and evolutionary lag. *Animal Behaviour* 59:823–840.

Howell, C. A., W. D. Dijak, and F. R. Thompson. 2007. Landscape context and selection for forest edge by breeding brown-headed cowbirds. *Landscape Ecology* 22:273–284.

Jackson, N. H., and D. D. Roby. 1992. Fecundity and egg-laying patterns of captive yearling brown-headed cowbirds. *Condor* 94:585–589.

Jaeckle, W. B., M. Kiefer, B. Childs, R. G. Harper, J. W. Rivers, and B. D. Peer. 2012. Comparison of eggshell porosity and estimated gas flux between the brown-headed cowbird and two common hosts. *Journal of Avian Biology* 43:486–490.

Jensen, W. E., and J. F. Cully. 2005. Geographic variation in brown-headed cowbird (*Molothrus ater*) parasitism on dickcissels (*Spiza americana*) in Great Plains tallgrass prairie. *Auk* 122:648–660.

Kalmbach, E. R. 1937. *Blackbirds of the Gulf Coast in relation to the rice crop with notes on their food habits and life history.* U.S. Bureau of Sport Fisheries and Wildlife, Denver Wildlife Research Center, Denver, CO.

Kattan, G. H. 1995. Mechanisms of short incubation period in brood-parasitic cowbirds. *Auk* 112:335–342.

Kilner, R. M. 2003. How selfish is a cowbird nestling? *Animal Behaviour* 66:569–576.

Kozlovich, D. R., R. W. Knapton, and J. C. Barlow.1996. Unsuitability of the house finch as a host of the brown-headed cowbird. *Condor* 98:253–258.

Kuehn, M. J., B. D. Peer, and S. I. Rothstein. 2014. Variation in host response to brood parasitism reflects evolutionary differences and not phenotypic plasticity. *Animal Behaviour* 88:21–28.

Lanyon, S. M. 1992. Interspecific brood parasitism in blackbirds (Icterinae): A phylogenetic perspective. *Science* 255:77–79.

Leitch, J. A., G. M. Linz, and J. F. Baltezore. 1997. Economics of cattail (*Typha* spp.) control to reduce blackbird damage to sunflower. *Agriculture, Ecosystems, and Environment* 65:141–149.

Lichtenstein, G., and S. G. Sealy. 1998. Nestling competition, rather than supernormal stimulus, explains the success of parasitic brown-headed cowbird chicks in yellow warbler nests. *Proceedings of the Royal Society of London B* 265:249–254.

Linz, G. M., and S. B. Bolin. 1982. Incidence of brown-headed cowbird parasitism on red-winged blackbirds. *Wilson Bulletin* 94:93–95.

Lowther, P. E. 1993. Brown-headed cowbird (*Molothrus* ater). No. 47. In *The birds of North America*, ed. P. G. Rodewald. Cornell Lab of Ornithology, Ithaca, NY. https://birdsna.org/Species-Account/bna/species/bnhcow (accessed November 12, 2016).

Lowther, P. E. 2015. *List of victims and hosts of the parasitic cowbirds (Molothrus). Field Museum version 02 June 2015.* http://www.fieldmuseum.org/sites/default/files/ plowther/2015/06/02/cowbird_hosts-02jun2015.pdf (accessed January 27, 2016).

Lustick, S. 1970. Energy requirements of molt in cowbirds. *Auk* 87:742–746.

Lyon, L. A., and D. F. Caccamise. 1981. Habitat selection by roosting blackbirds and starlings: Management implications. *Journal of Wildlife Management* 45:435–443.

Mann, C. C. 2005. *1491: New revelations of the Americas before Columbus.* Knopf, NY.

Mayfield, H. F. 1972. Third decennial census of Kirtland's warbler. *Auk* 89:263–268.

Meanley, B. 1971. *Blackbirds and the southern rice crop.* Resource Publication 100. U.S. Department of Interior, U.S. Fish and Wildlife Service. http://pubs.usgs.gov/rp/100/report.pdf (accessed June 28, 2016).

Mermoz, M. E., and J. F. Ornelas. 2004. Phylogenetic analysis of life-history adaptations in parasitic cowbirds. *Behavioral Ecology* 15:109–119.

Middleton, A. L. A. 1991. Failure of brown-headed cowbird parasitism in nests of the American goldfinch. *Journal of Field Ornithology* 62:200–203.

Morrison, M. L., and D. C. Hahn. 2002. Geographic variation in cowbird distribution, abundance, and parasitism. *Studies in Avian Biology* 25:65–72.

Morrison, M. L., L. S. Hall, S. K. Robinson, S. I. Rothstein, D. C. Hahn, and T. D. Rich, eds. 1999. Research and management of the brown-headed cowbird in western landscapes. *Studies in Avian Biology* 18.

National Audubon Society. 2010. *The Christmas bird count historical results.* http://www.christmasbirdcount.org (accessed November 12, 2016).

Neff, J. A., and B. Meanley. 1957. *Blackbirds and the Arkansas rice crop, Bulletin 584.* Agricultural Experiment Station, University of Arkansas, Fayetteville, AR.

Newton, I. 1998. *Population limitation in birds.* Academic Press, New York.

O'Loghlen, A. L., and S. I. Rothstein. 2003. Female preference for the songs of older males and the maintenance of dialects in brown-headed cowbirds (*Molothrus ater*). *Behavioral Ecology and Sociobiology* 53:102–109.

Ortego, B. 2000. Brown-headed cowbird population trends at a large winter roost in southwest Louisiana, 1974–1992. In *Ecology and management of cowbirds and their hosts: Studies in the conservation of North American passerine birds*, eds. J. N. M. Smith, T. L. Cook, S. I. Rothstein, S. K. Robinson, and S. G. Sealy, 58–62. University of Texas Press, Austin, TX.

Ortega, C. P. 1998. *Cowbirds and other brood parasites.* University of Arizona Press, Tucson, AZ.

Ortega C. P., J. F. Chace, and B. D. Peer. 2005. Management of cowbirds and their hosts: Balancing science, ethics, and mandates. *Ornithological Monographs* 57:1–114.

Ortega, C. P., and A. Cruz. 1991. A comparative study of cowbird parasitism in yellow-headed blackbirds and red-winged blackbirds. *Auk* 108:16–24.

Ortega, C. P., and A. Cruz. 1992. Gene flow of the *obscurus* race into the north-central Colorado population of brown-headed cowbirds. *Journal of Field Ornithology* 63:311–317.

Patten, M. A., D. L. Reinking, and D. H. Wolfe. 2011. Hierarchical cues in brood parasite nest selection. *Journal of Ornithology* 152:521–532.

Patten, M. A., E. Shochat, D. L. Reinking, D. H. Wolfe, and S. K. Sherrod. 2006. Habitat edge, land management, and rates of brood parasitism in tallgrass prairie. *Ecological Applications* 16:687–695.

Payne, R. B. 1976. The clutch size and numbers of eggs of brown-headed cowbirds: Effects of latitude and breeding season. *Condor* 78:337–342.

Peer, B. D. 2006. Egg destruction and egg removal by avian brood parasites: Adaptiveness and consequences. *Auk* 123:16–22.

Peer, B. D., and E. K. Bollinger. 1997. Explanations for the infrequent cowbird parasitism on common grackles. *Condor* 99:151–161.

Peer, B. D., H. J. Homan, G. M. Linz, and W. J. Bleier. 2003. Impact of blackbird damage to sunflower: Bioenergetic and economic models. *Ecological Applications* 13:248–256.

Peer B. D., M. J. Kuehn, S. I. Rothstein, and R. C. Fleischer. 2011a. Persistence of host defence behaviour in the absence of brood parasitism. *Biology Letters* 7:670–673.

Peer B.D., C. E. McIntosh, M. J. Kuehn, S. I. Rothstein, R. C. Fleischer. 2011b. Complex biogeographic history of shrikes and its implications for the evolution of defenses against avian brood parasitism. *Condor* 113:385–394.

Peer, B. D., J. W. Rivers, and S. I. Rothstein. 2013a. The brown-headed cowbird: North America's avian brood parasite. *Chinese Birds* 4:93–98.

Peer, B. D., J. W. Rivers, and S. I. Rothstein. 2013b. Cowbirds, conservation, and coevolution: Potential misconceptions and directions for future research. *Chinese Birds* 4:15–30.

Peer, B. D., S. K. Robinson, and J. R. Herkert. 2000. Egg rejection by cowbird hosts in grasslands. *Auk* 117:892–901.

Peer, B. D., S. I. Rothstein, K. S. Delaney, and R. C. Fleischer. 2007. Defence behaviour against brood parasitism is deeply rooted in mainland and island scrub-jays. *Animal Behaviour* 73:55–63.

Peer, B. D., and S. G. Sealy. 1999. Laying time of the bronzed cowbird. *Wilson Bulletin* 111:137–139.

Peer, B. D., and S. G. Sealy. 2004a. Correlates of egg rejection in hosts of the brown-headed cowbird. *Condor* 106:580–599.

Peer, B. D., and S. G. Sealy. 2004b. Fate of grackle (*Quiscalus*) defenses in the absence of brood parasitism: Implications for long-term brood parasite–host coevolution. *Auk* 121:1172–1186.

Pielou, E. C. 1991. *After the ice age: The return of life to glaciated North America*. University of Chicago Press, Chicago, IL.

Pierce, R. A. 1970. Bird depredations on rice and other grains in Arkansas. *Proceedings of the Bird Control Seminar* 5:101–109.

Post, W., and P. W. Sykes, Jr. 2011. Reproductive status of the shiny cowbird in North America. *Wilson Journal of Ornithology* 123:151–154.

Powell, A. F., F. K. Barker, S. M. Lanyon, K. J. Burns, J. Klicka, and I. J. Lovette. 2014. A comprehensive species-level molecular phylogeny of the New World blackbirds (Icteridae). *Molecular Phylogenetics and Evolution* 71:94–112.

Pyne, S. J. 1977. *Fire in America: A cultural history of wildland and rural fire*. University of Washington Press, Seattle, WA.

Richard, T. 2008. Confirmed occurrence and nesting of Kirtland's Warbler at CFB Petawawa, Ontario: A first for Canada. *Ontario Birds* 26:2–15.

Rivers, J. W. 2007. Nest mate size, but not short-term need, influences begging behavior of a generalist brood parasite. *Behavioral Ecology* 18:222–230.

Rivers, J. W., T. M. Loughin, and S. I. Rothstein. 2010. Brown-headed cowbird nestlings influence nestmate begging, but not parental feeding, in hosts of three distinct sizes. *Animal Behaviour* 79:107–116.

Rivers, J. W., and B. D. Peer. 2016. Relatedness constrains virulence in an obligate avian brood parasite. *Ornithological Science* 15:1–11.

Robertson, R. J., P. J. Weatherhead, F. J. S. Phelan, G. L. Holroyd, and N. Lester. 1978. On assessing the economic and ecological impact of winter blackbird flocks. *Journal of Wildlife Management* 42:53–60.

Robinson, S. K. 1992. Population dynamics of breeding Neotropical migrants in a fragmented Illinois landscape. In *Ecology and Conservation of Neotropical Migrant Landbirds*, eds. J. M. Hagan III and D. W. Johnston, 455–471. Smithsonian Institution Press, Washington, DC.

Robinson, S. K., S. I. Rothstein, M. C. Brittingham, L. J. Petit, and J. A. Grzybowski. 1995a. Ecology and behavior of cowbirds and their impact on host populations. In *Ecology and management of Neotropical migratory birds: A synthesis and review of critical issues*, eds. D. M. Finch and T. E. Martin, 428–460. Oxford University Press, New York.

Robinson, S. K., F. R. Thompson, T. M. Donovan, D. R. Whitehead, and J. Faaborg. 1995b. Regional forest fragmentation and the nesting success of migratory birds. *Science* 267:1987–1990.

Rosenberg, K. V., J. A. Kennedy, R. Dettmers, R. P. Ford, D. Reynolds, C. J. Beardmore, P. J. Blancher, et al. 2016. *Partners in Flight Landbird Conservation Plan: 2016 Revision for Canada and Continental United States*. Partners in Flight Science Committee. http://www.partnersinflight.org/ (accessed September 25, 2016).

Rothstein, S. I. 1975. An experimental and teleonomic investigation of avian brood parasitism. *Condor* 77:50–271.

Rothstein, S. I. 1976. Cowbird parasitism of the cedar waxwing and its evolutionary implications. *Auk* 93:498–509.

Rothstein, S. I. 1978. Geographical variation in the nestling coloration of parasitic cowbirds. *Auk* 95:152–160.

Rothstein, S. I. 1990. A model system for coevolution: Avian brood parasitism. *Annual Review of Ecology and Systematics* 21:481–508.

Rothstein, S. I. 1994. The cowbird's invasion of the far west: History, causes and consequences experienced by host species. *Studies in Avian Biology* 15:301–315.

Rothstein, S. I. 2001. Relic behaviors, coevolution and the retention versus loss of host defenses after episodes of avian brood parasitism. *Animal Behaviour* 61:95–107.

Rothstein, S. I., and T. L. Cook. 2000. Cowbird management, host population regulation and efforts to save endangered species: Introduction. In *Ecology and management of cowbirds and their hosts: Studies in the conservation of North American passerine birds*, eds. J. N. M. Smith, T. L. Cook, S. I. Rothstein, S. K. Robinson, and S. G. Sealy, 323–332. University of Texas Press, Austin, TX.

Rothstein, S. I., and R. C. Fleischer. 1987. Vocal dialects and their possible relation to honest status signaling in the brown-headed cowbird. *Condor* 89:1–23.

Rothstein, S. I., B. E. Kus, M. J. Whitfield, and S. J. Sferra. 2003. Recommendations for cowbird management in recovery efforts for the southwestern willow flycatcher. *Studies in Avian Biology* 26:157–167.

Rothstein, S. I., M. Patten, and R. C. Fleischer. 2002. Phylogeny, specialization, and brood parasite-host coevolution: Some possible pitfalls of parsimony. *Behavioral Ecology* 13:1–10.

Rothstein, S. I., and B. D. Peer. 2005. Conservation solutions for threatened and endangered cowbird (*Molothrus* spp.) hosts: Separating fact from fiction. *Ornithological Monographs* 57:98–114.

Rothstein, S. I., and S. K. Robinson. 1998. The evolution and ecology of avian brood parasitism: An overview. *In Parasitic birds and their hosts*, eds. S. I. Rothstein and S. K. Robinson, 3–58. Oxford University Press, London, UK.

Rothstein, S. I., J. Verner, and E. Stevens. 1980. Range expansion and diurnal changes in dispersion of the brown-headed cowbird in the Sierra Nevada. *Auk* 97:253–267.

Rothstein, S. I., J. Verner, and E. Stevens. 1984. Radio-tracking confirms a unique diurnal pattern of spatial occurrence in the parasitic brown-headed cowbird. *Ecology* 65:77–88.

Rothstein, S. I., D. A. Yokel, and R. C. Fleischer. 1986. Social dominance, mating and spacing systems, female fecundity, and vocal dialects in captive and free-ranging brown-headed cowbirds. *Current Ornithology* 3:127–185.

Sauer, J. R., D. K. Niven, J. E. Hines, D. J. Ziolkowski, Jr, K. L. Pardieck, J. E. Fallon, and W. A. Link. 2017. The North American Breeding Bird Survey, results and analysis 1966–2015. Version 2.07.2017 USGS Patuxent Wildlife Research Center, Laurel, MD. https://www.mbr-pwrc.usgs.gov/bbs/bbs.html (accessed April 30, 2017).

Saunders, C. A., P. Arcese, and K. D. O'Connor. 2003. Nest site characteristics in the song sparrow and parasitism by brown-headed cowbirds. *Wilson Bulletin* 115:24–28.

Scott, D. M. 1991. The time of day of egg laying by the brown-headed cowbird and other Icterines. *Canadian Journal of Zoology* 69:2093–2099.

Scott, D. M., and C. D. Ankney. 1980. Fecundity of the brown-headed cowbird in southern Ontario. *Auk* 97:677–683.

Scott, D. M., and D. Ankney. 1983. The laying cycle of brown-headed cowbirds: Passerine chicken? *Auk* 100:583–592.

Scott, D. M., P. J. Weatherhead, and C. D. Ankney. 1992. Egg-eating by female brown-headed cowbirds. *Condor* 94:579–584.

Sealy, S. G. 1992. Removal of yellow warbler eggs in association with cowbird parasitism. *Condor* 94:40–54.

Sealy, S. G. 1995. Burial of cowbird eggs by parasitized yellow warblers: An empirical and experimental study. *Animal Behaviour* 49:877–889.

Sealy, S. G., D. L. Neudorf, and D. P. Hill. 1995. Rapid laying by brown-headed cowbirds *Molothrus ater* and other parasitic birds. *Ibis* 137:76–84.

Searcy, W. A., and K. Yasukawa. 1995. *Polygyny and sexual selection in red-winged blackbirds.* Princeton University Press, Princeton, NJ.

Sherry, D. F., M. R. Forbes, M. Khurgel, and G. O. Ivy. 1993. Females have a larger hippocampus than males in the brood-parasitic brown-headed cowbird. *Proceedings of the National Academy of Sciences of the United States of America* 90:7839–7843.

Smith, J. N. M. 1981. Cowbird parasitism, host fitness, and age of the host female in an island song sparrow population. *Condor* 83:152–161.

Smith, J. N. M., and P. Arcese. 1994. Brown-headed cowbirds and an island population of song sparrows: A 16-year study. *Condor* 96:916–934.

Smith, J. N. M., T. L. Cook., S. I. Rothstein, S. K. Robinson., and S. G. Sealy. 2000. *Ecology and management of cowbirds and their hosts: Studies in the conservation of North American passerine birds.* University of Texas Press, Austin, TX.

Spaw, C. D., and S. Rohwer. 1987. A comparative study of eggshell thickness in cowbirds and other passerines. *Condor* 89:307–318.

Spottiswoode, C. N., K. F. Stryjewski, S. Quader, J.F.R. Colebrook-Robjent, and M. D. Sorenson. 2011. Ancient host specificity within a single species of brood parasitic bird. *Proceedings of the National Academy of Sciences of the United States of America* 43:17738–17742.

Strausberger, B. M. 1998. Temperature, egg mass, and incubation time: A comparison of brown-headed cowbirds and red-winged blackbirds. *Auk* 115:843–850.

Strausberger, B. M., and M. V. Ashley. 1997. Community-wide patterns of parasitism of a host "generalist" brood-parasitic cowbird. *Oecologia* 112:254–262.

Strausberger, B. M., and M. V. Ashley. 2003. Breeding biology of brood parasitic cowbirds characterized by parent-offspring and sib-group reconstruction. *Auk* 120:433–445.

Strausberger, B. M., and M. V. Ashley. 2005. Host use strategies of individual female brown-headed cowbirds *Molothrus ater* in a diverse avian community. *Journal of Avian Biology* 36:313–321.

Swan, D. C., L. Y. Zanette, and M. Clinchy. 2015. Brood parasites manipulate their hosts: Experimental evidence for the farming hypothesis. *Animal Behaviour* 105:29–35.

Teather, K. L., and R. J. Robertson. 1985. Female spacing patterns in brown-headed cowbirds. *Canadian Journal of Zoology* 63:218–222.

Teather, K. L., and R. J. Robertson. 1986. Pair bonds and factors influencing the diversity of mating systems in brown-headed cowbirds. *Condor* 88:63–69.

Terborgh, J. 1989. *Where have all the birds gone?* Princeton University Press, Princeton, NJ.

USDA APHIS. 2015. *Conflicts with People.* https://www.aphis.usda.gov/wps/portal/?1dmy&urile=wcm%3Apath%3A/aphis_content_library/sa_our_focus/sa_wildlife_damage/sa_operational_activities/sa_blackbirds/ct_conflicts (accessed July 20, 2016).

U.S. Department of Agriculture. 2015. *Wildlife damage blackbirds.* https://www.aphis.usda.gov/aphis/our-focus/wildlifedamage/operational-activities/sa_blackbirds/ct_blackbirds_european_starlings (accessed July 20, 2016).

U.S. Fish and Wildlife Service. 2002. *Southwestern willow flycatcher recovery plan.* Albuquerque, NM. Appendices A–O.

U.S. Fish and Wildlife Service. 2011. *Kirtland's Warbler 2011 Nesting Season Summary.* http://www.fws.gov/midwest/endangered/birds/Kirtland/kiwa-nest-sum.html (accessed July 20, 2016).

U.S. Fish and Wildlife Service. 2016. *Kirtland's Warbler Census Results: 1951, 1961, 1971 thru 2015.* https://www.fws.gov/midwest/endangered/birds/Kirtland/Kwpop.html (accessed July 20, 2016).

White, S. B., R. A. Dolbeer, and T. A. Bookhout. 1985. Ecology, bioenergetics, and agricultural impacts of a winter-roosting population of blackbirds and starlings. *Wildlife Monographs* 93:3–42.

Williams, R. E., and W. B. Jackson. 1981. Dietary comparisons of red-winged blackbirds, brown-headed cowbirds, and European starlings in north-central Ohio. *Ohio Journal of Science* 81:217–225.

Wilsey, C. B., J. J. Lawler, D. Cimprich, N. H. Schumaker. 2013. Dependence of the endangered Black-capped Vireo on sustained cowbird management. *Conserv. Biol.* 28:561–571.

Wilson, E. A. 1985. *Blackbird depredation on rice in southwestern Louisiana*. M.S. Thesis. Louisiana State University, Baton Rouge, LA.

Wilson, E. A., E. A. LeBoeuf, K. M. Weaver, and D. J. LeBlanc. 1990. The effect of planting date on blackbird damage to sprouting rice. *Louisiana Agriculture* 33:5–15.

Woodward, P. W. 1983. Behavioral ecology of fledgling brown-headed cowbirds and their hosts. *Condor* 85:151–163.

Woodward, P. W., and J. C. Woodward. 1979. Brown-headed cowbird parasitism on eastern bluebirds. *Wilson Bulletin* 91:321–322.

Woolfenden, B. E. 2000. *Demography and breeding behaviour of brown-headed cowbirds: An examination of host use, individual mating patterns and reproductive success using microsatellite DNA markers*. PhD thesis, McMaster University, Hamilton, CA.

Woolfenden, B. E., H. L. Gibbs, and S. G. Sealy. 2002. High opportunity for sexual selection in both sexes of an obligate brood parasitic bird, the brown-headed cowbird (*Molothrus ater*). *Behavioral Ecology and Sociobiology* 52:417–425.

Woolfenden, B. E., H. L. Gibbs, S. G. Sealy, and D. G. McMaster. 2003. Host use and fecundity of individual female brown-headed cowbirds. *Animal Behaviour* 66:95–106.

Yokel, D. A. 1986. The social organization of the brown-headed bowbird in the Owens Valley, California. In *Natural history of the White-Inyo Range, Eastern California and Western Nevada and high altitude physiology White Mountain Research Station Symposium Vol 1*, eds. C. A. Hall and D. J. Young, 164–172. Regents of the University of California, Los Angeles, CA.

Yokel, D. A. 1989. Intrasexual aggression and the mating behavior of brown-headed cowbirds: Their relation to population densities and sex ratios. *Condor* 91:43–51.

Zahavi, A. 1979. Parasitism and nest predation in parasitic cuckoos. *American Naturalist* 113:157–159.

CHAPTER 6

Effects of Habitat and Climate on Blackbird Populations

Greg M. Forcey
University of Florida
Gainesville, Florida

Wayne E. Thogmartin
Upper Midwest Environmental Sciences Center
Middleton, Wisconsin

CONTENTS

6.1 Breeding Season	104
6.1.1 Breeding Habitat	104
6.1.2 Nesting Microhabitat	105
6.1.3 Breeding Season Climate and Weather Associations	106
6.2 Winter Season	107
6.2.1 Winter Habitat	107
6.2.2 Winter Climate and Weather Associations	107
6.3 Migration Season	108
6.3.1 Migratory Habitat	108
6.3.2 Migratory Climate and Weather Associations	109
6.4 Landscape Effects	109
6.4.1 Landscape Change	111
6.5 Climate Change	112
6.6 Summary	113
References	114

Global biodiversity loss is proceeding at an accelerating pace (Newbold et al. 2015, 2016), in large part due to land use and climate change and associated spread of disease and nonnative species (Hobbs et al. 2006; Williams and Jackson 2007; Ellis 2011; Radeloff et al. 2015). Over the last century, the U.S. average temperature has increased 0.7°C–1.1°C, leading to an increased frost-free season, more frequent and intense heat waves, and increased frequency and intensity of winter storms; mean precipitation has increased, with increases in heavy downpours (Melillo et al. 2014). The dominant land uses in the United States are lands devoted to forest (272 million ha; 30%), pasture/range (249 million ha, 27%), and agriculture (165 million ha, 18%) (Economic Research Service 2011). Martinuzzi et al. (2015) projected changes in land use to the middle of the twenty-first century and found that at least 11% of the U.S. land cover (an area larger than Texas) was expected to change cover class (Figure 6.1). At the same time, mean temperature is expected to further increase 1.1°C–1.7°C by midcentury and 2.2°C–3.9°C by the end of the century (Melillo et al. 2014). In this age of unprecedented human-induced environmental change, understanding the relationships of species to the habitat and climatic conditions they experience is crucial to conservation and management. Improved understanding of relationships with habitat and climate will better inform management decisions designed to reduce crop depredation caused by blackbirds.

Hall et al. (1997) defined *habitat* as "the resources and conditions present in an area that produce occupancy by a given organism." In addition to vegetation associations, this definition encompasses other variables, such as the presence of food and water, all of which can influence blackbird occupancy of an area (Lowther 1993; Twedt and Crawford 1995; Yasukawa and Searcy 1995; Peer and Bollinger 1997). Here, we review the state of knowledge concerning habitat and climate associations for the red-winged blackbird (*Agelaius phoeniceus*), yellow-headed blackbird (*Xanthocephalus xanthocephalus*), common grackle (*Quiscalus quiscula*), and brown-headed cowbird (*Molothrus ater*) (collectively referred to as blackbirds [Icteridae]). We organize our review around the full annual cycle of blackbirds, beginning with responses to conditions faced during the breeding season in spring and summer, followed by the overwintering period and then the migratory connections between these seasons. We also consider the landscape effects of habitat in addition to those that can occur at the local scale, given their importance from a management perspective (Forcey 2006; Forcey et al. 2007, 2015).

Generally, blackbirds inhabit open habitats while avoiding deep forest or highly urbanized areas. All four blackbird species use agricultural habitat to some extent throughout the year, and surrounding croplands that contain abundant food resources can increase the suitability of nearby wetlands for nesting red-winged blackbirds and yellow-headed blackbirds (Creighton et al. 1997). Common grackles are closely associated with the presence of agricultural areas (especially in the Maritime provinces of Canada; Erskine 1971); however, even in these habitats, trees are often required for nesting. Grackles nest in a diverse array of substrate, including coniferous trees, but also in cattails (*Typha* spp.), rafters of open sheds, and even in the sticks below active nests of great-blue heron (*Ardea herodias*) (P. Weatherhead, personal observation). Like common grackle, brown-headed cowbirds show a similar preference for agriculture, with breeding bird atlas distribution maps showing a strong association with agricultural areas (Lowther 1993).

Climate affects blackbirds through (1) direct effects of temperature on energetics and behavior, (2) precipitation affecting wetland availability for habitat, and (3) temperature influencing the availability of forage. The energy requirements of birds, including blackbirds, are roughly linear or near-linear functions of ambient air temperature (Kendeigh 1944; Seibert 1949; Lewies and Dyer 1969). As such, variation in physiology (and associated mediation of physiology by behavior) should be expected with variation in temperature. Early work on this topic by Brenner (1966a) reported increased existence energy requirements with decreasing temperature, which he associated with hyperphagia rather than metabolic efficiency. White et al. (1985), following Kendeigh et al. (1977), described the relationship between metabolism and temperature as follows:

$$Metabolism\ (kcal\,/\,day) \propto 4.4W^{0.5224} - 0.1571W^{0.2427}T,$$

where W is the mass (g) and T is the ambient temperature (°C). As ambient temperature increases, existence metabolism decreases. Lewies and Dyer (1969, 297) reported similar findings for red-winged blackbirds but suggested the possibility of sex-related differences, which they attributed to "slightly different metabolic substrate being utilized by the two sexes." They reported differences in temperature–calorie relations between day and night—daytime relations of metabolism

EFFECTS OF HABITAT AND CLIMATE ON BLACKBIRD POPULATIONS

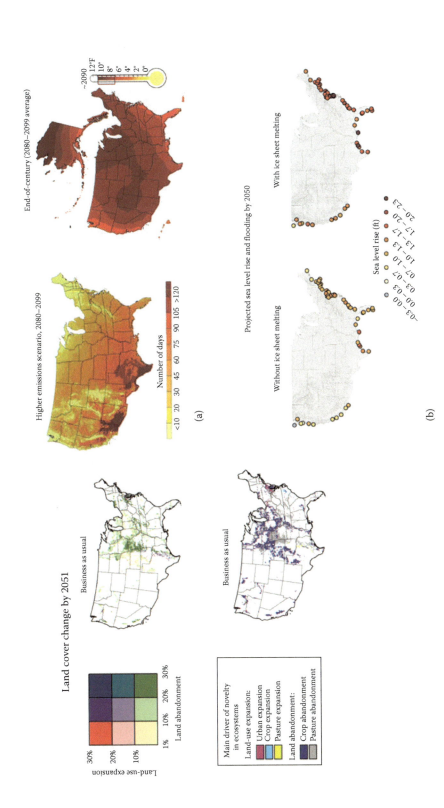

Figure 6.1 Land use and climate are expected to change considerably through the end of the century. Martinuzzi et al. (2015) predicted considerable crop abandonment along the periphery of Cornbelt and through the central United States, potentially affecting blackbird occurrence and nuisance behavior. At the same time, the number of days at 37.8°C (close to the lethal limit in blackbirds) is expected to increase considerably in the United States, especially in the southern United States (USGCRP 2009). The temperature is expected to increase considerably throughout the entirety of the United States (USGCRP 2009). With rising temperature, sea levels are expected to rise, potentially inundating coastal salt-marsh roosting areas for blackbirds. (Melillo et al. 2014.)

against temperature were curvilinear (e.g., for males, metabolic rate [O_2/(g × hr)] = 0.2 + 0.0047 air temperature [°C] – 0.000039 air temperature^2), whereas nighttime relations were strictly linear (e.g., for females, metabolic rate [O_2/(g × hr)] = 3.26 – 0.0188 air temperature [°C]). Yasukawa and Searcy (1995) similarly reported a curvilinear relationship between metabolism in red-winged blackbirds and daytime ambient temperature, as well as a linear relationship between metabolism and nocturnal temperature.

The upper lethal temperature in blackbirds is not well documented. Johnson and Cowan (1975) reported a mean lethal dose (LD_{50}) = 40° C in common starlings (*Sturnus vulgaris*), which should be approximately correct for blackbirds. At the opposite end of the temperature spectrum, reports on the lower lethal limit in blackbirds are unavailable. However, Kendeigh (1944, 1969) reported a lower lethal temperature of –35° C in the house sparrow (*Passer domesticus*) and –40° C in the evening grosbeak (*Coccothraustes vespertinus*); a similar temperature might be expected for blackbirds. The metabolic cost of thermoregulation in cold environs suppresses territorial advertisement and other social displays in red-winged blackbirds (Santee and Bakken 1987). Paladino (1989) speculated that the energetic cost of thermoregulation was highest in the northern portion of a species range, early in the breeding season, and early in the morning; as a result, he suggested the northern range limit in landbirds, including blackbirds, was likely a result of the inability of species to accommodate the cost of thermoregulation and the attendant mate attraction, territorial behavior, and other social behaviors, necessarily reducing fitness.

6.1 BREEDING SEASON

6.1.1 Breeding Habitat

Breeding habitat is critical for blackbirds during the spring and summer months, providing food and shelter for adult birds and their offspring. The red-winged blackbird is commonly regarded as a habitat generalist, breeding in a wide range of wetland and upland cover types (Yasukawa and Searcy 1995), though wetlands and hayfields are preferred (Clark and Weatherhead 1987). Wetlands dominated by cattail are especially preferred (Allen 1914; Robertson 1972; Twedt and Crawford 1995; Yasukawa and Searcy 1995; Linz et al. 1996; Linz and Homan 2011), although wetlands dominated by invasive vegetation such as purple loosestrife (*Lythrum salicaria*) are also habitable (Rawinski and Malecki 1984). Within a wetland, yellow-headed blackbirds are generally found in deep-water palustrine areas with extensive channeling, whereas red-winged blackbirds inhabit shallower locations with dense emergent vegetation; edge habitat between open water and emergent vegetation within a wetland is important for foraging by both species (Orians 1980; Schroeder 1982; Orians and Wittenberger 1991; Murkin et al. 1997; Turner and McCarty 1998; Naugle et al. 1999b).

In locations where the preferred habitat for red-winged blackbirds and yellow-headed blackbirds overlap, the former will often decrease or be forced to nest over shallower water because of interspecific competition with the latter (Orians and Willson 1964; Creighton et al. 1997). Although yellow-headed blackbirds are a more wetland-obligate species, red-winged blackbirds also can be found nesting in upland areas including shrub scrub (Whitt et al. 1999), croplands, sedge meadows, and overgrown fields (Yasukawa and Searcy 1995), with occupancy being positively related to the presence of native warm-season grasses (West et al. 2016). Despite the adaptability to upland habitat, red-winged blackbirds are vulnerable to cutting of hayfields during breeding (Vierling 2000), which can result in breeding territory abandonment (Albers 1978). Red-winged blackbirds have also been shown to decline in areas where small wetlands are lost due to tillage (Besser et al. 1984), suggesting that some upland habitat types are not ideal for this species if not in proximity to wetlands.

Both common grackles and brown-headed cowbirds have benefited from the conversion of forest habitat to agriculture and have greatly expanded their range since European settlement (Lowther 1993; Peer and Bollinger 1997). Unlike the preference for wetlands by red-winged blackbirds and yellow-headed blackbirds, both common grackles and brown-headed cowbirds prefer open areas with scattered trees, including both wetland and upland areas. Common grackles are a habitat generalist and breed in a variety of open woodland, forest edges, hammocks, swamps, marshes, and developed areas; they avoid deep mature forests (Peer and Bollinger 1997). For individual common grackles inhabiting wetlands, water depth has been shown to be a positive influence on common grackle abundance (Lariviere and Lepage 2000). Like common grackles, brown-headed cowbirds also prefer woody vegetation scattered among open areas, including forest edges, prairies, fields, pastures, orchards, and residential areas. They are especially fond of forest-field edges rather than strictly forests or fields (Brittingham and Temple 1983; Lowther 1993). Fragmentation of forests in eastern North America has benefited brown-headed cowbirds by making large areas accessible to a species that normally shuns deep forest (Lowther 1993).

6.1.2 Nesting Microhabitat

The preferred nesting substrate varies widely among the four blackbird species. Red-winged blackbird nest locations are likely the most variable and can be found in wetlands, uplands, and agricultural areas. Nests in wetlands are typically woven in between several vertical shoots or branches of cattail (*Typha*), bulrush (*Scirpus*), sedge (*Carex*), reed (*Phragmites*), or willow (*Salix*) and are placed 20–80 cm above water (Yasukawa and Searcy 1995). Red-winged blackbirds nesting early in the season will use old growth herbaceous vegetation from the previous year, whereas later-nesting birds will use new growth (Short 1985). Although deeper water is preferred for decreasing risk of predation (Searcy and Yasukawa 1995; Pribil and Picman 1997), increasing vegetation cover can increase nest predation by reducing the predator-mobbing efficiency of adults defending the nest (Picman 1980). Areas with deeper water may also contain higher densities of marsh wrens (*Cistothorus palustris*), which commonly destroy the eggs of other birds (Beletsky 1996). Upland nesting substrates include trees, buttonbush, blackberry, sedges, or grasses. Wheat, barley, alfalfa, or rice are used in agricultural habitats (Yasukawa and Searcy 1995).

Concomitant with the yellow-headed blackbird's specific wetland habitat requirements is specific microhabitat for nesting. Yellow-headed blackbird nests are almost always located over water and attached to robust emergent vegetation (cattail, bulrush, reeds, and willow) from the previous or current year (Schroeder 1982). Most nests are placed above deep water (>16 cm depth) and are over 15 cm above the water line (Weller and Spatcher 1965; Twedt and Crawford 1995).

Common grackle nests are often placed near water, in agricultural fields, or in residential areas (Peer and Bollinger 1997) and have been observed in wetlands (G. M. Linz, personal observation). Nest heights frequently range from 0.2 to 22.0 m above ground and are most often placed in conifers, deciduous trees, and shrubs between two or more upright branches (Maxwell et al. 1976; Peer and Bollinger 1997). Conifers are often the preferred choice of substrate because of the early spring nesting habitats of this species and the associated cover provided at this time. Common grackles nest either singly or with conspecifics in colonies of up to 10–18 nests (Peer and Bollinger 1997).

Because the brown-headed cowbird is a brood parasite, it does not build nests but rather relies on its hosts to raise its young. Over 220 host species have been recorded, with 144 species having successfully raised brown-headed cowbird young (Freidmann and Kiff 1985). Hosts span a wide range in size from 10 to 150 g (Lowther 1993). The most frequent host taxa include warblers, sparrows, and vireos; however, even some larger hosts such as the red-winged blackbird and eastern meadowlark are vulnerable (Freidmann 1963; Blankespoor et al. 1982; Linz and Bolin 1982; Lowther 1993). Female brown-headed cowbirds watch for nest-building or brooding activity to find prospective hosts (Norman and Robertson 1975). Hosts that are often selected appear to lay eggs smaller than their own, have an active nest with two or more eggs, and consist of a large closed nest or small open nest (King 1979).

6.1.3 Breeding Season Climate and Weather Associations

Forcey et al. (2015) studied red-winged blackbirds, yellow-headed blackbirds, and common grackles during breeding in the Prairie Pothole Region of the United States. They related North American Breeding Bird Survey counts to National Oceanic and Atmospheric Administration climatic variables (mean annual temperature, mean spring temperature, previous year spring mean temperature, total annual precipitation, and previous year total annual precipitation) using overdispersed Poisson mixed-effects regression. They reported that both red-winged and yellow-headed blackbird population counts were positively associated with precipitation in the previous year, indicating that a 1 standard deviation (SD) increase in precipitation in the previous year (207 mm) led to a 2.6% and 7.9% increase, respectively, in abundance. Common grackles were not associated with precipitation, however, but were strongly associated with warmer spring temperature, with a 1 SD increase in spring temperature (2.1°C), increasing abundance by 3.3%. Forcey et al. (2015) suggested that the reason for the differences between red-winged and yellow-headed blackbirds and common grackles was because common grackles are less wetland dependent than the other two species.

This regionally identified effect of precipitation on blackbird abundance has been elucidated by a number of field-level studies. Fletcher and Koford (2004) examined the consequences of rainfall variation in the Iowa portion of the Prairie Pothole Region on density and reproduction in red-winged and yellow-headed blackbirds. Their results indicated that yellow-headed blackbirds were markedly reduced in density and failed to reproduce during dry years; reproductive failure was principally due to easier access of nests by predators. Red-winged blackbirds, however, were largely immune to variation in precipitation, neither increasing nor decreasing in abundance or reproductive success. However, both species produced smaller clutches and nested later in dry years, which they suggested was a reflection of reduced food availability. In comparison, Brenner (1966b) reported severe drought-impacted reproductive success of red-winged blackbirds as a result of changes in vegetative composition (with drought causing a cattail-sedge marsh to become primarily a sedge marsh) and decreased insect biomass. Vierling (2000), however, studying red-winged blackbirds in a rural–suburban landscape in Colorado, reported a delay in egg laying in wet years (May 30) compared to a dry year (May 21), which she associated with increased predation. The most reasonable hypothesis for reconciling the contrasting findings of Vierling (2000) and Fletcher and Koford (2004) is that variation in climatic conditions induces variation in vegetative composition and insect biomass, which in turn causes deviation from an optimal laying date.

There are few studies examining the proximate consequences of temperature on blackbirds. Zimmerling and Ankney (2005) hypothesized that the warmer temperatures (2°–3°C difference) they observed in one of their three study years resulted in a shorter incubation period for nesting red-winged blackbirds in Ontario, Canada. However, they suggested that this effect of temperature was not directly related to egg development but rather through increased availability of insects, allowing incubating females to stay on eggs for longer periods of time. Similarly, Weatherhead (2005a) suggested that earlier nesting was associated with warmer spring weather and that the proximate factor driving this pattern was food abundance. Solar radiation is a major source of heat for open-cup nesters like blackbirds (Webb and King 1983), exacerbating the effects of ambient temperature. As a result, solar radiation influences parental nest site selection (Lloyd and Martin 2004) and nestling behavior in blackbirds (red-winged blackbirds, Choi and Bakken 1990; common grackle, Glassy and Amos 2009).

During the breeding season, the principal mechanisms by which climate influences blackbirds can be summarized as follows. Reduced precipitation may lead to altered timing of reproduction (Vierling 2000; Fletcher and Koford 2004) or reduced availability of nesting habitat, associated with reduced food availability (Brenner 1966b) or increased risk of nest predation (Robertson 1972; Shipley 1979; Fletcher and Koford 2004). At least in the case of Fletcher and Koford (2004), the top–down consequences of predation swamped the bottom–up consequences of food availability.

Increased precipitation also has the possibility of reducing the potential availability of breeding sites by lowering the ratio of emergent vegetation relative to open water (Lederer et al. 1975; Murkin et al. 1997). Increased temperature in spring affects the timing and duration of nesting, principally through increased availability of insect forage (Weatherhead 2005a; Zimmerling and Ankney 2005).

6.2 WINTER SEASON

6.2.1 Winter Habitat

Red-winged blackbirds form large roosts in the winter, with the largest occurring in the southeastern United States around major grain-producing areas (Mott 1984). Winter roosts are characterized by dense cover and include wetlands, deciduous thickets, coniferous stands, and sugarcane fields (Meanley 1965; Yasukawa and Searcy 1995; Miller et al. 2011). Yellow-headed blackbirds also occur in large flocks that are often sex-specific, with males spending the winter further north than females. Yellow-headed blackbirds roost in wetlands and prefer to forage in disturbed habitats during the winter, including agricultural fields and farmyards (Twedt and Crawford 1995). Common grackle winter roosts usually occur in dense conifer stands adjacent to woodlots and agricultural fields used for foraging. Agricultural fields include harvested cornfields, rice fields, and peanut fields; feedlots may also be used when snow prevents foraging in preferred habitats (White et al. 1985). Urban areas with high tree density can often have the largest common grackle roosts, at times exceeding 1 million birds during the winter (White et al. 1985; Peer and Bollinger 1997). Brown-headed cowbird winter roosts commonly occur in both deciduous and coniferous trees and will often occur in mixed-species flocks with other blackbirds, especially red-winged blackbirds (Lowther 1993).

6.2.2 Winter Climate and Weather Associations

Many species migrate to access resource-rich environments suited for breeding that are only seasonally available; at the conclusion of breeding, inclement conditions motivate migration to more benign environs. Brenner (1966a) suggested that the increased temperature in the wintering area, along with a propensity to gather in large flocks while roosting, allowed red-winged blackbirds to survive brief inclement winter conditions.

Strassburg et al. (2015) examined the relationships between species abundance and climatic variables for blackbirds and grackles overwintering in the south-central United States. These species included red-winged blackbird, common grackle, rusty blackbird (*Euphagus carolinus*), and Brewer's blackbird (*Euphagus cyanocephalus*). They related National Audubon Society Christmas Bird Count data to National Oceanic and Atmospheric Administration climatic variables (annual and winter minimum, mean, and maximum temperature, as well as total precipitation) using overdispersed Poisson mixed-effects regression. Relations of wintering blackbirds to climate were not consistent among species. Strassburg et al. (2015) reported negative associations of rusty blackbird abundance with minimum winter temperature, whereas Brewer's blackbird abundance was positively associated with annual minimum temperature and negatively associated with annual precipitation. The abundance of both red-winged blackbird and common grackle was positively associated with annual precipitation.

White et al. (1985) observed the relative constancy of red-winged blackbird, brown-headed cowbird, and common grackle abundance during mild winters in Tennessee but considerable variability during the following two severe winters. They reported significant associations between mean weekly population estimates and several weather variables, including mean weekly temperature ($r = 0.80$), change in weekly temperature ($r = 0.75$), and snow cover ($r = -0.79$). Blackbirds increased their use of cattle feedlots as snow cover increased over the course of the winter, particularly when snow was >2.5 cm, which confirmed observations by Besser et al. (1968).

Over a 21-year period, Weatherhead (2005b) studied the effects of large-scale climate phenomena on red-winged blackbird population trajectory in Ontario, Canada. He found that an "unprecedented" positive phase of the North Atlantic Oscillation, a climate phenomenon with a weather-governing role for every month in the year, was associated with a nearly 50% change in harem size. This positive phase of the North Atlantic Oscillation was associated with warmer, wetter, stormier winters in the southeastern United States, where Ontario red-winged blackbirds winter (Dolbeer 1982). Coupled with a strong positive association between change in harem size and male return rate, Weatherhead (2005b) suggested that this stormier winter weather led to poorer survival and, therefore, declining abundance of red-winged blackbirds in Ontario.

Weatherhead (2005b) and Strassburg et al. (2015) conducted correlational studies examining the consequences of climatic processes operating over large scales. The proximal mechanisms influencing blackbirds in winter were only partly revealed. One mechanism by which winter weather may influence blackbirds is through their diet. Stewart (1978a) reported the movement of common grackles in response to forage being made unavailable by snow. Additionally, winter storms have been reported to kill blackbirds, grackles, and starlings (MacReynolds 1917; Forbush, cited in Stewart 1978b), often as a result of severe fluctuations in temperature leading to freezing precipitation (Odum and Pitelka 1939). Francis (1976) found that dense winter roost aggregations in pine trees in Kentucky were as much as 2°C warmer as compared to the surrounding area, elevated in part by the metabolism of roosting birds. In high wind conditions, this benefit dissipated, and areas where trees were thinned afforded no protection from wind. Francis (1976) speculated that in the densest aggregations, birds at the top of roosts protected those below from precipitation (but not feces).

Birds may not be able to avoid the quick onset of deleterious climatic conditions (Newton 1998). Brenner and Malin (1965) examined the timespan roosting red-winged blackbirds can survive if they are prevented from foraging and found that the birds can live 2.88 days (range = 2.84–2.94). Blackbirds actively searching but unable to find food because of, for instance, heavy snowfall, would be expected to perish sooner.

6.3 MIGRATION SEASON

6.3.1 Migratory Habitat

Dolbeer (1978) divided the life cycle of red-winged blackbirds into seasons, noting that spring migration was February 21 through April 24, whereas fall migration occurred October 16 through December 9. During migration, blackbirds will commonly form mixed-species flocks with as many as 500,000 to 1,000,000 birds and commonly associate with agricultural habitats for foraging. Harvested agricultural fields (particularly sunflower and corn fields) have been shown to be important stopover habitats for blackbirds during spring migration in the Prairie Pothole Region of the upper midwestern United States and Canada (Clark et al. 1986; Homan et al. 2004; Galle et al. 2009), whereas other crops such as soybean have shown low use by blackbirds during migration (Hagy et al. 2008).

Red-winged blackbirds, yellow-headed blackbirds, common grackles, and brown-headed cowbirds are often observed in the same flocks (Peer and Bollinger 1997; G. M. Linz, personal observation), whereas yellow-headed blackbirds will also often form sex-specific flocks of conspecifics (Twedt and Crawford 1995). Red-winged blackbirds and yellow-headed blackbirds prefer emergent wetland vegetation for roosting, and forage in agricultural areas such as harvested grain fields, plowed fields, meadows, and pastures (Twedt and Crawford 1995). Homan et al. (2006) found that the maximum roost size of mixed blackbird flocks during spring in east-central South Dakota was correlated with emergent wetland area and possibly wetland basin area. Common grackles commonly roost near agricultural fields during migration in a variety of vegetation, including urban treed areas, hardwood thickets, coniferous plantations, and emergent marsh vegetation (Peer and

Bollinger 1997). Brown-headed cowbirds have a preference for early successional forest with high densities of red maple (*Acer rubrum*) and sweetgum (*Liquidambar styraciflua*) and accompanying closed canopies (Lyon and Caccamise 1981).

6.3.2 Migratory Climate and Weather Associations

As compared to breeding and overwintering studies of blackbirds, there are relatively few studies of the effects of climatic conditions on migrating blackbirds. Savard et al. (2011) monitored the fall migration of rusty blackbirds at the mouth of the Saguenay River on the north shore of the St. Lawrence River estuary in Quebec, Canada, over a 15-year period. They found that rusty blackbird numbers were positively correlated with annual and winter North Atlantic Oscillation indices and negatively related to summer (June–August) precipitation. They suggested these climatic factors contributed through food web processes to cyclic variation in abundance. Work of a similar nature is not available for other blackbird species. However, Weatherhead (2005b) examined the consequences of the North Atlantic Oscillation on breeding in red-winged blackbirds and reported no effect of this climate process on initiation of egg laying, which might suggest that these large-scale climate processes have little effect on arrival time. However, low winter values of the North Atlantic Oscillation expanded the breeding season, which might lead to an alteration in the timing of fall migration.

6.4 LANDSCAPE EFFECTS

In addition to local-scale habitat factors, surrounding landscape-level habitat variables can influence blackbird species composition and abundance (Forcey 2006; Forcey et al. 2007, 2015). Given that habitat effects at the landscape level can either be similar or different from those found at the local scale (Thogmartin 2007), it is important to consider multiple spatial scales when evaluating habitat influences (Thogmartin and Knutson 2007). Tozer et al. (2010) demonstrated the importance of studying habitat influences at multiple spatial scales, as the effects can vary depending on whether abundance, nest success, or productivity is the variable of interest. Saab (1999) even found that landscape variables were more influential on the distribution and occurrence of some bird species than smaller-scale habitat effects. Additionally, a variable at one scale can have a different effect at another scale (e.g., Pribil and Picman 1997; Thogmartin 2007). Multiple studies have examined landscape-level habitat effects on blackbird species, and results have shown varying degrees of landscape-level habitat influences on blackbirds, depending on the species, the scale of the analysis, and whether the variable described landscape composition (habitat-specific percent coverage) or landscape configuration (arrangement of habitats in the landscape) (Pribil and Picman 1997).

Forcey (2006) and Forcey et al. (2007, 2015) developed multi–state/province habitat and climate models for blackbird species in the Prairie Potholes Region and determined that both land use and climate variables were influential on blackbirds, though effects on the habitat generalists (e.g., red-winged blackbird and common grackle) were harder to elucidate compared to habitat specialists (e.g., yellow-headed blackbird). Fairbairn and Dinsmore (2001) developed models to estimate the densities of red-winged blackbirds, yellow-headed blackbirds, and common grackles in Iowa wetlands based on habitat variables and discovered that the composition and configuration of habitats in the landscape were influential for blackbirds. Schafer (1996) developed models for predicting the nest success of red-winged blackbirds and yellow-headed blackbirds and found that nest success for both species was affected by distance to shore, water depth, and nest height. Finally, Strassburg et al. (2015) evaluated landscape habitat effects on the winter populations of blackbirds and found dissimilar relationships compared to habitat associations found during the breeding season.

During breeding, red-winged blackbird abundance is positively related to the amount of wetland in the landscape at smaller landscape scales; a 1 SD increase in the percentage of wetland at the 1,000 ha scale increased red-winged blackbird abundance by 2.9% (Forcey et al. 2015). Although total wetland area is influential for red-winged blackbirds, individual wetland size does not appear to have an effect (Tozer et al. 2010). During winter, landscape-scale wetland variables did not have a strong influence on red-winged blackbird abundance (Strassburg et al. 2015). In comparison, the yellow-headed blackbird's close ties with wetland habitat for nesting were evident, as their abundance showed strong positive relationships with the wetland area at multiple landscape scales (Naugle et al. 2001; Forcey et al. 2015). A 1 SD increase in the percentage of wetland in the landscape at the 1,000 ha (2.7%), 10,000 ha (3.4%), and 100,000 ha (7.2%) scales increased yellow-headed blackbird abundance by 12%, 11%, and 82.5%, respectively (Forcey et al. 2015). Naugle et al. (1999b) even noted that wetland area was the only landscape variable of importance for yellow-headed blackbirds and that other habitat influences occurred at the local scale. This is likely due to female yellow-headed blackbirds' close ties to the immediate nesting area, as they usually do not exploit resources away from the nest wetland (Twedt and Crawford 1995).

The influence of wetland area on common grackle abundance during breeding is less certain. Although Fairbairn and Dinsmore (2001) found wetland area to be an important predictor of common grackle abundance, Tozer et al. (2010), Forcey et al. (2015), and Strassburg et al. (2015) did not find a similar relationship. This finding is likely because common grackles are an adaptable habitat generalist and are not restricted to specific cover types (Peer and Bollinger 1997), which makes any habitat relationships harder to elucidate.

Woody vegetation can have a negative impact on blackbirds, which tend to prefer more open habitats. Previous studies have generally suggested that red-winged blackbirds and yellow-headed blackbirds are negatively associated with woody vegetation, whereas common grackles and brown-headed cowbirds show some positive associations. Naugle et al. (1999a) and Naugle et al. (2001) showed that encroachment of woody vegetation around prairie wetlands can have a negative impact on red-winged blackbird abundance, and West et al. (2016) noted a negative relationship between red-winged blackbird occupancy and forested cover at a 250-m scale. Forcey et al. (2015) found a similar negative relationship between forest edge density and yellow-headed blackbird abundance at the two smallest landscape scales (1,000 and 10,000 ha), although this relationship was not present at the largest landscape scale (100,000 ha). A 1 SD increase in the percentage of forest at the 1,000 ha (11.6%) and 10,000 ha (11.5%) scales decreased yellow-headed blackbird abundance by 7.7% and 6.7%, respectively (Forcey et al. 2015). In addition, Naugle et al. (1999a) did not find any significant associations between yellow-headed blackbirds and woody vegetation, which suggests uncertainty of this effect on this species with respect to scale. Common grackles are frequently associated with woody vegetation (Peer and Bollinger 1997), and Naugle et al. (1999a) recorded common grackles in 60% of wetlands surrounded by >75% tree cover. Although these findings were consistent, Forcey et al. (2015) found a negative relationship with common grackles and woody vegetation at the 100,000 ha landscape scale and no definitive relationships at the 1,000 and 10,000 ha scales. A 1 SD decrease in percentage of forest in the landscape at the 100,000 ha scale (10.9%) decreased common grackle abundance by 6.5% (Forcey et al. 2015). These conflicting results are likely because 1) common grackles respond differently to woody vegetation at landscape scales than at more local scales and 2) land cover data used in Forcey et al. (2015) may only represent larger continuous areas of forest, overlooking small patches of trees surrounding wetlands that common grackles may utilize. Brown-headed cowbirds seem to prefer woody vegetation, as host nests are more frequently parasitized around wetlands with woody vegetation compared to restored wetlands without this habitat (Delphey and Dinsmore 1993). This is likely because woody vegetation often provides good visibility of the surrounding area, which females will use to locate hosts and their associated nests (Lowther 1993). Brown-headed cowbirds have also been shown to be affected by landscape habitat variables including shrub cover, but other variables were not strongly influential. This suggests that brown-headed cowbirds are responding to other factors such as host densities rather than specific landscape features (Jacobs et al. 2012).

In addition to the influences of specific types of vegetation and habitat, the arrangement of those habitats in the landscape can also be influential to blackbirds, but this relationship is not consistent among studies (e.g., Clark and Weatherhead 1986; Fairbairn and Dinsmore 2001; Forcey et al. 2015). Landscapes with high habitat diversity have been shown to positively influence red-winged blackbird nest success in agricultural areas, compared to large homogenous expanses of crop fields (Schafer 1996). This is likely due to surrounding croplands containing abundant food resources for blackbirds that can increase the suitability of nearby wetlands for nesting (Creighton et al. 1997). Clark and Weatherhead (1986) found the density of male red-winged blackbirds to be strongly related to the mixing of breeding habitat (hayfields) with feeding habitat (cropland); abundance was less when either habitat became very abundant and habitat diversity decreased. Fairbairn and Dinsmore (2001) noted that the density of red-winged blackbirds was positively influenced by the wetland perimeter–area ratio, whereas yellow-headed blackbirds were negatively associated with this variable. Areas with higher perimeter–area ratios have more edge habitat and more potential for wet meadow and prairie vegetation, which likely explains the positive association of red-winged blackbirds preferring edge and the negative association of yellow-headed blackbirds preferring deeper water wetlands (Twedt and Crawford 1995; Yasukawa and Searcy 1995).

The influence of habitat patch size on blackbirds is inconclusive. Herkert (1994) discovered red-winged blackbirds more often in small habitat patches than in large ones. In comparison, Helzer and Jelinski (1999) found red-winged blackbirds more often in large patches than in small ones. Schafer (1996) noted no relationships between patch size and nest success rates for yellow-headed blackbirds but found that red-winged blackbird nest success rates were higher in smaller patches. Fairbairn and Dinsmore (2001) noted red-winged blackbirds to be positively related to the number of smaller wetland patches, whereas yellow-headed blackbirds avoided smaller patches of wetland. Forcey et al. (2015) found different relationships between red-winged blackbird abundance and patch richness density, suggesting that importance of habitat heterogeneity may differ among scales.

6.4.1 Landscape Change

Given the strong habitat effects associated with blackbirds, changes in those variables can have a concomitant effect on blackbird abundance. Both changes in agricultural practices and the creation of the Conservation Reserve Program (CRP) resulted in substantial changes in land use over the last several decades. The rapid expansion of agriculture in the twentieth century and the clearing of forests in the eastern United States largely resulted in blackbirds expanding their range and abundance (Dolbeer and Stehn 1983). While this expansion was initially favorable to blackbirds, the North American Breeding Bird Survey showed all four blackbird species in slight decline throughout their range, with the exception of yellow-headed blackbirds in the central United States (Sauer et al. 2014). Recent declines in blackbird species associated with agriculture are likely due to increased mechanization, reduced crop complexity, earlier mowing of hay (which can cause nesting failures), and increased use of chemical fertilizers, herbicides, and insecticides (Blackwell and Dolbeer 2001; Weatherhead 2005b).

The CRP, originally designed to remove marginal agricultural lands from production and replace them with perennial vegetation, served two purposes: 1) reduce erosion and stream sedimentation and 2) enhance fish and wildlife habitat (Johnson and Igl 1995). Although Sauer et al. (2014) noted overall declines in red-winged blackbird and brown-headed cowbird numbers throughout their range in North America, densities of these species were approximately 10 and 6 times higher, respectively, in CRP land compared to cropland in North Dakota (Johnson and Igl 1995). This result suggests that although agricultural habitat plays an important role in the life history of blackbirds, management of natural habitat has the potential to be more influential for regulating blackbird populations. Although red-winged blackbirds are more abundant in CRP habitat, other research suggests that nesting in CRP results in negative population growth rates, largely related to poor fecundity, and thus the effect on this species is deleterious (McCoy et al. 1999). Future CRP enrollments are unknown, but recent high crop prices have encouraged farmers to put CRP lands back into production, reducing the total amount of wildlife habitat available from this program (Stubbs 2014). If this current trend continues, reduction in CRP acreage could reduce the amount of available breeding habitat for blackbird species that utilize grassland areas. However, given the finding by McCoy et al. (1999) that CRP acts as a sink for red-winged blackbirds, reductions in CRP may lead to increased abundance for this species.

6.5 CLIMATE CHANGE

The Intergovernmental Panel on Climate Change (IPCC) stated in its 2014 assessment report that "warming of the climate system is unequivocal, and since the 1950s, many of the observed changes are unprecedented over decades to millennia. The atmosphere and ocean have warmed, the amounts of snow and ice have diminished, and sea level has risen" (IPCC 2014). Blackbirds, especially yellow-headed blackbirds, are likely adapted to highly unstable interannual precipitation leading to inconsistent wetland conditions (Beletsky and Orians 1994). As such, we might expect them to be particularly able to accommodate changing climatic conditions. Wilson (2009) found that red-winged blackbirds were the earliest arrivals among the species he studied and suggested that this species is immune to changes in leaf phenology caused by climatic variation, principally because they are able to forage on seeds that may tide them over until leaf out and subsequent insect emergence. Given the relative insensitivity of red-winged blackbirds to leaf phenology, he suggested that red-winged blackbirds would not provide the most sensitive indicator for climate change. Nevertheless, evidence suggests they have responded to changes in the climatic environment. Ledneva et al. (2004) reported a strongly significant association between observations of first spring activity in Massachusetts of red-winged blackbirds versus mean monthly temperature. They reported an earlier observation over time of 2.54 days per degree change in temperature (°C).

Bateman et al. (2016) examined how past changes in climate affected current potential breeding distributions of species in the conterminous United States. They calculated the bioclimatic velocity of potential breeding distributions, which describe the pace and direction of change in species distribution over the past 60 years. Using the species-habitat modeling software MaxEnt (Elith et al. 2011), they related Global Biodiversity Information Facility occurrence records for species in the breeding season (April–July) against monthly total precipitation and temperature maxima and minima data from the PRISM dataset (4-km resolution, PRISM Climate Group, Oregon State University, http://prism.oregonstate.edu). These monthly climate data were aggregated into eight BIOCLIM variables (mean annual temperature [°C], temperature seasonality [standard deviation × 100], maximum temperature of the warmest period [°C], minimum temperature of the coldest period [°C], annual precipitation [mm], precipitation in the wettest quarter [mm], precipitation in the driest quarter [mm], and precipitation seasonality [coefficient of variation]) for three time periods: 6, 12, and 36 months prior. Their results indicated that red-winged and rusty blackbirds expanded their ranges principally eastward, common grackles expanded southeastward, brown-headed cowbirds expanded westward, and Brewer's and yellow-headed blackbirds expanded northwestward. The differences in species response may be due to different favorable climatic conditions. Rusty blackbirds were most highly associated with the mean temperature of the previous 6 months and to a lesser extent temperature seasonality in the previous 3 years. Brewer's and yellow-headed blackbirds, conversely, responded most strongly to precipitation in the wettest quarter over the previous 3 years. Brown-headed cowbirds were most associated with temperature seasonality over the previous 3 years.

Torti and Dunn (2005) examined the effects of long-term changes in temperature on laying dates and clutch size in red-winged blackbirds, based upon a 50-year record of nest data. As temperature warmed between 1950 and 2000, red-winged blackbirds nested 7.5 days earlier, which they suggested was correlated to larger clutch sizes (their results, however, did not show a relationship between temperature and clutch size).

There are broad latitudinal responses of blackbirds to climatic conditions that may be expected to change as climate changes. Brenner and Hayes (1985), studying northern (Ohio) and southern (Florida) populations of red-winged blackbirds, found individuals from the northern population required significantly less energy, perhaps because of the insulation afforded by the larger body size of northern individuals. The longer day length experienced by southern individuals, rather than temperature, was associated with increased water intake by southern individuals compared to northern individuals. As temperature warms, we might expect larger body size to be selected against and commensurate increases in water intake.

As noted earlier, Weatherhead (2005b) examined red-winged blackbird abundance in association with the North Atlantic Oscillation. With respect to changing climate, Weatherhead (2005b) speculated that under a warming atmosphere, a persistent positive phase of the North Atlantic Oscillation could lead to warmer, wetter, stormier winters in the southeastern United States, leading to the possibility of increased winter mortality. In turn, associated population declines could lead to changes in blackbird sociality, including increasing propensity of monogamy.

6.6 SUMMARY

Blackbird species are affected by both habitat and climate effects at local and landscape scales throughout their range. All species tend to avoid deep forested and heavily urbanized areas while being able to survive in a range of other habitats including wetlands, uplands, and agricultural areas. Red-winged blackbirds, common grackles, and brown-headed cowbirds are habitat generalists and can inhabit multiple habitat types, whereas yellow-headed blackbirds are strongly tied to the presence of wetland habitat, especially during breeding. Weather variables such as temperature can affect food supplies, and precipitation can affect the amount of wetland habitat in the landscape, which is important for multiple blackbird species. Increased understanding of the habitat relationships of blackbirds can better inform management decisions so that resources can be focused on the habitats and locations providing the biggest impact. While weather variables cannot be managed, understanding the influences of weather variables on blackbirds may allow us to predict population growth or decline and adjust management practices accordingly. Improved understanding of habitat influences and incorporating weather information into predictive models may allow managers to better focus their resources on locations that will be optimally beneficial for a given management objective. This is particularly important with respect to management of blackbird populations due to agricultural depredation.

At present, it is entirely unclear to what extent the synergistic effects of a changing climate and land use will have on blackbird populations. Lethal levels of temperature are likely to become more common in the southern portion of the United States as the century progresses, pushing breeding season distribution northward. However, this northward movement can only be possible if suitable conditions exist. Precipitation forecasts are mixed, but most suggest increased precipitation in the central United States, which should be a boon to blackbirds and the habitats they require. Nevertheless, anthropogenic land change will ultimately determine the amount of suitable habitat for blackbirds.

Improved understanding of the combined effect of climate and habitat is necessary. Previous work has examined the consequences of multiple climate and habitat factors, but how these effects interact to accommodate blackbird life history will be useful in formulating management and conservation actions. For instance, nest placement with respect to vegetative cover and solar radiation subsequently influences bird behavior. These sorts of synergistic insights should be the focus of next-generation research on blackbird habitat and climate associations.

REFERENCES

Albers, P. H. 1978. Habitat selection by breeding red-winged blackbirds. *Wilson Bulletin* 90:619–634.

Allen, A. A. 1914. The red-winged blackbird: A study in the ecology of a cat-tail marsh. *Proceedings of the Linnean Society of New York* 24–25:43–128.

Bateman, B. L., A. M. Pidgeon, V. C. Radeloff, J. VanDerWal, W. E. Thogmartin, S. J. Vavrus, and P. J. Heglund. 2016. The pace of past climate change versus potential bird distributions and land use in the U.S. *Global Change Biology* 22:1130–1144.

Beletsky, L. 1996. *The red-winged blackbird: The biology of a strongly polygynous songbird.* Academic Press, London, UK.

Beletsky, L. D., and G. H. Orians. 1994. Site fidelity and territorial movements of males in a rapidly declining population of yellow-headed blackbirds. *Behavioral Ecology and Sociobiology* 4:257–265.

Besser, J. F., J. W. DeGrazio, J. L. Guarino, D. F. Mott, D. L. Otis, B. R. Besser, and C. E. Knittle. 1984. Decline in breeding red-winged blackbirds in the Dakotas, 1965–1981. *Journal of Field Ornithology* 55:435–443.

Besser, J. F., W. C. Royall, Jr., and J. W. DeGrazio. 1968. Costs of wintering starlings and red-winged blackbirds at feedlots. *Journal of Wildlife Management* 32:179–180.

Blackwell, B. F., and R. A. Dolbeer. 2001. Decline of the red-winged blackbird population in Ohio correlated to changes in agriculture (1965–1996). *Journal of Wildlife Management* 65:661–667.

Blankespoor, G. W., J. Oolman, and C. Uthe. 1982. Eggshell strength and cowbird parasitism of red-winged blackbirds. *Auk* 99:363–365.

Brenner, F. J. 1966a. Energy requirements of the red-winged blackbird. *Wilson Bulletin* 78:111–120.

Brenner, F. J. 1966b. The influence of drought on reproduction in a breeding population of red-winged blackbirds. *American Midland Naturalist* 76:201–210.

Brenner, F. J., and T. M. Hayes. 1985. Photoperiodic and temperature influences on metabolism and water intake in two populations of the red-winged blackbird. *American Midland Naturalist* 113:325–333.

Brenner, F. J., and W. F. Malin. 1965. Metabolism and survival time of the red-winged blackbird. *Wilson Bulletin* 77:282–289.

Brittingham, M. C., and S. A. Temple. 1983. Have cowbirds caused forest songbirds to decline? *BioScience* 33:31–35.

Choi, I. H., and G. S. Bakken. 1990. Begging response in nestling red-winged blackbirds (*Agelaius phoeniceus*): Effect of body temperature. *Physiological Zoology* 63:965–986.

Clark, R. G., and P. J. Weatherhead. 1986. The effect of fine-scale variations in agricultural land-use on the abundance of red-winged blackbirds. *Canadian Journal of Zoology* 64:1951–1955.

Clark, R. G., and P. J. Weatherhead. 1987. Influence of population size on habitat use by territorial male red-winged blackbirds in agricultural landscapes. *Auk* 104:311–315.

Clark, R. G., P. J. Weatherhead, H. Greenwood, and R. D. Titman. 1986. Numerical responses of red-winged blackbird populations to changes in regional land-use patterns. *Canadian Journal of Zoology* 64:1944–1950.

Creighton, J. H., R. D. Sayler, J. E. Tabor, and M. J. Monda. 1997. Effects of wetland excavation on avian communities in eastern Washington. *Wetlands* 17:216–227.

Delphey, P. J., and J. J. Dinsmore. 1993. Breeding bird communities of recently restored and natural prairie potholes. *Wetlands* 13:200–206.

Dolbeer, R. A. 1978. Movement and migration patterns of red-winged blackbirds: A continental overview. *Bird-Banding* 49:17–34.

Dolbeer, R. A. 1982. Migration patterns for sex and age classes of blackbirds and starlings. *Journal of Field Ornithology* 53:28–46.

Dolbeer, R. A., and R. A. Stehn. 1983. Population status of blackbirds and starlings in North America, 1966–81. *Eastern Wildlife Damage Control Conference* 1:51–61.

Economic Research Service. 2011. *Major land uses.* http://www.ers.usda.gov/data-products/major-land-uses/.aspx (accessed November 30, 2016).

Elith, J., S. J. Phillips, T. Hastie, M. Dudík, Y. E. Chee, and C. J. Yates. 2011. A statistical explanation of MaxEnt for ecologists. *Diversity and Distributions* 17:43–57.

Ellis, E. C. 2011. Anthropogenic transformation of the terrestrial biosphere. *Philosophical Transactions of the Royal Society A* 369:1010–1035.

Erskine, A. J. 1971. Some new perspectives on the breeding ecology of common grackles. *Wilson Bulletin* 83:352–370.

Fairbairn, S. E., and J. J. Dinsmore. 2001. Local and landscape-level influences on wetland bird communities of the prairie pothole region of Iowa, USA. *Wetlands* 21:41–47.

Fletcher, R. J., Jr., and R. R. Koford. 2004. Consequences of rainfall variation for breeding wetland blackbirds. *Canadian Journal of Zoology* 82:1316–1325.

Forcey, G. 2006. *Landscape-level influences on wetland birds in the prairie pothole region of the United States and Canada*. PhD Dissertation. North Dakota State University, Fargo, ND.

Forcey, G. M., G. M. Linz, W. E. Thogmartin, and W. J. Bleier. 2007. Influence of land use and climate on wetland breeding birds in the Prairie Pothole region of Canada. *Canadian Journal of Zoology* 85:421–436.

Forcey, G. M., W. E. Thogmartin, G. M. Linz, P. C. McKann, and S. M. Crimmins. 2015. Spatially explicit modeling of blackbird abundance in the Prairie Pothole Region. *Journal of Wildlife Management* 79:1022–1033.

Francis, W. J. 1976. Micrometeorology of a blackbird roost. *Journal of Wildlife Management* 40:132–136.

Freidmann, H. 1963. *Host relations of the parasitic cowbirds*. U.S. National Museum Bulletin Number 223. Smithsonian Institution, Washington, DC.

Freidmann, H., and L. Kiff. 1985. The parasitic cowbirds and their hosts. *Western Foundation of Vertebrate Zoology* 2:225–302.

Galle, A. M., G. M. Linz, H. J. Homan, and W. J. Bleier. 2009. Avian use of harvested crop fields in North Dakota during spring migration. *Western North American Naturalist* 69:491–500.

Glassy, B., and M. Amos. 2009. Shade seeking by common grackle (*Quiscalus quiscula*) nestlings at the scale of the nanoclimate. *Journal of Thermal Biology* 34:76–80.

Hagy, H. M., G. M. Linz, and W. J. Bleier. 2008. Optimizing the use of decoy plots for blackbird control in commercial sunflower. *Crop Protection* 27:1442–1447.

Hall, L. S., P. R. Krausman, and M. L. Morrison. 1997. The habitat concept and a plea for standard terminology. *Wildlife Society Bulletin* 25:173–182.

Helzer, C. J., and D. E. Jelinski. 1999. The relative importance of patch area and perimeter-area ratio to grassland breeding birds. *Ecological Applications* 9:1448–1458.

Herkert, J. R. 1994. The effects of habitat fragmentation on Midwestern grassland bird communities. *Ecological Applications* 4:461–471.

Hobbs R. J., S. Arico, J. Aronson, J. S. Baron, P. Bridgewater, V. A. Cramer, P. R. Epstein, et al. 2006. Novel ecosystems: Theoretical and management aspects of the new ecological world order. *Global Ecology and Biogeography* 15:1–7.

Homan, H. J., G. M. Linz, R. M. Engeman, and L. B. Penry. 2004. Spring dispersal patterns of red-winged blackbirds, *Agelaius phoeniceus*, staging in eastern South Dakota. *Canadian Field-Naturalist* 118:201–209.

Homan, H. J., R. S. Sawin, G. M. Linz, and W. J. Bleier. 2006. Habitat characteristics of spring blackbird roosts in east-central South Dakota. *Prairie Naturalist* 38:183–193.

Intergovernmental Panel on Climate Change (IPCC). 2014. Climate Change 2014: Synthesis report. *Contribution of Working Groups I, II and III to the Fifth Assessment Report of the Intergovernmental Panel on Climate Change*, eds. R. K. Pachauri and L. A. Meyer. Intergovernmental Panel on Climate Change, Geneva, Switzerland.

Jacobs, R. B., F. R. Thompson, R. R. Koford, F. A. La Sorte, H. D. Woodward, and J. A. Fitzgerald. 2012. Habitat and landscape effects on abundance of Missouri's grassland birds. *Journal of Wildlife Management* 76:372–381.

Johnson, D. H., and L. D. Igl. 1995. Contributions of the conservation reserve program to populations of breeding birds in North Dakota. *Wilson Bulletin* 107:709–718.

Johnson, S. R., and I. M. Cowan. 1975. The energy cycle and thermal tolerance of the starlings (Aves, Sturnidae) in North America. *Canadian Journal of Zoology* 53:55–68.

Kendeigh, S. C. 1944. Effect of air temperature on the rate of energy metabolism in the English sparrow. *Journal of Experimental Zoology* 96:1–16.

Kendeigh, S. C. 1969. Energy responses of birds to their thermal environments. *Wilson Bulletin* 81:441–449.

Kendeigh, S. C., V. R. Dolnik, and V. M. Gavrilov. 1977. Avian energetics. In *Granivorous birds in ecosystems*, eds. J. Pinowski and S. C. Kendeigh, 127–204. Cambridge University Press, Cambridge, UK.

King, A. P. 1979. *Variables affecting parasitism in the North American cowbird (Molothrus ater)*. PhD Dissertation. Cornell University, Ithaca, NY.

Lariviere, S., and M. Lepage. 2000. Effect of a water-level increase on use by birds of a lakeshore fen in Quebec. *Canadian Field-Naturalist* 114:694–696.

Lederer, R. J., W. S. Mazen, and P. J. Metropulos. 1975. Population fluctuation in a yellow-headed blackbird marsh. *Western Birds* 6:1–6.

Ledneva, A., A. J. Miller-Rushing, R. B. Primack, and C. Imbres. 2004. Climate change as reflected in a naturalist's diary, Middleborough, Massachusetts. *Wilson Bulletin* 116:224–231.

Lewies, R. W., and M. I. Dyer. 1969. Respiratory metabolism of the red-winged blackbird in relation to ambient temperature. *Condor* 71:291–298.

Linz, G. M., D. C. Blixt, D. L. Bergman, and W. J. Bleier. 1996. Responses of red-winged blackbirds, yellow-headed blackbirds and marsh wrens to glyphosate-induced alterations in cattail density. *Journal of Field Ornithology* 67:167–176.

Linz, G. M., and S. B. Bolin. 1982. Incidence of brown-headed cowbird parasitism on red-winged blackbirds. *Wilson Bulletin* 94:93–95.

Linz, G. M., and H. J. Homan. 2011. Use of glyphosate for managing invasive cattail (*Typha* spp.) to disperse blackbird (Icteridae) roosts. *Crop Protection* 30:98–104.

Lloyd, J., and T. E. Martin. 2004. Nest-site preference and maternal effects on offspring growth. *Behavioural Ecology* 15:816–823.

Lowther, P. E. 1993. Brown-headed cowbird (*Molothrus ater*). No. 47. *The birds of North America*, ed. P. G. Rodewald. Cornell Lab of Ornithology, Ithaca, NY. https://birdsna.org/Species-Account/bna/species/bnhcow (accessed November 12, 2016).

Lyon, L. A., and D. F. Caccamise. 1981. Habitat selection by roosting blackbirds and starlings: Management implications. *Journal of Wildlife Management* 45:435–443.

MacReynolds, G. 1917. A Pennsylvania starling roost. *Auk* 34:338–340.

Martinuzzi, S., G. I. Gavier-Pizarro, A. E. Lugo, and V. C. Radeloff. 2015. Future land-use changes and the potential for novelty in ecosystems of the United States. *Ecosystems* 18:1332–1342.

Maxwell, G. R., J. M. Nocilly, and R. I. Shearer. 1976. Observations at a cavity nest of the common grackle and an analysis of grackle nesting sites. *Wilson Bulletin* 88:505–507.

McCoy, T. D., M. R. Ryan, E. W. Kurzejeski, and L. W. Burger, Jr. 1999. Conservation reserve program: Source or sink habitat for grassland birds in Missouri? *Journal of Wildlife Management* 63:530–538.

Meanley, B. 1965. The roosting behavior of the red-winged blackbird in the southern United States. *Wilson Bulletin* 77:217–228.

Melillo, J. M., T. C. Richmond, and G. W. Yohe, eds. 2014. *Climate change impacts in the United States: The third national climate assessment*. U.S. Global Change Research Program. http://www.globalchange.gov/browse/reports/climate-change-impacts-united-states-third-national-climate-assessment-0 (accessed November 24, 2016).

Miller, M. W., E. V. Pearlstine, R. M. Dorazio, and F. J. Mazzotti. 2011. Occupancy and abundance of wintering birds in a dynamic agricultural landscape. *Journal of Wildlife Management* 75:836–847.

Mott, D. F. 1984. Research on winter roosting blackbirds and starlings in the southeastern United States. *Vertebrate Pest Conference* 11:183–187.

Murkin, H. R., E. J. Murkin, and J. P. Ball. 1997. Avian habitat selection and prairie wetland dynamics: A 10-year experiment. *Ecological Applications* 7:1144–1159.

Naugle, D. E., K. F. Higgins, and S. M. Nusser. 1999a. Effects of woody vegetation on prairie wetland birds. *Canadian Field Naturalist* 113:487–492.

Naugle, D. E., K. F. Higgins, S. M. Nusser, and W. C. Johnson. 1999b. Scale-dependent habitat use in three species of prairie wetland birds. *Landscape Ecology* 14:267–276.

Naugle, D. E., R. R. Johnson, M. E. Estey, and K. F. Higgins. 2001. A landscape approach to conserving wetland bird habitat in the prairie pothole region of eastern South Dakota. *Wetlands* 21:1–17.

Newbold, T., L. N. Hudson, A. P. Arnell, S. Contu, A. De Palma, S. Ferrier, S. L. Hill, et al. 2016. Has land use pushed terrestrial biodiversity beyond the planetary boundary? A global assessment. *Science* 353:288–291.

Newbold, T., L. N. Hudson, S. L. L. Hill, S. Contu, I. Lysenko, R.A. Senior, L. Borger, et al. 2015. Global effects of land use on local terrestrial biodiversity. *Nature* 520:45–50.

Newton, I. 1998. *Population limitation in birds*. Academic Press, San Diego, CA.

Norman, R. F., and R. J. Robertson. 1975. Nest-searching behavior in the brown-headed cowbird. *Auk* 92:610–611.

Odum, E. P., and F. A. Pitelka. 1939. Storm mortality in a winter starling roost. *Auk* 56:451–455.

Orians, G. H., and M. F. Willson. 1964. Interspecific territories of birds. *Ecology* 45:736–744.

Orians, G. H. 1980. *Some adaptations of marsh-nesting blackbirds*. Princeton University Press, Princeton, NJ.

Orians, G. H., and J. Wittenberger. 1991. Spatial and temporal scales in habitat selection. *American Naturalist* 137:S29–S49.

Paladino, F. V. 1989. Constraints of bioenergetics on avian population dynamics. *Physiological Zoology* 62:410–428.

Peer, B. D., and E. K. Bollinger. 1997. Common grackle (*Quiscalus quiscula*). No. 271. *The birds of North America*, ed. P. G. Rodewald. Cornell Lab of Ornithology, Ithaca, NY. https://birdsna.org/Species-Account/bna/species/rewbla (accessed September 25, 2016).

Picman, J. 1980. Impact of marsh wrens on reproductive strategy of red-winged blackbirds. *Canadian Journal of Zoology* 58:337–350.

Pribil, S., and J. Picman. 1997. The importance of using the proper methodology and spatial scale in the study of habitat selection by birds. *Canadian Journal of Zoology* 75:1835–1844.

Radeloff, V. C., J. W. Williams, B. L. Bateman, K. D. Burke, S. K. Karter, E. S. Childress, K. J. Cromwell, et al. 2015. The rise of novelty in ecosystems. *Ecological Applications* 25:2051–2068.

Rawinski, T. J., and R. A. Malecki. 1984. Ecological relationships among purple loosestrife, cattail, and wildlife at the Montezuma National Wildlife Refuge. *New York Fish and Game Journal* 31:81–87.

Robertson, R. J. 1972. Optimal niche space of the red-winged blackbird (*Agelaius phoeniceus*). I. Nesting success in marsh and upland habitat. *Canadian Journal of Zoology* 50:247–263.

Saab, V. 1999. Importance of spatial scale to habitat use by breeding birds in riparian forests. *Auk* 110:37–48.

Santee, W., and G. Bakken. 1987. Social displays in red-winged blackbirds (*Agelaius phoeniceus*): Sensitivity to thermoregulatory costs. *Auk* 104:413–420.

Sauer, J. R., J. E. Hines, J. E. Fallon, K. L. Pardieck, D. J. Ziolkowski, Jr., and W. A. Link. 2014. *The North American Breeding Bird Survey, Results and Analysis 1966–2013. Version 01.30.2015*. U.S. Geological Survey, Patuxent Wildlife Research Center, Laurel, MD.

Savard, J, P. L., C. Melanie, and D. Bruno. 2011. Exploratory analysis of correlates of the abundance of rusty blackbirds (*Euphagus carolinus*) during fall migration. *Ecoscience* 18:402–408.

Schafer, J. L. 1996. *A comparison of blackbird reproductive success in natural and restored Iowa wetlands*. MS Thesis. Iowa State University, Ames, IA.

Schroeder, R. L. 1982. *Habitat suitability index models: Yellow-headed blackbird*. U.S. Department of the Interior, US Fish and Wildlife Service FWS/OBS-82110.26. U.S. Fish and Wildlife Service, U.S. Department of Interior, Washington, DC. https://pubs.er.usgs.gov/publication/fwsobs82_10_26 (accessed November 24, 2016).

Searcy, W. A., and K. Yasukawa. 1995. *Polygyny and sexual selection in red-winged blackbirds*. Princeton University Press, Princeton, NJ.

Seibert, H. C. 1949. Differences between migrant and non-migrant birds in food and water intake at various temperatures and photoperiods. *Auk* 66:128–153.

Shipley, F. S. 1979. Predation on red-winged blackbird eggs and nestlings. *Wilson Bulletin* 91:426–433.

Short, H. L. 1985. *Habitat suitability index models: Red-winged blackbird*. Biological Report 82(10.95). U.S. Fish and Wildlife Service, U.S. Department of Interior, Washington, DC. http://www.nwrc.usgs.gov/wdb/pub/hsi/hsi-095.pdf (accessed November 12, 2016).

Stewart, P. A. 1978a. Possible weather-related southward movements of common grackles in early January. *Bird-Banding* 49:79–80.

Stewart, P. A. 1978b. Weather-related mortality of blackbirds and starlings in a Kentucky roosting congregation. *Wilson Bulletin* 90:655–656.

Strassburg, M., G. M. Linz, S. M. Crimmins, and W. E. Thogmartin. 2015. Habitat associations of native blackbirds and invasive European starlings wintering in the southeastern United States. *Human-Wildlife Interactions* 9:171–179.

Stubbs, M. 2014. *Conservation Reserve Program (CRP): Status and issues*. Congressional Research Service 7-5700 R42783, Washington, DC. http://nationalaglawcenter.org/wp-content/uploads/assets/crs/R42783.pdf (accessed November 24, 2016).

Thogmartin, W. E. 2007. Letter: Effects at the landscape scale may constrain habitat relations at finer scales. *Avian Conservation and Ecology* 2:6.

Thogmartin, W. E., and M. G. Knutson. 2007. Scaling local species-habitat relations to the larger landscape with a hierarchical spatial count model. *Landscape Ecology* 22:61–75.

Torti, V. M., and P. O. Dunn. 2005. Variable effects of climate change on six species of North American birds. *Oecologia* 145:486–495.

Tozer, D. C., E. Nol, and K. F. Abraham. 2010. Effects of local and landscape-scale habitat variables on abundance and reproductive success of wetland birds. *Wetlands Ecology and Management* 18:679–693.

Turner, A. M., and J. P. McCarty. 1998. Resource availability, breeding site selection, and reproductive success of red-winged blackbirds. *Oecologia* 113:140–146.

Twedt, D. J., and R. D. Crawford. 1995. Yellow-headed blackbird (*Xanthocephalus xanthocephalus*). No. 192. *The birds of North America*, ed. P. G. Rodewald. Cornell Laboratory of Ornithology, Ithaca, NY. https://birdsna.org/Species-Account/bna/species/yehbla (accessed June 23, 2016).

USGCRP. 2009. *Global climate change impacts in the United States*. eds. T. R. Karl, J. M. Melillo, and T. C. Peterson, Cambridge University Press, New York, NY. http://www.globalchange.gov/browse/reports/global-climate-change-impacts-united-states (accessed November 24, 2016).

Vierling, K. T. 2000. Source and sink habitats of red-winged blackbirds in a rural/suburban landscape. *Ecological Applications* 10:1211–1218.

Weatherhead, P. J. 2005a. Effects of climate variation on timing of nesting, reproductive success, and offspring sex ratios of red-winged blackbirds. *Oecologia* 144:168–175.

Weatherhead, P. J. 2005b. Long-term decline in a red-winged blackbird population: Ecological causes and sexual selection consequences. *Proceedings of the Royal Society of London: Biological Sciences* 272:2313–2317.

Webb, D. R., and J. R. King. 1983. Heat-transfer relations of avian nestlings. *Journal of Thermal Biology* 8:301–310.

Weller, M. W., and C. S. Spatcher. 1965. *Role of habitat in the distribution and abundance of marsh birds. Special Report No. 43*. Ames, IA. http://publications.iowa.gov/21788/1/Role%20of%20habitat%20in%20the%20distibution%20and%20abundance%20of%20marsh%20birds.pdf (accessed November 24, 2016).

West, A. S., P. D. Keyser, C. M. Lituma, D. A. Buehler, R. D. Applegate, and J. Morgan. 2016. Grasslands bird occupancy of native warm-season grass. *Journal of Wildlife Management* 80:1081–1090.

White, S. B., R. A. Dolbeer, and T. A. Bookhout. 1985. Ecology, bioenergetics, and agricultural impacts of a winter-roosting population of blackbirds and starlings. *Wildlife Monographs* 93:1–42.

Whitt, M. B., H. H. Prince, and R. R. Cox. 1999. Avian use of purple loosestrife dominated habitat relative to other vegetation types in a Lake Huron wetland complex. *Wilson Bulletin* 111:105–114.

Williams, J. W., and S. T. Jackson. 2007. Novel climates, no-analog communities, and ecological surprises. *Frontiers in Ecology and the Environment* 5:475–482.

Wilson, W. H., Jr. 2009. Variability of arrival dates of Maine migratory breeding birds: Implications for detecting climate change. *Northeastern Naturalist* 16:443–454.

Yasukawa, K., and W. A. Searcy. 1995. Red-winged blackbird (*Agelaius phoeniceus*). No. 184. *The Birds of North America*, ed. P. G. Rodewald. Cornell Lab of Ornithology, Ithaca, NY. https://birdsna.org/Species-Account/bna/species/rewbla (accessed September 25, 2016).

Zimmerling, J. R., and C. D. Ankney. 2005. Variation in incubation patterns of red-winged blackbirds nesting at lagoons and ponds in eastern Ontario. *Wilson Bulletin* 117:280–290.

CHAPTER 7

Dynamics and Management of Blackbird Populations

Richard A. Dolbeer
Wildlife Services
Sandusky, Ohio

CONTENTS

7.1 The Annual Cycle of the Red-Winged Blackbird Population 121
7.2 Population Control Efforts 123
    7.2.1 Miscellaneous Early Attempts, 1950s Through 1960s 123
        7.2.1.1 Dynamiting Roosts 123
        7.2.1.2 Poison Baits 124
        7.2.1.3 Floodlight Traps 124
        7.2.1.4 Decoy Traps 124
    7.2.2 Major Organized Efforts, 1970s through 2000s 125
        7.2.2.1 Winter Roost Spraying with Surfactant PA-14 125
        7.2.2.2 Ripening Sunflowers in the Dakotas 126
        7.2.2.3 Sprouting Rice in Louisiana 127
        7.2.2.4 Trapping Cowbirds to Reduce Nest Parasitism 127
7.3 Conclusions 130
References 130

Whenever birds have threatened agricultural crops, the natural response of farmers has been to attempt to reduce the depredating populations (Dolbeer 1986). Laws were established as early as 1424 in Europe and 1667 in North America to encourage the killing of rooks (*Corvus frugilegus*) and blackbirds (Icteridae), respectively, to protect grain crops (Dolbeer 1980; Wright et al. 1980). During the nineteenth century, initial attempts to establish laws in North America protecting overexploited species such as the passenger pigeon (*Ectopistes migratorius*) were often thwarted by agricultural groups concerned about bird depredations (Schorger 1973). Large-scale efforts to reduce populations of weaver finches, primarily quelea (*Quelea quelea*), to protect agricultural crops in Africa were attempted from the 1950s through the 1980s, in which hundreds of millions of birds were killed annually with explosives, flamethrowers, and toxic sprays (Ward 1979; Bruggers and Elliott 1989). In North Africa, millions of European starlings (*Sturnus vulgaris*) were killed in the late 1950s to protect olive groves by the application of the insecticide parathion to winter roosts (Bub 1980).

Today in North America, three members of the blackbird family—red-winged blackbirds (*Agelaius phoeniceus*), common grackles (*Quiscalus quiscula*), and brown-headed cowbirds (*Molothrus ater*)—are dominant components of the avifauna, all being in the top 10 most abundant birds based on Breeding Bird Survey data, 1997–2006 (Table 7.1, Figure 7.1). Red-winged blackbirds alone represented 21% of the total relative abundance of the top 10 species. These three blackbird species, both independently or in mixed flocks that frequently include the introduced and equally abundant European starling, often conflict with agriculture, pose zoonotic disease threats

Table 7.1 Relative Abundance of the 10 Most Frequently Recorded Species on the North American Breeding Bird Survey (BBS), 1998–2007[a]

Rank	Species	Number of BBS Routes Recorded[a]	Mean Number/ BBS Route[b]	Relative Continental Abundance[c]
1	Red-winged blackbird	3,452	31.20	107,710
2	American robin	3,500	24.54	85,906
3	European starling	3,281	18.51	60,745
4	Mourning dove	3,476	16.86	58,606
5	American crow	3,199	15.17	48,532
6	Common grackle	2,661	12.43	33,082
7	House sparrow	2,861	11.53	32,990
8	Brown-headed cowbird	3,449	7.93	27,340
9	Chipping sparrow	2,920	8.63	25,211
10	Western meadowlark	1,491	16.23	24,198

[a] Number of BBS routes with acceptable data on which the species was detected in 1998–2007 out of 4,003 total routes with acceptable data (Blancher et al. 2013; Partners in Flight Science Committee 2013).
[b] Mean number of birds detected per route per year for all routes in Bird Conservation Regions within provinces, states, and territories where species was detected (Partners in Flight Science Committee 2013; North American Bird Conservation Initiative 2016).
[c] Number of BBS routes in which species was recorded times mean number per BBS route.

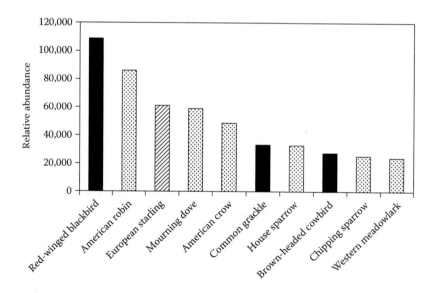

Figure 7.1 The list of top 10 bird species in North America, based on relative abundance from Breeding Bird Survey routes, 1998–2007, includes three species of blackbirds (Table 7.1). These three species, along with the European starling, often forage and roost together, making the group a dominant component of the avifauna of North America.

DYNAMICS AND MANAGEMENT OF BLACKBIRD POPULATIONS

and risks to aviation safety, and sometimes negatively impact endangered species. Thus, it is not surprising that considerable effort has been expended in attempts to reduce populations locally, regionally, and nationally.

SIDEBAR 7.1   THE GRACKLE

The grackle's voice is less than mellow,
His heart is black, his eye is yellow,
He bullies more attractive birds
With hoodlum deeds and vulgar words,
And should a human interfere,
Attacks that human in the rear.
I cannot help but deem the grackle
An ornithological debacle.

Ogden Nash (1902–1971).

(Courtesy of Heath Hagy, North Dakota State University.)

7.1 THE ANNUAL CYCLE OF THE RED-WINGED BLACKBIRD POPULATION

Before describing various attempts that have been made to control blackbird populations, it is critical to understand the sheer numbers of birds involved and the predictable but dramatic fluctuation in numbers during the annual cycle. This discussion will use the red-winged blackbird as an example, but the findings are applicable to common grackles, brown-headed cowbirds, and most other passerine species.

Although accurate estimates of the numbers of the various species of blackbirds in North America are impossible to obtain, the numbers are certainly in the hundreds of millions. Red-winged blackbirds comprised 204 million (38%) of the 537 million blackbirds and starlings reported in roosts in the United States during the winter of 1974–1975, the last time a comprehensive winter roost survey was done (Meanley 1975; Meanley and Royal 1976). Common grackles and brown-headed cowbirds comprised 118 million (22%) and 97 million (18%) of the total roosting population, respectively. Dolbeer (2002) calculated the total red-winged blackbird population at the start of the nesting season (mid-April) to be about 170 million, based on converting indices of abundance from the North American Breeding Bird Survey to density estimates for the 92 ecological strata in North America (Dolbeer et al. 1976). Bird Life International (2016) currently lists the estimated population size for red-winged blackbirds (time of year not specified) at 210 million. Partners in Flight Science Committee (2013; updated by Rosenberg et al. 2016) estimated the breeding population of red-winged blackbirds, brown-headed cowbirds, and common grackles at 150 million, 120 million, and 69 million, respectively.

In this discussion, I assume that 170 million is a reasonable population estimate for the North American population of red-winged blackbirds at the start of the nesting season, the low point in the annual cycle. Based on a population model that incorporates five age classes and mean population parameters from the literature (Table 7.2; Dolbeer et al. 1976; Dolbeer 1998), the annual population cycle can be simulated. Obviously, conditions will vary from year to year and region to region, but this simulation provides a depiction of the mean or typical annual cycle for the North American population of red-winged blackbirds.

Based on the model and parameter values, the population increases 92% from the low of 170 million in mid-April to a peak of 328 million in mid-June (Figure 7.2). From mid-June until the following mid-April, approximately 158 million red-winged blackbirds naturally die from various sources as the population declines to the low point in the annual cycle of 170 million at the start of the next nesting season. During this 10-month period, this represents, on average, about 525,000 red-winged blackbirds dying per day.

Table 7.2 Population Parameter Values Used in Simulation of Annual Cycle of Red-Winged Blackbird Population in North America[a]

Population Parameter	Definition	Parameter Value
JSR	Juvenile (age 0–1) survival rate	0.4
ASR	Adult (>age 1) survival rate (annual)	0.56
ESR	Egg survival rate (egg laying to fledging/weaning)	0.5
EPRA	Eggs per reproducing adult	2.4
FFR1	Fraction of females breeding in age class 1	0.8
FFR2	Fraction of females breeding in age class 2	1.0
FFR3	Fraction of females breeding in age class 3	1.0
FFR4	Fraction of females breeding in age class 4	1.0
FFR5	Fraction of females breeding in age class 5+	1.0

[a] An initial population of 170 million birds at start of nesting season in mid-April. See Dolbeer (1998) for description of model. The initial age composition of population at start of breeding season for birds 1, 2, 3, 4, and 5+ years old was set at 44%, 25%, 14%, 8%, and 10%, respectively.

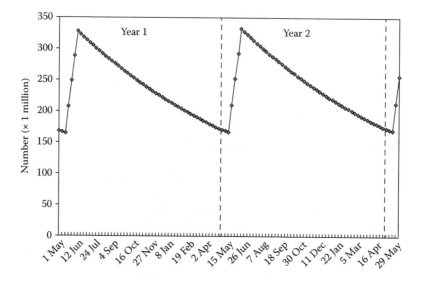

Figure 7.2 Simulation of two annual cycles of the North American red-winged blackbird population, assuming an initial population of about 170 million birds at the start of the nesting season and population parameters listed in Table 7.2. (Adapted from Dolbeer et al. 1976; Dolbeer 1998.)

If one combines red-winged blackbirds, brown-headed cowbirds, and common grackles, the North American population is likely at least 350 million at the start of the nesting season in mid-April and increases to 675 million by mid-June. Thus, at least 325 million birds of these three species die naturally each year between mid-June and mid-April (about 1.1 million/day).

Compensatory factors likely increase the survival and reproductive rates of birds remaining after any control effort (Tanner 1966; Dolbeer 1998), especially for relatively short-lived species such as blackbirds (Péron 2013). Thus, many millions of birds would need to be killed before any detectable decline in the populations of these abundant species would be manifested in subsequent years, as long as nesting and foraging habitats were available. The large size of populations and the magnitude of natural annual turnover in numbers, combined with these compensatory factors and

DYNAMICS AND MANAGEMENT OF BLACKBIRD POPULATIONS   123

the dynamics of migration in which northern breeding populations disperse over wide areas of the southern United States in winter (Dolbeer 1982), present major challenges in attempts to control populations of these blackbird species, as described below.

7.2 POPULATION CONTROL EFFORTS

7.2.1 Miscellaneous Early Attempts, 1950s through 1960s

Although individual farmers had been shooting, trapping, and attempting to poison blackbirds in North America since the 1600s (Meanley 1971; Dolbeer 1980), it was not until the 1950s that organized attempts at reducing populations were implemented (Figure 7.3).

7.2.1.1 Dynamiting Roosts

From 1934 to 1945, 127 American crow (*Corvus brachyrhynchos*) roosts were dynamited in Oklahoma in winter to reduce predation on waterfowl eggs and damage to grain crops. An estimated 3.8 million crows were killed (averaging about 30,000 per operation), but no evidence was obtained to indicate the explosions influenced total population levels, agricultural damage, or waterfowl production during the 12-year period (Hanson 1946). In spite of these negative results, experimental dynamite bombing of winter roosts of blackbirds was conducted in Arkansas from 1951 to 1953 in an attempt to reduce damage to sprouting and ripening rice (*Oryza sativa*, Neff and Meanley 1952; Meanley 1971).

Figure 7.3   Sign commonly posted at the edges of ripening corn fields near marshes along Lake Erie in north-central Ohio during the 1960s, where large populations of red-winged blackbirds roosted in late summer and fall. (Courtesy of R.A. Dolbeer.)

In one 6-ha roost containing an estimated 20 million blackbirds, 100 "dynamite-shot" bombs were strung in the trees. Only about 200,000 birds were killed (1% of the population in the roost). This approach was abandoned for various obvious reasons: labor, expense, hazards involved, large rate of crippling, limited sites where dynamite could be used, and lack of effectiveness in solving problems.

7.2.1.2 Poison Baits

During the 1950s, biologists with the U.S. Fish and Wildlife Service experimented with strychnine-treated baits in fields to control red-winged blackbird populations damaging ripening corn (*Zea mays*) in Delaware, Florida, and Ohio (Mitchell 1953, 1955; Snyder 1961). Strychnine poisoning of red-winged blackbirds and brown-headed cowbirds to protect sprouting and ripening rice was attempted in Arkansas (Neff and Meanley 1957). These efforts were abandoned because blackbirds generally avoided the baits and few birds were killed. There was also concern about exposure of bait to nontarget birds. However, baiting with other toxicants was revived in the 1980s in attempts to manage blackbird populations (see Sections 7.2.2.2 and 7.2.2.3).

7.2.1.3 Floodlight Traps

The floodlight trap, developed by the U.S. Fish and Wildlife Service in the 1950s, was a large funnel of netting that tapered back to an opening in a tent that could be zippered shut. Floodlights behind the tent were directed at the roost. The funnel mouth, 10–15 m high by 30 m wide, was supported by poles and placed as close as possible to the edge of the woodlot where blackbirds and starlings were roosting at night in winter. After dark, the lights were turned on and several people entered the opposite side of the roost to flush birds toward the lights. Captured birds were euthanized by injection of hydrogen sulfide or carbon monoxide into the tent; in some cases birds were banded and released. From 1957 to 1962, floodlight traps were used at winter roosts in various states in the eastern United States on 101 nights (Mitchell 1963; Meanley 1971). The largest catch made in a single night was 120,000 European starlings and blackbirds at a woodlot roost in Arkansas in January 1961; only six operations caught over 20,000 birds. The mean capture per operation was 6,700 birds (4,100 European starlings and 2,600 blackbirds, mainly common grackles and brown-headed cowbirds). Most efforts caught fewer than 10,000 birds. The last known use of a floodlight trap was at a blackbird roost in West Tennessee in 1978 to capture birds for marking (White et al. 1985; E.K. Bollinger, personal communication). The effort expended to erect and operate the traps and the lack of measurable effects in reducing overall populations or agricultural damage were the main factors for abandoning this method.

7.2.1.4 Decoy Traps

Decoy traps are spacious cages (typically 2 m high and at least 3 m by 3 m in area) in which a few live blackbirds are placed along with grain and water. "Decoyed" birds gain entrance by folding their wings and dropping into the trap through openings in the top. In 1963, a program using 20 U.S. Fish and Wildlife Service traps and four farmer-built traps was implemented in Arkansas to reduce damage to ripening rice (Meanley 1971). About 56,000 blackbirds, mainly brown-headed cowbirds, were captured and euthanized (mean = 39 birds/trap/day). A similar program was implemented in the St. Lawrence Valley of Quebec in the 1970s to evaluate decoy traps to reduce blackbird damage to ripening corn. This evaluation indicated that damage actually increased in the vicinity of traps because the number of birds attracted to the nearby crops exceeded the numbers removed (Weatherhead et al. 1980; Weatherhead 1982). Decoy traps have been used successfully to reduce cowbird parasitism (see Section 7.2.2.4) and may have some utility in protecting fruit crops from house finches (*Haemorhous mexicanus*) and European starlings (Elliot 1964; Larsen and Mott 1970; Conover et al. 2006) in limited areas. However, their use to reduce agricultural damage in large acreages of grain crops and to solve other conflicts caused by blackbird populations has not been successful. Decoy traps are labor-intensive and simply capture too few birds to make a difference, given the millions of blackbirds present during the annual population cycle.

7.2.2 Major Organized Efforts, 1970s through 2000s

7.2.2.1 Winter Roost Spraying with Surfactant PA-14

The surfactant PA-14, registered with the U.S. Environmental Protection Agency (EPA) in 1973 by the Animal Damage Control (ADC) program of the U.S. Department of the Interior's Fish and Wildlife Service (now Wildlife Services in the U.S. Department of Agriculture), was the first attempt at large-scale killing of blackbird populations in North America. PA-14 was used for 19 years (1974–1992) for lethal control of winter-roosting blackbirds and European starlings in the United States (U.S. Department of Interior 1976; Dolbeer et al. 1995). The goal was primarily to reduce concentrations of roosting birds near urban areas because of the noise, fecal accumulation, general nuisance, and disease threat (Figure 7.4) (Garner 1978; White et al. 1985). There was also a hope that these population reductions in winter would reduce agricultural damage at nearby feedlots and to grain crops in the northern parts of the United States in subsequent summers. Finally, some ornithologists speculated that the removal of cowbirds from winter roosts might be beneficial for reducing parasitism of endangered songbirds in subsequent summers (Griffith and Griffith 2000; Ortego 2000; Rothstein and Cook 2000).

PA-14 was applied to winter roosts at night by aircraft when rain was imminent or by fire hoses or irrigation systems. The surfactant allowed water to penetrate feathers, inducing death by hypothermia if temperatures fell below about 5°C. There were 83 roosts encompassing 178 ha treated

Figure 7.4 A mixed flock of red-winged blackbirds, common grackles, and brown-headed cowbirds descends at sunset into a thicket of *Phragmites* sp. along the shore of Lake Erie near Huron, Ohio, on November 21, 2016. This fall roost contained an estimated 100,000 birds. (Courtesy of R. A. Dolbeer.)

with 33,300 L of PA-14 from 1974 to 1992, primarily in Tennessee and Kentucky (Dolbeer et al. 1995). An estimated 38.2 million birds (48% common grackles, 30% European starlings, 13% red-winged blackbirds, and 9% brown-headed cowbirds) were killed, an average of 2.0 million per year. The annual kill represented <1.3% of the estimated national winter population of blackbirds and starlings.

Dolbeer et al. (1995) found no evidence from North American Breeding Bird Survey data that PA-14 applications caused declines in regional breeding population in subsequent summers. Although PA-14 applications sometimes caused treated roosts to break up for that winter, municipalities had to deal with the disposal of large numbers of dead birds. New roosts often formed in the same or subsequent winters in nearby locations. This was demonstrated in January 1977, when 1.1 million blackbird and starlings (96% of population in roost) were killed by a PA-14 application in Tennessee (White et al. 1985). Within 2 weeks, five new roosts totaling over 1 million birds had formed within 21 km of the treated roost. In 1992, the ADC program withdrew the registration of PA-14 because of the costs required to provide additional EPA-requested data and the expense and general lack of efficacy of roost spraying in solving problems.

7.2.2.2 Ripening Sunflowers in the Dakotas

Starting in the late 1970s, at the time PA-14 was being applied to winter blackbird roosts in the southern United States, researchers initiated efforts lasting over three decades to discover an environmentally safe and cost-beneficial strategy for managing blackbird populations responsible for damaging ripening sunflower (*Helianthus annuus*) in the Prairie Pothole Region (PPR) of the northern Great Plains (Linz et al. 2011; Linz 2013; Linz and Hanzel 2015). Fall-migrating blackbirds in the PPR, composed mainly of red-winged blackbirds, common grackles, and yellow-headed blackbirds (*Xanthocephalus xanthocephalus*), numbered about 75 million (Peer et al. 2003). Efforts were expended to control birds from this population at winter, spring, and late-summer roosts.

For winter roosts, the toxicants DRC-2698 and DRC-1337, compounds that had a supposed advantage over PA-14 in not requiring water or cold weather to be effective, were sprayed on three roosts in Mississippi and Arkansas in 1979–1988. Few birds apparently were killed in the three spraying operations, and this approach was abandoned (LeFebvre et al. 1979, 1980; Heisterberg et al. 1990; Linz 2013).

For springtime, an approach tried in the 1990s was an evaluation of baiting sites with brown rice treated with DRC-1339 toxicant (a compound closely related to DRC-1337 and DRC 2698) in corn stubble near the spring migration roosts of blackbirds in eastern South Dakota. An estimated 230,000 blackbirds were killed at two roost sites in 1995 (Barras 1996). Considerable effort was expended over the next decade to evaluate the exposure to and impact on nontarget granivorous birds from spring baiting for blackbird control (Linz 2013). A population model developed by Blackwell et al. (2003) indicated that even with the removal of up to 2 million red-winged blackbirds annually during spring migration, any minor benefits in reduced damage to sunflower in late summer would be outweighed at least 2:1 by the costs of spring baiting. In the end, the conclusion was reached that spring baiting was not a practical means of reducing blackbird populations enough to reduce sunflower damage in late summer and fall, especially given the concern over exposure of toxic bait to nontarget birds (Linz 2013).

For late summer, efforts expended from the 1980s through the 2000s included spraying a roost in North Dakota with DRC-2698 and placing baits with the toxicant DRC-2698 or DRC-1339 in sunflower fields or on platforms in sunflower fields. A maximum estimated mortality of 31,000 blackbirds occurred during the baiting of three fields with DRC-2698 in 1985, and some mortality of nontarget birds was detected. These efforts were terminated in 2008 (Linz 2013).

Linz (2013) summarized these research efforts with blackbirds and sunflower by stating the following:

> Three decades of research by my research team and others has not resulted in an environmentally-safe and cost-beneficial method of using DRC-1339 and related compounds for reducing local, regional and

national populations of blackbirds doing or about to do damage to ripening sunflower. None of the strategies including baiting during winter in the southern U.S., baiting at local spring roost sites in eastern South Dakota and baiting to reduce local blackbird populations that were damaging ripening crops is currently used because of logistical difficulties, cost-effectiveness, environmental risks and societal concerns.

The fundamental issue was that the numbers of birds needed to be killed was simply too great to make these programs practical.

Instead of focusing on blackbird population management, Linz (2013) recommended that the best options for farmers to reduce blackbird damage to sunflower were primarily habitat and crop management strategies. These included use of herbicides to thin dense cattail stands to disperse large roost concentrations of blackbirds near sunflower fields in late summer, using a plant desiccant to accelerate fall harvest (minimizing exposure of the ripe crop in field), planting decoy crops so that blackbirds have alternate feeding sites when visual and auditory repellents are used to scare the birds from ripening fields, synchronizing planting time of sunflower with neighbors to eliminate a mix of early or late-maturing sunflower crops, and leaving stubble after harvest to serve as alternate food sites.

7.2.2.3 Sprouting Rice in Louisiana

The tendency of blackbirds to form communal roosts in the rice-growing areas of southwestern Louisiana during the winter and early spring and to travel and feed in large flocks often results in locally serious damage to the sprouting rice crop. Red-winged blackbirds, brown-headed cowbirds, and common grackles are the primary blackbird species responsible for causing damage to sprouting rice (Meanley 1971; Cummings et al. 2005). Glahn and Wilson (1992) evaluated the use of DRC-1339–treated brown rice placed in "staging areas" (e.g., stubble fields, levee roads, and open grass sites where the birds gather in late afternoon before entering nearby roosts to spend the night) in 1989 and 1990 for reducing blackbird populations in Louisiana at the time rice was planted and sprouting (mainly mid-February to mid-March). Treatment included 3,487 kg and 3,071 kg of bait in 1989 and 1990, respectively, with an estimated 70% of bait eaten. Blackbird mortality was estimated at 1.3 and 2.7 million birds for the respective years. The baiting significantly reduced the number of blackbirds at a nearby roost and growers subjectively reported a reduction in losses of >80%. The authors concluded that the baiting program was cost-effective for reducing blackbird damage to sprouting rice, although no objective measures of damage reduction were made.

Based primarily on this work, the U.S. Department of Agriculture's Wildlife Services carried out a spring baiting program with DRC-1339 in the rice-growing area of southwest Louisiana from 2007 to 2015 (U.S. Department of Agriculture 2015, 2016). The mean number of birds killed per year was 314,000 red-winged blackbirds, 56,000 common grackles, and 482,000 brown-headed cowbirds (Figure 7.5). For all species except cowbirds, this mortality represented <1% of annual mortality for these species in the estimated population of the Mississippi Flyway (Peer et al. 2003); for cowbirds it was 1.9%–2.7%. Based on the annual cycle of these birds (as depicted for red-winged blackbirds in Figure 7.2), this low level of mortality has likely not had any effect on the overall population in the flyway. The fact that the number of birds being removed each year has not shown a trend of decline also indicates this poisoning program is not having a significant effect on the overall population. Because the removal is occurring at the exact time and place where damage occurs, some reduction in damage to sprouting rice is likely achieved. The degree of reduction and cost–benefit ratio are unknown.

7.2.2.4 Trapping Cowbirds to Reduce Nest Parasitism

During the 1960s, the total nesting population of Kirtland's warblers (*Setophaga kirtlandii*), confined to early growth jack pine (*Pinus banksiana*) habitat in northern lower Michigan, declined

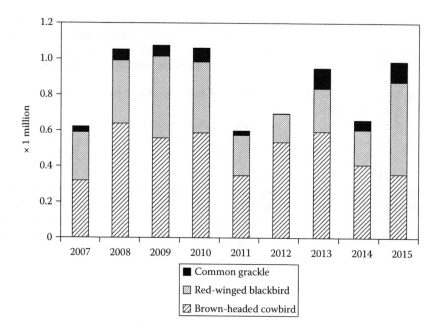

Figure 7.5 Number of blackbirds killed by baiting with DRC-1339 during spring in Louisiana to reduce damage to sprouting rice, 2007–2015.

from about 500 nesting pairs (based on counts of singing males) to only 200 nesting pairs in 1971 (U.S. Fish and Wildlife Service 2016). Walkinshaw (1972) identified brown-headed cowbirds, a species that spread into Michigan from the west as land was cleared for agriculture, as an important factor in the decline; 24% of nests examined from 1931 to 1955 were parasitized and 69% in 1957–1971. Generally, no warblers fledged from parasitized nests and overall nest success (<1 bird fledged/nest) in the 1960s was felt to be too low to sustain the population.

Beginning in 1972, a program was initiated to remove cowbirds during the nesting season (U.S. Department of Interior 2015). In the first year, 15 decoy traps were operated in the nesting area of the Kirtland's warbler from mid-April through June. Each trap had 10 cowbirds as "decoys"; traps were checked daily and all captured cowbirds were removed and euthanized. The program expanded to 40 traps in 1978 and ranged from 40 to 70 traps (mean = 50) through 2015 as the nesting habitat of the Kirtland's warbler expanded. The program resulted in 160,000 cowbirds removed between 1972 and 2015 (maximum of 7,500 in 1990, mean = 3,631/year). In addition to the cowbird removal, prescribed burning was implemented on National Forest Service land in the 1970s to expand the nesting habitat of the early growth, dense, jack pine forest preferred by Kirtland's warblers.

The trapping program had an immediate impact, as cowbird parasitism was reduced to only 6% of nests in 1972 and averaged 3.4% of nests from 1972 to 1981 (Kelly and DeCapita 1982; DeCapita 2000). The Kirtland's warbler population stabilized in the 1970s and 1980s at about 200 nesting pairs. Beginning about 1990, as a result of the cowbird trapping and emerging availability of dense jack pine habitat, the population began increasing, reaching about 2,500 nesting pairs (5,000 adults) in 2015 (U.S. Fish and Wildlife Service 2016). Birds dispersing from the core breeding area in the lower peninsula of Michigan have expanded the nesting range to areas of jack pine in the upper peninsula of Michigan and in Wisconsin.

The annual removal of several thousand cowbirds in an 11-county area of Michigan has obviously not impacted the North American population of 120 million or more birds. In fact, this removal in Michigan represents only about 1% of the 430,000 cowbirds removed each spring in Louisiana (2007–2015). The program in Louisiana, as documented in Section 7.2.2.3, has not

DYNAMICS AND MANAGEMENT OF BLACKBIRD POPULATIONS

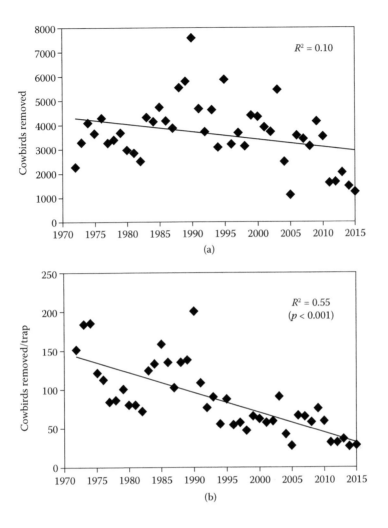

Figure 7.6 Number of brown-headed cowbirds removed (a) and number removed per trap (b) from April through June in the Kirtland's warbler nesting area of the lower peninsula of Michigan, 1972–2015. (Data from U.S. Department of Interior 2015.)

impacted returning populations to date. However, there is some indication that 43 consecutive years of removal has had some impact on the population that returns to this part of Michigan each year. The number trapped in 2011–2015 (mean = 1,606/year) was only 41% of the annual mean of 3,850 for 1972–2010 (Figure 7.6). There was a highly significant ($p < 0.001$) decline in the number of cowbirds removed per trap, from 1972 to 2015. There may be other factors at work, such as cowbirds developing trap shyness or habitat changes. Regardless, by removing the cowbirds immediately before and during the nesting season, the program has shown success in at least temporarily reducing the population available to parasitize Kirtland's warbler nests each year. Although there is some debate as to the importance of cowbird removal in the recovery of this species (Rothstein and Cook 2000), the evidence is clear that the removal greatly reduced nest parasitism and increased fledging success. I conclude this intervention was an important factor in the recovery of the Kirtland's warbler, especially in the 1970s when the population dipped below 200 nesting pairs.

A similar, but smaller-scale, trapping operation was carried out at Camp Pendleton, California, from 1983 to 1996 to reduce brown-headed cowbird parasitism of least Bell's vireo (*Vireo bellii pusillus*) nests (Griffith and Griffith 2000). The program removed about 425 cowbirds per year, which effectively eliminated parasitism. The vireo population increased during this period, likely as a result of the increased number of fledglings per nest. There was no evidence that cowbird removal influenced the number of cowbirds present in subsequent years.

7.3 CONCLUSIONS

Because of the large size of populations and the magnitude of the annual turnover in numbers, the major efforts to permanently reduce populations of blackbirds in North America through trapping, dynamiting, roost spraying, and poisoning generally have not been successful, even when millions of birds have been killed. Localized, temporary reductions in populations at the time damage occurs, notably with brown-headed cowbirds in the limited nesting area of Kirtland's warblers in Michigan, have provided positive results. The temporary, localized reduction in blackbird populations by poisoning birds in spring roosts in southwestern Louisiana at the time rice is sprouting may also provide some reduction in damage, although cost–benefit data are lacking. These programs provide only temporary reductions and must be repeated yearly.

Thus, as long as suitable habitat exists for the various blackbird species to nest and forage, efforts to reduce the overall population regionally or nationally are not achievable without massive control programs that would have to kill many millions of birds on an annual basis. Such efforts likely would result in negative impacts on populations of nontarget birds, especially grassland species whose numbers are already declining because of loss of habitat and other factors (Brennan and Kuvlesky 2005). Such massive control programs would also negate the positive attributes of blackbirds, such as consumption of insects and weed seeds and their role as an abundant prey base for many predatory species.

To generalize from the recommendations made by Linz (2013) regarding blackbirds and ripening sunflower (see Section 7.2.2.2) and by Dolbeer (1986) and Linz et al. (2015) for various bird–human conflicts, the resolution of conflicts with blackbirds should focus much more on integrated habitat and crop management strategies and less on attempts at blackbird population management. These strategies may include the avoidance of vulnerable crops in areas of high blackbird populations, use of bird-resistant varieties of vulnerable crops, management of roosting vegetation near vulnerable crops or in urban areas to disperse birds, planting decoy crops, and leaving crop residue in harvested fields to provide alternate feeding sites when birds are dispersed with frightening devices from ripening fields being protected. Blackbirds are dominant, natural components of the avian community of North America; our goal should be to coexist and not to eradicate.

REFERENCES

Barras, A. E. 1996. Evaluation of spring baiting with an avicide, DRC-1339, and habitat preferences of migratory blackbirds. MS Thesis. North Dakota State University, Fargo, ND.

Bird Life International. 2016. *Data zone.* http://www.birdlife.org/datazone/species/factsheet/22724191 (accessed June 15, 2016).

Blackwell, B. F., E. Huszar, G. M. Linz, and R. A. Dolbeer. 2003. Lethal control of red-winged blackbirds to manage damage to sunflower: An economic evaluation. *Journal of Wildlife Management* 67:818–828.

Blancher, P. J., K. V. Rosenberg, A. O. Panjabi, B. Altman, A. R. Couturier, W. E. Thogmartin, and the Partners in Flight Science Committee. 2013. *Handbook to the partners in flight population estimates database, Version 2.0.* PIF Technical Series Number 6. http://rmbo.org/pifpopestimates/downloads/Handbook%20to%20the%20PIF%20Population%20Estimates%20Database%20Version%202.0.pdf (accessed March 3, 2017).

Brennan, L. A., and W. P. Kuvlesky, Jr. 2005. Invited paper: North American grassland birds: An unfolding conservation crisis? *Journal of Wildlife Management* 69:1–13.

Bruggers, R. L., and C. C. H. Elliott, Editors. 1989. *Quelea quelea: Africa's bird pest*. Oxford University Press, Oxford, UK.

Bub, H. 1980. Fundumstande beringter Stare (*Sturnus vulgaris*). *Ornithologische Mitteilungen* 32:242–245.

Conover, M. R., G. S. Kania, and R. A. Dolbeer. 2006. Use of decoy traps to protect blueberries from juvenile European starlings. *Human-Wildlife Conflicts* 1:265–270.

Cummings, J. L., S. A. Shwiff, and S. K. Tupper. 2005. Economic impacts of blackbird damage to the rice industry. *Wildlife Damage Management Conference* 11:317–322.

DeCapita, M. E. 2000. Brown-headed cowbird control on Kirtland's warbler nesting areas. In *Ecology and management of cowbirds and their hosts*, eds. Smith, J. N. M., T. L. Cook, S. I. Rothstein, S. K. Robinson, and S. G. Sealy, 333–341. University of Texas Press, Austin, TX.

Dolbeer, R. A. 1980. *Blackbirds and corn in Ohio*. U.S. Department of the Interior Fish and Wildlife Service Resource Publication 136. Washington, DC.

Dolbeer, R. A. 1982. Migration patterns for age and sex classes of blackbirds and starlings. *Journal of Field Ornithology* 53:28–46.

Dolbeer, R. A. 1986. Current status and potential of lethal means for reducing bird damage in agriculture. Vol. 1. *ACTA XIX Congress of International Ornithological Congress*, 474–483. University of Ottawa Press, Ottawa, Canada.

Dolbeer, R. A. 1998. Population dynamics: The foundation of wildlife damage management for the 21st century. *Vertebrate Pest Conference* 18:2–11.

Dolbeer, R. A. 2002. Population dynamics of red-winged blackbirds, one of the most abundant birds in North America. In *Management of North American Blackbirds: Special Symposium of The Wildlife Society Ninth Annual Conference*, ed. Linz, G. M., 110. National Wildlife Research Center, Fort Collins, CO.

Dolbeer, R. A., C. R. Ingram, and J. L. Seubert. 1976. Modeling as a management tool for assessing the impact of blackbird control measures. *Vertebrate Pest Conference* 7:35–45.

Dolbeer, R. A., D. F. Mott, and J. L. Belant. 1995. Blackbirds and starlings killed at winter roosts from PA-14 applications: Implications for regional population management. *Eastern Wildlife Damage Management Conference* 7:77–86.

Elliot, H. N. 1964. Starlings in the Pacific Northwest. *Vertebrate Pest Conference* 2:29–39.

Garner, K. M. 1978. Management of blackbird and starling winter roost problems in Kentucky and Tennessee. *Vertebrate Pest Conference* 8:54–59.

Glahn, J. F., and E. A. Wilson. 1992. Effectiveness of DRC-1339 baiting for reducing blackbird damage to sprouting rice. *Eastern Damage Control Conference* 5:117–123.

Griffith, J. T., and J. C. Griffith. 2000. Cowbird control and the endangered least Bell's vireo: A management success story. In *Ecology and management of cowbirds and their hosts*, eds. Smith, J. N. M., T. L. Cook, S. I. Rothstein, S. K. Robinson, and S. G. Sealy, 343–356. University of Texas Press, Austin, TX.

Hanson, H. G. 1946. *Crow center of the United States*. Oklahoma Game and Fish News, March 4–7. Oklahoma Game and Fish Department, Oklahoma, OK.

Heisterberg, J. F., J. L. Cummings, G. M. Linz, C. E. Knittle, T. W. Seamans, and P. P. Woronecki. 1990. Field trial of a CPT-avicide aerial spray. *Vertebrate Pest Conference* 14:350–356.

Kelly, S. T., and M. E. DeCapita. 1982. Cowbird control and its effect on Kirtland's warbler reproductive success. *Wilson Bulletin* 94:363–365.

Larsen, K. H., and D. F. Mott. 1970. House finch removal from a western Oregon blueberry planting. *The Murrelet* 51:15–16.

Lefebvre, P. W., R. A. Dolbeer, N. R. Holler, G. A. Hood, C. E. Knittle, R. E. Matteson, E. W. Schafer, Jr., and R. A. Stehn. 1980. *Second field trial of CAT as a blackbird/starling roost toxicant*. Unpublished Bird Damage Research Report Number 145. U.S. Fish and Wildlife Service, Denver Wildlife Research Center, Denver, CO.

Lefebvre, P. W., N. R. Holler, R. E. Matteson, E. W. Schafer, Jr., and D. J. Cunningham. 1979. Development status of N-(3-chloro-4-methylphenyl) acetamide as a candidate black-bird/starling roost toxicant. *Bird Control Seminar* 8:65–70.

Linz, G. 2013. Blackbird population management to protect sunflower: A history. *Wildlife Damage Management Conference* 15:42–53.

Linz, G. M., E. H. Bucher, S. B. Canavelli, E. Rodriguez, and M. L. Avery. 2015. Limitations of population suppression for protecting crops from bird depredation: A review. *Crop Protection* 76:46–52.

Linz, G. M, and J. J. Hanzel. 2015. Sunflower and bird pests. In *Sunflower: Chemistry, production, processing, and utilization*, eds. Force, E. M., N. T. Dunford, and J. J. Salas, 175–186. AOCS Press, Urbana, IL.

Linz, G. M., H. J. Homan, S. W. Werner, H. M. Hagy, and W. J. Bleier. 2011. Assessment of bird management strategies to protect sunflower. *BioScience* 61:960–970.

Meanley, B. 1971. *Blackbirds and the southern rice crop*. U.S. Department of Interior, U.S. Fish and Wildlife Service, Resource Publication 100. http://pubs.usgs.gov/rp/100/report.pdf (accessed June 28, 2016).

Meanley, B. 1975. *Distribution and abundance of blackbird and starling roosts in the United States*. Progress Report July1974–June 1975, Work Units P-F-25.1 and P-F-25.2. U.S. Fish and Wildlife Service, Patuxent Wildlife Research Center, Laurel, MD.

Meanley, B., and W. C. Royall, Jr. 1976. Nationwide estimates of blackbirds and starlings. *Bird Control Seminar* 7:39–40.

Mitchell, R. T. 1953. *Corn depredation by blackbirds in the Lower Delaware River Valley*. Unpublished report, 1–19. U.S. Fish and Wildlife Service, Bureau of Sport Fisheries and Wildlife, Patuxent Research Refuge, Laurel, MD.

Mitchell, R. T. 1955. *Investigations of blackbird depredations to corn in the Florida Everglades area*. Fourth Quarter Progress Report, 1–27. U.S. Fish and Wildlife Service, Bureau of Sport Fisheries and Wildlife, Patuxent Research Refuge, Laurel, MD.

Mitchell, R. T. 1963. *The floodlight trap: A device for capturing large numbers of blackbirds and starlings at roosts*. Special Scientific Report-Wildlife Number 77. U.S. Fish and Wildlife Service. Washington, DC.

Neff, J. A., and B. Meanley. 1952. *Experimental bombing tests in an Arkansas blackbird roost, 1952*. Special report. U.S. Fish and Wildlife Service, Bureau of Sport Fisheries and Wildlife, Denver Wildlife Research Center, Denver, CO.

Neff, J. A., and B. Meanley. 1957. *Blackbirds and the Arkansas rice crop*. Arkansas Agricultural Experiment Station Bulletin 584. Fayetteville, AR.

North American Bird Conservation Initiative. 2016. *Bird conservation regions*. http://www.nabci-us.org/bcrs.htm (accessed June 24, 2016).

Ortego, B. 2000. Brown-headed cowbird population trends at a large winter roost in southwest Louisiana. In *Ecology and management of cowbirds and their hosts*, eds. Smith, J. N. M., T. L. Cook, S. I. Rothstein, S. K. Robinson, and S. G. Sealy, 58–62. University of Texas Press, Austin, TX.

Partners in Flight Science Committee. 2013. *Population estimates database, version 2013*. http://rmbo.org/pifpopestimates (accessed June 24, 2016).

Peer, B. D., H. J. Homan, G. M. Linz, and W. J. Bleier. 2003. Impact of blackbird damage to sunflower: Bioenergetic and economic models. *Ecological Applications* 13:248–256.

Péron, G. 2013. Compensation and additivity of anthropogenic mortality: Life-history effects and review of methods. *Journal of Animal Ecology* 82:408–417.

Schorger, A. W. 1973. *The passenger pigeon: Its natural history and extinction*. University of Oklahoma Press, Norman, OK.

Snyder, D. B. 1961. Strychnine as a potential control for red-winged blackbirds. *Journal of Wildlife Management* 25:96–99.

Rosenberg, K. V., J. A. Kennedy, R. Dettmers, R. P. Ford, D. Reynolds, C. J. Beardmore, P. J. Blancher, et al. 2016. *Partners in Flight Landbird Conservation Plan: 2016 Revision for Canada and Continental United States*. Partners in Flight Science Committee. http://www.partnersinflight.org/ (accessed September 25, 2016).

Rothstein, S. I., and T. L. Cook. 2000. Part V. Introduction. In *Ecology and management of cowbirds and their hosts*, eds. Smith, J. N. M., T. L. Cook, S. I. Rothstein, S. K. Robinson, and S. G. Sealy, 323–332. University of Texas Press, Austin, TX.

Tanner, J. T. 1966. Effects of population density on growth rates of animal populations. *Ecology* 47:733–745.

U.S. Department of Agriculture. 2015. *Environmental assessment: Managing blackbird damage to sprouting rice in southwestern Louisiana*. U.S. Department of Agriculture Wildlife Services, Washington, DC.

U.S. Department of Agriculture. 2016. *Wildlife services program data reports, 2007–2014*. https://www.aphis.usda.gov/aphis/ourfocus/wildlifedamage/sa_reports/sa_pdrs/ct_pdr_home_2014 (accessed June 24, 2016).

U.S. Department of the Interior. 1976. *Use of compound PA-14 avian stressing agent for control of blackbirds and starlings at winter roosts*. Final Environmental Impact Statement. U.S. Fish and Wildlife Service, Washington, DC.

U.S. Department of the Interior. 2015. *2015 brown-headed cowbird control in Kirtland's warbler nesting areas, northern Lower Michigan*. Fish and Wildlife Service, East Lansing Field Office, East Lansing, MI.

U.S. Fish and Wildlife Service. 2016. *Kirtland's warbler census results: 1951, 1961, 1971 thru 2015*. https://www.fws.gov/midwest/endangered/birds/Kirtland/Kwpop.html (accessed June 24, 2016).

Walkinshaw, L. H. 1972. Kirtland's warbler – Endangered. *American Birds* 26:3–9.

Ward, P. 1979. Rational strategies for the control of queleas and other migrant bird pests in Africa. *Philosophical Transactions, Royal Society of London B* 287:289–300.

Weatherhead, P. J. 1982. Assessment, understanding and management of blackbird-agriculture interactions in eastern Canada. *Vertebrate Pest Conference* 10:193–196.

Weatherhead, P. J., H. Greenwood, S. H. Tinker, and J. R. Bider. 1980. Decoy traps and the control of blackbird populations. *Phytoprotection* 61:65–71.

White, S. B., R. A. Dolbeer, and T. A. Bookhout. 1985. Ecology, bioenergetics, and agricultural impact of a winter-roosting population of blackbirds and starlings. *Wildlife Monographs* 93:1–42.

Wright, E. N., L. R. Inglis, and C. J. Feare, Editors. 1980. *Bird problems in agriculture*. British Crop Protection Council, Croydon, UK.

CHAPTER 8

Chemical Repellents

Scott J. Werner
National Wildlife Research Center
Fort Collins, Colorado

Michael L. Avery
National Wildlife Research Center
Gainesville, Florida

CONTENTS

8.1	Avian Repellent Testing in North America	136
8.2	Anthraquinone	137
8.3	Methiocarb	144
8.4	Aminopyridine, or Avitrol	145
8.5	Methyl Anthranilate and Dimethyl Anthranilate	145
8.6	Registered Insecticides, Fungicides, and Insect Repellents	150
	8.6.1 Registered Insecticides	150
	8.6.2 Registered Fungicides	151
	8.6.3 Insect Repellents	152
8.7	Other Plant Derivatives	152
8.8	Other Candidate Repellents	154
8.9	Suggested Future Research	155
References		155

Agricultural depredations caused by blackbirds can be managed with various lethal and nonlethal methods, including chemical repellents. For many people, nonlethal chemical repellents represent an appealing approach to managing crop depredation because the depredating birds are targeted but not killed; they are just inconvenienced. An effective repellent application can cause the crop-depredating birds to leave their present feeding site and seek food elsewhere. Where the birds go to feed is immaterial to the producer as long as the birds leave the producer's field. Thus, an effective repellent application will not likely affect the overall size of the blackbird population, but it may reduce the population associated with depredation and thereby reduce losses within the treated field. As a consequence, nearby crop fields might incur greater damage unless appropriate crop protection measures are employed.

Blackbirds flock to fields of rice, sunflower, corn, and other crops because these sites represent accessible sources of abundant and energy-rich food that is obtainable with relatively little effort. Agricultural crops are especially important to young birds and, in the late summer and fall, newly fledged birds constitute a large portion of many depredating blackbird flocks. Crop fields can provide ideal feeding situations for blackbirds learning to fend for themselves. Ever-increasing alteration of the natural landscape to accommodate expansion of human activities makes it increasingly difficult for blackbirds to find natural sources of food. Field crops are powerful attractions to blackbirds, and depredating birds are not easily dissuaded. The potential benefits of feeding on the crop are great, so there must be a commensurately high potential cost to the birds to discourage them.

Increasing the cost to the depredating birds translates to increasing the amount of time and energy required to feed on the crop. The more time a blackbird spends acquiring nutritional resources, the less time it can spend on other essential life activities such as territorial defense, mate acquisition, predator vigilance, and so on. There is substantial incentive to feed efficiently. If it becomes difficult for a bird to maintain a certain rate of energy intake by feeding on the crop, then the bird will likely look for other sources of food. Thus, the net effect of applying a chemical repellent to the crop may be to lower the value of the crop to the bird by reducing its rate of energy intake. The availability of nearby alternate food sources may dissuade depredating blackbirds from repellent-treated fields.

The challenge for researchers is to identify a chemical compound that can be formulated and applied to a crop so as to make that crop so unpalatable, or render its immediate feeding environment so unsuitable, that blackbirds will be unable to feed there efficiently. The development, registration and eventual field application of the chemical repellent must be accomplished within the context of numerous constraints imposed by economics, human health and safety concerns, and environmental regulations.

Although published investigations regarding the research and development of chemical repellents date back to the 1830s, worldwide few wildlife repellents are presently registered for agricultural applications. Repellents and other nonlethal management techniques are important components of integrated pest management strategies, so it is therefore useful to review our current understanding of chemical repellents relevant to blackbirds. In this chapter, we review previous research regarding the use of nonlethal chemical repellents for blackbird damage management, provide detailed information regarding several repellent compounds of particular relevance to the ecology and management of North American blackbirds, and suggest prospects for future repellent research and development.

8.1 AVIAN REPELLENT TESTING IN NORTH AMERICA

Native Americans used extracts from plants such as hellebore (*Veratrum* spp.) to protect seeded corn from depredations by "starlings, crows, and other birds" (Benson 1996). Godman (1833) described the efforts of farmers in Maryland to stave off crow depredations to newly planted corn. One method involved coating corn seed with a mixture of grease, tar, and slaked lime (calcium hydroxide). Crows encountering seed planted with this coating "quickly left it for some less carefully managed grounds, where pains had not been taken to make all the corn so nauseous and bitter" (Godman 1833, 109).

Commercial bird repellents such as Pestex, Cock Robin, and Corbin (unknown active ingredients) were sold in the United States during the 1930s (Neff and Meanley 1956). The first U.S. patent for an avian repellent was issued to Franz Heckmanns and Marianne Meisenheimer in 1944. This U.S. use patent (No. 2,339,335) covered anthraquinone (CAS No. 84-65-1) and several related quinones as bird-repellent seed treatments. In 1945, Michael Arnold obtained a U.S. use patent (No. 2,372,046) for mixtures of sulfur nitride (CAS No. 64885-69-4) and iminosulfur as fungicides and bird repellents (Neff and Meanley 1956). Numerous other chemicals were added to corn seed for avian repellency during the 1940s, including sulfur (CAS No. 7704-34-9), nicotine dust (CAS No. 54-11-5), Bordeaux dust (copper sulfate with lime), cryolite (sodium aluminum fluoride; CAS No. 15096-52-3), anthraquinone, benzene hexachloride (CAS No. 58-89-9), naphthalene (CAS No. 91-20-3), dinitronaphthalene

(CAS No. 605-71-0), dinitrophenol (CAS No. 51-28-5), trinitrophenol (CAS No. 88-89-1), dinitrocresol (CAS No. 8071-51-0), mercaptobenzothiazole (CAS No. 149-30-4), aloes, sulfur, iron sulfate (CAS No. 7720-78-7), red ochre (iron oxide containing unhydrated hematite; CAS No. 76774-74-8), and tar (coal and pine) derivatives (Neff and Meanley 1956).

The first systematic investigations of blackbird repellents were conducted by Johnson Neff, Brooke Meanley, and Ronald Brunton in eastern Arkansas rice fields from 1951 to 1954 (Denver Wildlife Research Laboratory, U.S. Fish and Wildlife Service). Subsequent investigations by this group were conducted at the Denver Federal Center and in the vicinity of Alexandria, Louisiana, in 1955–1956. Neff and Meanley (1957) summarized their cage and small-scale field evaluations of more than 25 compounds as blackbird repellents (Table 8.1). Of these compounds, good blackbird repellency was observed for actidione (CAS No. 66-81-9), anthracene (CAS No. 120-12-7), anthraquinone, Arasan (thiram; CAS No. 137-26-8), benzanthrone (CAS No. 82-05-3), dinitroanthraquinone (CAS No. 129-39-5), orthophos (parathion), phenanthraquinone (CAS No. 84-11-7), sucrose octaacetate (CAS No. 126-14-7), and zinc dimethyl dithiocarbamate cyclohexamine (Neff and Meanley 1957). According to Neff and Meanley (1956), a good repellent tends to drive away, ward off, and/or create aversion through some odious or distasteful nature, and "the definition seems to restrict the reaction largely to the senses of taste, touch or smell."

Neff et al. (1957) summarized their basic field testing of more than 10 compounds as candidate blackbird repellents (Table 8.1). Of these compounds, anthraquinone, Arasan, dinitroanthraquinone, quinizarine (CAS No. 128-80-3), tetramethylthiuram disulfide (CAS No. 205-286-2), and thiram provided good repellency. The first peer-reviewed investigation of avian repellents was published in 1958. Abbott (1958) concluded that anthraquinone, Morkit (a.i., 9,10-anthraquinone), quinizarine, and Arasan all effectively repelled common grackles (*Quiscalus quiscula*) from eastern white pine seeds (*Pinus strobus*).

Starr et al. (1964) identified a clear need for a quantitative method for reliably comparing one chemical against another, including concentration–effect measurements. Starr et al. (1964) comparatively evaluated more than 10 chemical repellents and reported R_{50} values among these compounds (i.e., the concentration of a chemical required to repel 50% of the test birds under given test conditions) for red-winged blackbirds (*Agelaius phoeniceus*). Good blackbird repellency was observed for 1,1-iminodianthraquinone; 1,3,5-trinitrobenzene aniline complex (CAS No. 3101-79-9); 1-hydroxy-2-pyridine thione disulfide; anthraquinone; benzanthrone; N,N-diethyl-3-methyl-benzamide (CAS No. 134-62-3); coumaphos (CAS No. 56-72-4); 3-methyl-4-(methylthio)phenol methylcarbamate (CAS No. 3566-00-5); carbaryl (CAS No. 63-25-2); 4-thiazolidinone, 3-(4-chlorophenyl)-5-methyl-2-thioxo- (CAS No. 6012-92-6); *n*-dodecylguanidine acetate; and tetramethylthiuram disulfide (Table 8.1).

8.2 ANTHRAQUINONE

Among 162 publications regarding chemical repellents from 1956 to 2016, the greatest number of publications per chemical were associated with anthraquinone. Quinones are distributed throughout plant and invertebrate animal taxa (Thomson 1987). Anthraquinone compounds, mostly found in plants, constitute the largest group of natural quinones (Sherburne 1972). The functions of these compounds are not well understood, but one of them, emodin (1-3-8-trihydroxy-6-methyl-anthraquinone), is a potent avian antifeedant (Sherburne 1972). Many anthraquinones that occur in invertebrates might have predator defense functions (Hilker and Köpf 1994). Anthraquinones are primarily used in industrial dyes and in bleaching pulp for papermaking, but one compound, 9,10-anthraquinone (i.e., anthraquinone; Figure 8.1), holds particular interest for wildlife managers as an avian feeding deterrent.

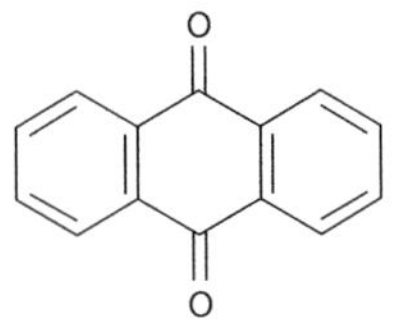

Figure 8.1 Chemical structure of 9,10-anthraquinone.

Table 8.1 Nonlethal Chemical Repellents

Chemical (Active Ingredient; N = 119)	Tested Matrix	Tested Concentrations (%)	Reported Efficacy	Reference
1,1-Dianthrimide	Rice seed	2	Poor repellency	Neff and Meanley 1957
1,1-Iminodianthraquinone	Milo and rice seed	0.20	Good repellency	Starr et al. 1964
1,3,5-Trinitrobenzene aniline complex	Milo and rice seed	0.9–1.1	Good repellency	Starr et al. 1964
1,4-Naphthalenedione	Milo and rice seed	1	Good repellency	Schafer and Jacobson 1983
1-Amino-1,3-dibrom anthraquinone	Rice seed	2	No repellency	Neff and Meanley 1957
1-Amino-2,4-dibrom anthraquinone	Rice seed	2	Poor repellency	Neff and Meanley 1957
1-Amino-4-hydroxy anthraquinone	Rice seed	2	Poor repellency	Neff and Meanley 1957
1-Azetidinecarbothioic acid		1	Good repellency	Schafer et al. 1986
1-Chloro-9,10-anthracenedione		1	Good repellency	Schafer and Jacobson 1983
1-Hydroxy-2-pyridine thione disulfide	Milo and rice seed	0.1–0.3	Good repellency	Starr et al. 1964
1-Pyrrolidinecarbothioic acid		1	Good repellency	Schafer et al. 1986
2-Chloroanthraquinone	Rice seed	2	Poor repellency	Neff and Meanley 1957
2-Methyl-a,a-diphenyl-1-pyrrolidinebutyramide	Rice seed	0.1	Good repellency	Schafer and Brunton 1971
3-(4-Chlorophenyl)-5-methyl-2-thioxo-4-thiazolidinone	Milo and rice seed	0.02–0.11	Good repellency	Starr et al. 1964
3-Methyl-4-(methylthio)phenol methylcarbamate	Milo and rice seed	0.004–0.02	Good repellency	Starr et al. 1964
4-Thiazolidinone,3-(4-chlorophenyl)-5-methyl-2-thioxo-	Milo and rice seed	0.02–0.11	Good repellency	Starr et al. 1964
4,8-Diamino anthrarufin	Rice seed	2	No repellency	Neff and Meanley 1957
Actidione	Rice seed	2	Good repellency	Neff and Meanley 1957
Activated charcoal (carbon black)	Rice seed	1–4	Good repellency	Neff et al. 1957; Belant 1997b
Allegiance® FL (metalaxyl)	Rice seed		Poor repellency	Werner et al. 2010
Aluminum pigment	Rice seed	Unknown	Poor repellency	Neff et al. 1957
Anthocyanins	Sunflower meal	0.5–5	Good repellency	Mason et al. 1989b
Anthracene	Rice seed	2	Good repellency	Neff and Meanley 1956, 1957; Avery et al. 1997
Anthraquinone	Rice seed	0.05–2	Good repellency	Neff and Meanley 1956, 1957; Neff et al. 1957; Abbott 1958; Wright 1962; Starr et al. 1964; Avery et al. 1997; DeLiberto and Werner 2016
Anthrone	Rice seed	0.05–0.25	Good repellency	Avery et al. 1997

(Continued)

CHEMICAL REPELLENTS

Table 8.1 (Continued) Nonlethal Chemical Repellents

Chemical (Active Ingredient; N = 119)	Tested Matrix	Tested Concentrations (%)	Reported Efficacy	Reference
Aprocarb	Rice seed	0.04	Good repellency	Schafer et al. 1983
Apron XL® LS (see text)	Rice seed		Poor repellency	Werner et al. 2008b
Arasan (thiram)	Rice seed	2	Good repellency	Neff and Meanley 1957; Neff et al. 1957; Abbott 1958
Asana® XL (esfenvalerate)	Sunflower seed		Moderate repellency	Linz et al. 2006
Aza-Direct® (azadirachtin)	Rice seed		Poor repellency	Werner et al. 2008a
Bay 22408		0.05	Good repellency	Schafer et al. 1983
Bay 32651		0.02	Good repellency	Schafer et al. 1983
Bay 38920		0.05	Good repellency	Schafer et al. 1983
Baythroid® 2 (cyfluthrin)	Sunflower seed		Poor repellency	Linz et al. 2006
Benzathrone	Rice seed	0.3–2	Good repellency	Neff and Meanley 1957; Starr et al. 1964
Beta amino anthraquinone	Rice seed	2	Poor repellency	Neff and Meanley 1957
Caffeine	Rice seed	0.25–2	Good repellency	Avery et al. 2005; Werner et al. 2007
Capsaicin		0.001–1	Poor repellency	Mason and Maruniak 1983; Mason et al. 1991b
Carbaryl	Rice seed	0.1–0.2	Good repellency	Starr et al. 1964
Chlor benzanthrone	Rice seed	2	Poor repellency	Neff and Meanley 1957
Cinnamamide	Rice seed	0.8	Good repellency	Gill et al. 1994
Cinnamyl derivatives		0.2–3.4	Good repellency	Avery and Decker 1992; Jakubas et al. 1992
Cobalt™ (chlorpyrifos)	Sunflower seed		Good repellency	Werner et al. 2010
Coniferyl derivatives		0.25–3.2	Good repellency	Jakubas et al. 1992
Copper-8-quinolinolate	Rice seed	2	Good repellency	Neff and Meanley 1957
Copper-8-quinolinolate	Rice seed	2	Poor repellency	Neff and Meanley 1957
Coumaphos	Milo and rice seed	0.002–0.02	Good repellency	Starr et al. 1964; Schafer et al. 1983
Diazinon		0.02	Good repellency	Schafer et al. 1983
Di-brom benzathrone	Rice seed	2	Poor repellency	Neff and Meanley 1957
DID 95		0.06	Good repellency	Schafer et al. 1983

(Continued)

Table 8.1 (Continued) Nonlethal Chemical Repellents

Chemical (Active Ingredient; N = 119)	Tested Matrix	Tested Concentrations (%)	Reported Efficacy	Reference
Diketone	Adult lace bugs	≈2.8 μg per bug	Good repellency	Mason et al. 1991c
Dimethyl anthranilate	Livestock feed	0.28–1	Good repellency	Mason et al. 1985; Glahn et al. 1989; Mason et al. 1991a
Dinitroanthraquinone	Milo and maize	10	Good repellency	Neff et al. 1957
Dithane (mancozeb)	Rice seed	0.1–1	Poor repellency	Avery and Decker 1991
Dividend Extreme® (difenoconazole)	Rice seed		Poor repellency	Werner et al. 2008b
Dolomitic lime ([4-(methylthio)-3,5-xylyl N-methyl-carbamate])	Millet	1–4	Good repellency	Belant 1997b
Dursban	Rice seed	0.1	Good repellency	Schafer and Brunton 1971
(E)-1,2,4-trimethoxy-5-(1-propenyl)benzene		1	Good repellency	Schafer and Jacobson 1983
Endosodulfan 3EC® (endosulfan)	Sunflower seed		Moderate repellency	Linz et al. 2006
Endura® (boscalid)	Sunflower seed		Poor repellency	Linz et al. 2006
Ethyl cinnamate	Rice seed	0.05–1	Moderate repellency	Avery and Decker 1992
Fensulfothion		0.001	Good repellency	Schafer et al. 1983
Fipronil	Rice seed	0.03–0.05	Poor repellency	Avery et al. 1998
Flock Buster	Sunflower seed		Poor repellency	Werner et al. 2010
Gander Gone (citrus terpenes)	Rice seed		Poor repellency	Werner et al. 2008a
GWN-4770 (flutolanil)	Rice seed		Good repellency	Werner et al. 2008a
Hercules AC-5727		0.02	Good repellency	Schafer et al. 1983
Hydrochromone	Adult lace bugs	≈2.8 μg per bug	Good repellency	Mason et al. 1991c
Imidacloprid	Rice seed	0.06–0.25	Good repellency	Avery et al. 1993b, 1994
Isosafrole		1	Good repellency	Schafer and Jacobson 1983
Karate® with Zeon Technology™ (l-cyhalothrin)	Rice seed		Poor repellency	Werner et al. 2008b
Kocide SD (copper hydroxide)	Rice seed	0.01–1	Moderate repellency	Avery and Decker 1991
Lime	Millet and corn	6.25–25	Good repellency	Neff and Meanley 1956; Belant 1997c; Clark and Belant 1998
Lindane	Pea & corn seed		Good repellency	Neff and Meanley 1956
Lorsban® 4E (chlorpyrifos)	Sunflower seed		Good repellency	Linz et al. 2006
Mangone	Millet	0.001–0.1	Poor repellency	Belant 1997a
Maxim® 4FS (fludioxonil)	Rice seed		Poor repellency	Werner et al. 2008b

(Continued)

CHEMICAL REPELLENTS

Table 8.1 (Continued) Nonlethal Chemical Repellents

Chemical (Active Ingredient; N = 119)	Tested Matrix	Tested Concentrations (%)	Reported Efficacy	Reference
Methiocarb & Mesurol	Rice, corn, and fruit	0.1–1	Good repellency	Schafer and Brunton 1971; Guarino et al. 1974; Stone et al. 1974; Woronecki et al. 1981; Mason 1989; Avery and Decker 1991; Avery et al. 1993a
Methyl (1-(2-pyridinyl) ethylidene), hydrazinecarbodithioate		1	Good repellency	Schafer et al. 1986
Methyl anthranilate	Feed, rice seed	0.1–2.5	Good repellency	Mason et al. 1991a, 1993; Avery et al. 1995; Werner et al. 2005
Methyl cinnamate	Rice seed	0.005–1	Good repellency	Avery and Decker 1992
Mexacarbate		0.04	Good repellency	Schafer et al. 1983
Mistron (talc)	Milo and maize	10	Poor repellency	Neff et al. 1957
Morkit (9,10-anthraquinone)	White pine seed		Good repellency	Abbott 1958
Mustang® Maxx (zeta cypermethrin)	Sunflower seed		Poor repellency	Linz et al. 2006
n-Dodecylguanidine acetate	Milo and rice seed	0.4–0.6	Good repellency	Starr et al. 1964
N,N-diethyl-3-methyl-benzamide	Milo and rice seed	0.8–2.3	Good repellency	Starr et al. 1964
Naftalofos		0.02	Good repellency	Schafer et al. 1983
Narlene		0.03	Good repellency	Schafer et al. 1983
Nicotine sulfate	Rice seed	0.1	Good repellency	Neff and Meanley 1956; Schafer and Brunton 1971
Nitrobenzene potassium sulfonate	Rice seed	2	No repellency	Neff and Meanley 1957
Nutra-lite	Millet	1–4	Moderate repellency	Belant 1997b
Ortho benzoyl benzoic acid	Rice seed	2	Poor repellency	Neff and Meanley 1957
Orthophos (parathion)	Rice seed	2	Good repellency	Neff and Meanley 1957
Panoctine (guazatine acetates)	Rice seed	0.01–1	Poor repellency	Avery and Decker 1991
Pennyroyal oil	Rice seed	0.1–1	Good repellency	Avery et al. 1996
Phenanthraquinone	Rice seed	2	Good repellency	Neff and Meanley 1957
Phenanthrene	Rice seed	2	Poor repellency	Neff and Meanley 1957

(Continued)

Table 8.1 (Continued) Nonlethal Chemical Repellents

Chemical (Active Ingredient; N = 119)	Tested Matrix	Tested Concentrations (%)	Reported Efficacy	Reference
Phygon (2,3-dichloro-1,4 naphthoquinone)	Rice seed	2	Good repellency	Neff and Meanley 1957; Neff et al. 1957
Polyphenols (sorghum)	Sorghum		Good repellency	Bullard et al. 1980
Pulegone & d-pulegone	Rice and millet	0.001–1	Good repellency	Mason 1990; Avery et al. 1996; Belant 1997a
Quadris® (azoxystrobin)	Rice seed		Poor repellency	Werner et al. 2008a
Quinizarine (1,4-dihydroxyanthraquinone)	Southern pine seed	"High levels"	Effective	Neff and Meanley 1956; Neff et al. 1957; Abbott 1958
RE 5305		0.03	Good repellency	Schafer et al. 1983
Safrole		1	Moderate repellency	Schafer and Jacobson 1983
Scout X-Tra® (tralomethrin)	Sunflower seed		Moderate repellency	Linz et al. 2006
Sevin® (carbaryl [1-naphthyl methylcarbamate)	Sweet corn		Decreased insects	Woronecki et al. 1981
Spergon (benzoquinone)	Rice seed	2	Good repellency	Neff and Meanley 1957; Neff et al. 1957
Sucrose octaacetate	Rice seed	2	Good repellency	Neff and Meanley 1957
Sulfotepp		0.06	Good repellency	Schafer et al. 1983
Tilt® (propiconazole)	Rice seed		Moderate repellency	Werner et al. 2008b
Thiram (tetramethylthiuram disulphide)	Milo, rice, and corn	0.01–10	Varied repellency	Neff and Meanley 1956; Neff et al. 1957; Wright 1962; Starr et al. 1964; Avery and Decker 1991; Werner et al. 2010
Trans-asarone		1	Moderate repellency	Schafer and Jacobson 1983
Trilex® (trifloxystrobin)	Rice seed		Poor repellency	Werner et al. 2010
Turpentine	Sunflower seed	0.13–5	Varied repellency	Neff and Meanley 1956; Mason and Bonwell 1993
Vitavax® 200 (thiram, carboxin)	Rice seed		Poor repellency	Werner et al. 2010
Warrior® T (lambda cyhalothrin)	Sunflower seed		Moderate repellency	Linz et al. 2006
White quartz	Millet	1–4	Moderate repellency	Belant 1997b
Zinc dimethyl dithiocarbamate cyclohexamine	Rice seed	5	Good repellency	Neff and Meanley 1957

CHEMICAL REPELLENTS

The mode of action of anthraquinone as an avian repellent is unknown, but its postingestive effects are likely responsible for subsequent feeding repellency (Avery et al. 1997, 1998a).

The Denver Wildlife Research Laboratory initiated an extensive study of blackbird depredation in eastern Arkansas rice fields in 1949. This study included a comparative evaluation of chemical repellents in 1951–1954. By 1952, anthraquinone was identified as the gold standard for blackbird repellents and was used in each subsequent screening test for comparison with other candidate repellents (Neff et al. 1957). Rice seeds were treated with 0.125%–2% anthraquinone, and reproducible repellency was observed at 0.5%–2% anthraquinone during 2–7-choice assays (Neff et al. 1957). Starr et al. (1964) estimated R_{50} values of 0.13%, 0.26%, and 0.49% anthraquinone (wt/wt) for red-winged blackbirds in captivity. Wright (1962) also evaluated anthraquinone as an avian repellent for the protection of germinating corn (Table 8.1).

DeLiberto and Werner (2016) reviewed the uses of anthraquinone as a chemical repellent, perch deterrent, insecticide, and feeding deterrent in many wild birds and some mammals, insects, and fishes. This thorough review highlighted 111 publications (1943–2016) regarding anthraquinone applications for international pest management and agricultural crop protection. Criteria for evaluation of effective chemical repellents include efficacy, potential for wildlife hazard, phytotoxicity, and environmental persistence. As a biopesticide, anthraquinone often meets these criteria of efficacy for the nonlethal management of agricultural depredation caused by pest wildlife (DeLiberto and Werner 2016). In January 2016, the U.S. Environmental Protection Agency (U.S. EPA) issued a national registration for anthraquinone-based seed treatments and the protection of newly planted rice from blackbird depredation (i.e., AV-1011® rice seed treatment; Arkion Life Sciences, New Castle, DE; Table 8.2). Additional research and development of foliar anthraquinone-based repellents are ongoing for the protection of ripening crops.

Table 8.2 Products Registered by the U.S. Environmental Protection Agency as Nonlethal Chemical Repellents for Blackbirds

Product Name	Target Species	EPA Registration No.	Active Ingredient	Registrant
AV-1011® Rice Seed Treatment (restricted use product)	Blackbirds	69969-4	9,10-Anthraquinone (50%)	Arkion Life Sciences, LLC, New Castle, DE
Avian Control® (unclassified registration)	Starlings, gulls (Larinae), blackbirds (Icteridae), rock doves, cliff swallows, house swallows, American crows, house finches, geese (Anserinae), mute swans, and coots	88889-1	Methyl anthranilate (20%)	Avian Enterprises, LLC, Jupiter, FL
Bird Shield® Repellent Concentrate (unclassified registration)	Blackbirds, cedar waxwings, crows, finches, geese, jays, magpies, pigeons, ravens, robins, sparrows, starlings, and woodpeckers	66550-1	Methyl anthranilate (26.4%)	Bird Shield Repellent Corp., Pullman, WA
Rejex-it® TP-40 (unclassified registration)	Starlings, gulls (Larinae), blackbirds (Icteridae), rock doves, cliff swallows, house swallows, American crows, house finches, geese (Anserinae), mute swans, coots, woodpeckers, and Sapsuckers	91897-1	Methyl anthranilate (40%)	Avian Enterprises, LLC, Jupiter, FL

8.3 METHIOCARB

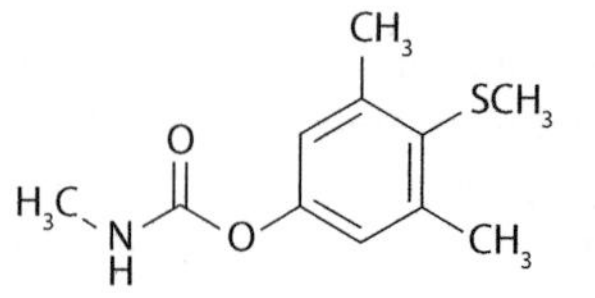

Methiocarb (3,5-dimethyl-4-[methylthio]phenyl methylcarbamate; Figure 8.2) is a carbamate pesticide that was originally developed by Bayer Chemical scientists in Germany as an insecticide. Testing soon revealed that methiocarb (CAS No. 2032-65-7) had great promise for use as a bird repellent (Hermann and Kolbe 1971). Because methiocarb is a carbamate, it inhibits acetylcholinesterase at synapses in the nervous system. However, unlike most cholinesterase-inhibiting compounds, the effects of methiocarb are rapidly reversible, and cholinesterase disruption is only transitory. Affected birds exhibit a range of symptoms, including retching, vomiting, and temporary paralysis. The onset of symptoms and their severity are dependent on the dose received. Typically, vomiting starts within 10 minutes of ingestion of treated food. Some affected birds become immobilized within 30 minutes of consuming an appropriate dose, but they are fully recovered 30 minutes later. Birds feeding on methiocarb-treated food present no sign of irritation or that the chemical tastes bad. Treated food is readily accepted, and feeding activity diminishes only as the bird starts to detect the physiological effects of the chemical.

Figure 8.2 Chemical structure of methiocarb (3,5-dimethyl-4-[methylthio]phenyl methylcarbamate).

Schafer and Brunton (1971) suggested that "the most productive area of research for alleviating bird damage in the past decade has been the development of chemical agents to kill, immobilize, stupefy, and repel destructive species. Since all birds have beneficial qualities, and most are protected by law, the most potentially useful compounds are nontoxic repellents." These authors were the first to publish laboratory efficacy data regarding the repellency of methiocarb in a peer-reviewed journal (Table 8.1). From 1961 to 1971, the Denver Wildlife Research Center screened 724 compounds in a search for safe and effective avian repellents. Of these candidate repellents, 679 were rejected from further consideration because of insufficient repellency in red-winged blackbirds. Twenty-four of the remaining 45 compounds were too toxic to red-winged blackbirds, nine were too toxic to rats, and six were too phytotoxic to corn seeds. Of the remaining six compounds, 2-methyl-α,α-dephenyl-1-pyrrolidone butyramide and methiocarb yielded acceptable R_{50} and LD_{50} (i.e., median lethal dose) values for red-winged blackbirds, common grackles, brown-headed cowbirds (*Molothrus ater*), and tricolored blackbirds (*Agelaius tricolor*; Schafer and Brunton 1971).

In the United States, methiocarb was evaluated extensively as a bird repellent for numerous crops. The Mesurol® 75% seed treatment formulation was very effective in protecting newly sown rice seed from blackbird depredations (Holler et al. 1982). A 0.5% methiocarb hopper-box treatment reduced blackbird and pheasant (*Phasianus colchicus*) damage to seeded corn by 96% and 74%, respectively (Ingram et al. 1973). For fruit crops, application of an aqueous suspension of methiocarb, formulated as 75% wettable powder, reduced bird damage 65.6% in sweet cherries and 62.2% in tart cherries (Guarino et al. 1974). A similar degree of efficacy occurred in aviary and field applications of methiocarb on blueberries (Stone et al. 1974; Avery et al. 1993a).

In addition to methiocarb applications for the protection of plant agriculture, Woronecki et al. (1981) suggested that their application of Mesurol (a.i., methiocarb) to sweet corn fields reduced insect numbers and blackbird activity within treated fields. The correlation between insect populations and reduced bird damage after the chemical treatment supports the hypothesis that cornfields are made less attractive to blackbirds by the reduction of insects (Woronecki et al. 1981).

Years of field use in a variety of applications demonstrated that methiocarb could be applied effectively and safely to control bird depredations to crops (Dolbeer et al. 1994). Many studies were also conducted to identify means to lower application rates, thereby reducing costs and potential residues, without sacrificing efficacy (e.g., Avery 1989; Mason 1989; Nelms and Avery 1997).

CHEMICAL REPELLENTS

Nevertheless, because of concerns for human health and safety, registrations for methiocarb and other carbamate pesticides applied to food crops were discontinued in the early 1990s by the U.S. EPA. As of 2016, methiocarb is registered for U.S. uses as an insecticide, miticide, and molluscicide for control of certain insects and mollusks on ornamentals (Mesurol® 75-W; Gowan Company, Yuma, AZ) but not as an avian repellent.

8.4 AMINOPYRIDINE, OR AVITROL

In contrast to nonlethal chemical repellents, aminopyridine (CAS No. 504-24-5) is an organic compound that is used as a poison with flock-alarming properties under the trade name of Avitrol (Figure 8.3; 0.5%–1% bird control bait; Avitrol Corporation, Tulsa, OK). The greatest industrial application of 4-aminopyridine is as a precursor to the human pharmaceutical pinacidil, which affects potassium ion channels. Avitrol is applied as a chemically treated bait on corn chop, whole corn, and mixed grains to repel blackbirds, rock pigeons (*Columba livia*), house sparrows (*Passer domesticus*), and European starlings (*Sturnus vulgaris*) from noncrop areas. The reaction of Avitrol-treated birds frightens other members of the flock so that they leave the treated area. Presumably, after one such experience, the frightened birds do not return to the site. Birds that react and alarm a flock usually die. In experimental evaluations of Avitrol in corn and sunflower fields, however, the compound was not proven to be consistently effective (DeGrazio et al. 1971; Dolbeer et al. 1976; Stickley et al. 1976; Somers et al. 1981). Avitrol is currently registered by the U.S. EPA as a restricted use pesticide for the control of pest birds (e.g., blackbirds, sparrows, starlings, pigeons, and crows) from a given noncrop location.

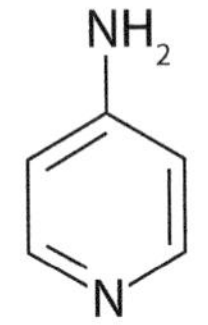

Figure 8.3 Chemical structure of 4-aminopyridine.

8.5 METHYL ANTHRANILATE AND DIMETHYL ANTHRANILATE

Methyl anthranilate (CAS No. 134-20-3) and dimethyl anthranilate (CAS No. 85-91-6) are esters of anthranilic acid (Figure 8.4). Methyl anthranilate is approved by the U.S. Food and Drug Administration as a grape flavoring for human consumption (e.g., candy, soft drinks, chewing gum, pharmaceuticals, and nicotine products). Methyl anthranilate is also used in modern perfumes, as a component of various essential oils, and as a synthesized aroma chemical. Methyl anthranilate occurs naturally in Concord grapes and other *Vitis labrusca* grapes, as well as bergamot, black locust, jasmine, lemon, mandarin orange, orange, strawberry, wisteria, and ylang ylang.

Although palatable to mammals, methyl and dimethyl anthranilate are irritants to birds primarily because they trigger pain receptors in the avian trigeminal nerve (Mason et al. 1989a; Table 8.3). Unlike with illness-inducing repellents such as anthraquinone and methiocarb, birds contacting methyl or dimethyl anthranilate are immediately affected. The concentration of the chemical exposure, the availability of alternative food, and the bird's level of hunger interact to determine the degree of irritation it will tolerate to continue feeding on the treated food.

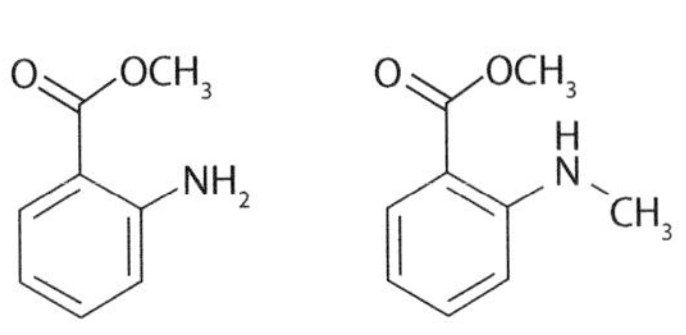

Mason et al. (1985) evaluated the field efficacy of dimethyl anthranilate as an avian repellent for livestock feed. Dimethyl anthranilate reduced the consumption of treated livestock feed by blackbirds and European starlings, and this compound may be useful as a feed additive to reduce avian depredation of

Figure 8.4 Chemical structure of methyl anthranilate (left) and dimethyl anthranilate (right).

Table 8.3 Patents Filed at the U.S. Patent and Trademark Office for Nonlethal Chemical Repellents and Blackbirds

Publication Date	Publication Number	Application Number	Patent Name	Active Ingredient(s)	Inventor(s)
01/03/1961	US2967128 A	n/a	Bird repellent	Methyl ortho-*N*-methylaminobenzoate; methyl anthranilate; ethyl anthranilate; phenyl ethyl anthranilate; methyl anthranilate; or dimethyl benzyl carbinyl acetate	Morley R. Kare
07/17/1962	US3044930 A	n/a	N-oxides of heterocyclic nitrogen compounds as bird and rodent repellents	N-oxides of heterocyclic nitrogen-containing compounds	Kenneth E. Cantrel, Lyle D. Goodhue
08/28/1962	US3051617 A	n/a	Bird repellent	Whole anise seed, crushed and finely divided, oil of anise, pure anise extract, and a light weight oil acting as a vehicle for the above	Alma F. Mann
04/19/1966	US3247060 A	n/a	Methods for controlling birds with halogenated-4-lower alkyl aniline and nitrobenzene compounds	3-Iodo-4-methylaniline hydrochloride; 3-bromo-4-methylaniline hydrochloride; 3-chloro-4-methyl aniline sulfate; 3-chloro-4-methylaniline hydrochloride; 3-bromo-4-methylaniline; 3-bromo-4-methylaniline; 3-iodo-4-methylnitrobenzene; or 3-bromo-4-methylnitrobenzene	Waletzky Emanuel, Kantor Sidney
03/18/1969	US3433873 A	n/a	Compositions and methods for controlling birds	4-Formamidopyridine; 4-acetamidopyridine; 4-propionamidopyridine; 3-acetamidopyridine; 2-acetamidopyridine; or 3-formamidopyridine	Andrew J. Reinert, Ralph P. Williams
10/28/1969	US3475539 A	n/a	2,2-Bis(chloromethyl)-1,3-propanediol cyclic sulfite as a bird management agent	2,2-Bis(chloromethyl)-1,3-propanediol cyclic sulfite	Kenneth E. Cantrel, Raymond L. Cobb, Andrew J. Reinert
01/27/1970	US3492407 A	n/a	Pest repelling compositions and methods of use	Halophenyl-substituted guanidines	Bertram Anders, Gunther Hermann, Rudolf Hiltmann, Englebert Kuhle, Klaus Sasse, Hartmund Wolleber
05/16/1972	US3663692 A	n/a	Methods of bird control	Caffeine; lithium carbonate; lithium chloride; procainamide hydrochloride; phenmetrazine hydrochloride; or trifluoperazine dihydrochloride	Morley R. Kare

(Continued)

CHEMICAL REPELLENTS

Table 8.3 (Continued) Patents Filed at the U.S. Patent and Trademark Office for Nonlethal Chemical Repellents and Blackbirds

Publication Date	Publication Number	Application Number	Patent Name	Active Ingredient(s)	Inventor(s)
06/19/1984	US4455304 A	US 06/369,984	Composition for repelling birds	Dried capsicum pepper and dried garlic	Kourken Yaralian
04/16/1985	US4511579 A	US 06/549,747	Pest repellant	Trialkylphenyl alkylcarbamates	George L. Rotramel, Daniel P. Veilleux, Joseph L. Allen
09/15/1987	US4693889 A	US 06/806,877	Bird-repellent composition	Polyisobutylene	Michael T. Chirchirillo, Terrance Cannan
12/13/1988	US4790990 A	US 06/892,188	Mammalian livestock feed, mammalian livestock feed additive, and methods for using same	Dimethyl anthranilate	J. Russell Mason, Morley R. Kare, Dorf A. DeRovira
12/19/1989	US4888173 A	US 07/062,219	Anthocyanin bird repellents	Anthocyanins, including enocyanin and those extracted from Neagra de Cluj sunflower seeds	James R. Mason, Michael A. Adams
10/29/1991	US5061478 A	US 07/488,982	Sprayable bird and animal pest repellant composition containing a tacky polyolefin and methods for the preparation and use thereof	Tacky polyolefin, including tacky polypropylene, tacky polyisobutylene, or tacky polybutene	Eitan Yarkony, Yair Yarkony
03/23/1993	US5196451 A	US 07/793,292	Avian control	3,5-Dimethoxycinnamic acid, or a carboxylic ester or carboxylate salt thereof	Peter W. Greig-Smith, Michael F. Wilson
03/22/1994	US5296226 A	US 07/954,952	Bird-repellent compositions	Benzoic derivative of esters of anthranilic acid, phenylacetic acid, or dimethyl benzyl carbonyl acetate	Leonard R. Askham
11/14/1995	US5466674 A	US 08/274,408	Bird aversion compounds	Methyl anthranilate; methyl phenyl acetate; ethyl phenyl acetate; ortho-amino acetophenone; 2-amino-4,5-dimethyl acetophenone; veratroyl amine; dimethyl anthranilate; cinnamic aldehyde; cinnamamide; cinnamic acid; and combinations thereof	Marvin F. Preiser, Peter F. Vogt
08/27/1996	US5549902 A	US 08/358,462	Bird aversion compounds	Methyl anthranilate; methyl phenyl acetate; ethyl phenyl acetate; ortho-amino acetophenone; 2-amino-4,5-dimethyl acetophenone; veratroyl amine; dimethyl anthranilate; cinnamic aldehyde; cinnamamide; cinnamic acid; and combinations thereof	Marvin F. Preiser, Peter F. Vogt

(Continued)

Table 8.3 (Continued) Patents Filed at the U.S. Patent and Trademark Office for Nonlethal Chemical Repellents and Blackbirds

Publication Date	Publication Number	Application Number	Patent Name	Active Ingredient(s)	Inventor(s)
09/30/1997	US5672352 A	US 08/236,350	Methods of identifying the avian repellent effects of a compound and methods of repelling birds from materials susceptible to consumption by birds	Aromatic core structure characterized by one of the following core ring structures ##STR2## wherein R1, R1′, or R1″ is an electron-donating group and R2 is an electron-withdrawing group or a neutral group which does not substantially hinder electron donation to the core ring structure by R1	Larry Clark, J. Russell Mason, Pankaj S. Shah, Richard A. Dolbeer
08/11/1998	US5792468 A	US 08/818,676	Lime feeding repellent	Lime	Jerrold L. Belant, Richard A. Dolbeer
03/23/1999	US5885604 A	US 08/918,800	Method for protecting seeds from birds	Polycyclic quinone or precursor thereof	Kenneth E. Ballinger, Jr.
02/15/2000	US6024971 A	US 08/834,585	Water fog for repelling birds	Anthranilates	Thomas J. Nachtman, John H. Hull, Larry Clark
12/11/2001	US6328986 B1	US 09/549,637	Method of deterring birds from plant and structural surfaces	9,10-Anthraquinone	Kenneth E. Ballinger, Jr.
08/25/2005	US20050186237 A1	US 11/016,569	Bird repellent	Anthraquinone and a visual cue; anthraquinone and d-pulegone; or anthraquinone, a visual cue, and d-pulegone; wherein the visual cue is a blue or green dye with a lowered relative reflective wavelength in the range from 500–700 nm	Tim Day, Lindsay Matthews
08/02/2007	US20070178127 A1	US 11/343,396	Agrochemical bird repellent and method	Flutolanil	Nina Wilson
09/15/2015	US9131678 B1	US 13/755,671	Ultraviolet strategy for avian repellency	Anthraquinone and titanium (IV) oxide, trisiloxanes, or siloxanes	Scott J. Werner
06/09/2016	US20160157477 A1	US 14/910,099	Use of visual cues to enhance bird-repellent compositions	Polycyclic quinones and titanium (IV) oxides (TiO_2), trisiloxanes, siloxanes, UV-B absorbent agents, UV-A absorbent agents, $CaCO_3$, $MgCO_3$, carbon black, or ZnO	Kenneth E. Ballinger, Jr., Scott J. Werner

Note: n/a, not applicable.

livestock feed without primary or secondary hazards to nontarget birds (Mason et al. 1985). Glahn et al. (1989) investigated the repellency of dimethyl anthranilate that had been encapsulated in a food grade starch at experimental feedlots. Compared to the pretreatment phase, when 22.7 kg of untreated poultry pellets were consumed by blackbirds and starlings, the consumption of treated feed was nearly eliminated (range = 0–0.01 kg) during the treatment. Thus, 1% dimethyl anthranilate in livestock feed appears to provide a practical bird repellent for the protection of livestock feed from avian depredation (Glahn et al. 1989).

Mason et al. (1991a) evaluated the effectiveness of methyl anthranilate as a bird-repellent additive for livestock feed (Table 8.1). Although red-winged blackbirds were repelled by layer crumbles treated with 1% methyl anthranilate, consumption returned to baseline levels by treatment day 3 (Mason et al. 1991a). Mason et al. (1993) evaluated methyl anthranilate–treated pelleted baits for mitigating the risks of granular pesticide formulations for nontarget birds. The addition of methyl anthranilate decreased the consumption of pelleted baits by brown-headed cowbirds under laboratory and field conditions (Mason et al. 1993).

Avery et al. (1995) evaluated a formulation of methyl anthranilate in aviary and field tests to assess its potential as an avian feeding deterrent for rice seed. Methyl anthranilate suppressed rice consumption at 1%–2.5% (wt/wt). Controlled field trials showed that seed loss from plots containing 1.7% methyl anthranilate treatments averaged 27% and 34% compared to losses on untreated plots that averaged 52% and 73%. Thus, Avery et al. (1995) concluded that methyl anthranilate has potential in the management of blackbird damage to rice, particularly if methyl anthranilate residues on rice seed can be prolonged throughout the period of needed protection from blackbird depredation.

Werner et al. (2005) evaluated Bird Shield™ (a.i., 26.4% methyl anthranilate) as a blackbird repellent in ripening rice and ripening sunflower fields. The repellent was aerially applied by fixed-wing aircraft at the manufacturer-recommended label rate and volume (1.17 L Bird Shield/ha and 46.7 L/ha, respectively); one field received 200% of the label rate (Figure 8.5). No difference was observed in average bird activity (birds/min) between treated and untreated rice fields over the 3-day post-treatment period. Reversed-phase liquid chromatography was used to quantify methyl anthranilate residues in treated fields. The maximum concentration of methyl anthranilate in rice samples was 4.71 µg/g. This concentration was below reported threshold values that irritate birds (i.e., 80,000 µg/g). One sunflower field from each of six pairs was selected for two aerial applications of Bird Shield at the label-recommended rate of ~1 week apart. The remaining six fields served as untreated controls. Daily bird counts, starting on the first day of application and continuing for 5–7 days after the second application, showed similar numbers of blackbirds within treated and untreated sunflower fields. No difference in sunflower damage was observed within treated and control fields prior and subsequent to the treatment. Werner et al. (2005) therefore concluded that Bird Shield was not effective for repelling blackbirds from ripening rice or ripening sunflower fields.

Figure 8.5 Aerial application of a methyl anthranilate-based repellent on a ripening rice field in southeastern Missouri. (Werner et al. 2005.)

Several methyl anthranilate–based repellents are commercially available in the United States (Table 8.2). For example, Avex (Corvus Repellent Inc., Greeley, CO) is a new-generation methyl anthranilate–based bird repellent. Avian Control® (Avian Enterprises LLC, Jupiter, FL) is registered by the U.S. EPA for use on numerous crops to prevent damage from foraging birds. Bird Shield (Bird-X, Inc., Chicago, IL) is registered for several agricultural uses (e.g., blueberries, pome and stone fruits, cereal grains, sunflowers, table grapes) and residential uses (outdoor recreational structures, decorative non–fish-bearing bodies of water, turf and ornamentals). Bird Stop® (Bird-X, Inc., Chicago, IL) creates an invisible barrier that irritates birds' trigeminal system. EcoBird 4.0® (Roth Chemical Company, Overland Park, KS) is a methyl anthranilate–based bird repellent used for the humane and effective dispersal of pest birds in open spaces.

8.6 REGISTERED INSECTICIDES, FUNGICIDES, AND INSECT REPELLENTS

The effectiveness of wildlife repellents for agricultural crop protection is not only dependent upon their safety and efficacy considerations but also their cost. The cost of developing agricultural pesticides, including wildlife repellents, includes the cost of registering the pesticide through the U.S. EPA. The cost associated with the registration of a new active ingredient as an agricultural pesticide in the United States was estimated to be $7.8 million (Eisemann et al. 2011). For comparison, the cost to register an existing food-use pesticide as a wildlife repellent (i.e., additional use) was $732,976. Thus, much repellent research has been focused on evaluating the repellent efficacy of pesticides that are already registered for agricultural applications. We summarized the registered insecticides, registered fungicides, and insect repellents that have been evaluated as blackbird repellents.

8.6.1 Registered Insecticides

Woronecki et al. (1981) suggested that their application of Sevin® insecticide (Tessenderlo Kerley, Inc., Phoenix, AZ) to sweet corn fields reduced insect numbers and blackbird activity within treated fields (Table 8.1). Avery et al. (1993b) observed good repellency among red-winged blackbirds and brown-headed cowbirds offered rice seeds treated with 0.062% and 0.187% imidacloprid (CAS No. 138261-41-3; wt/wt). In an independent test, red-winged blackbirds avoided rice seed treated with 0.0833% and 0.25% imidacloprid (Avery et al. 1994). When applied to wheat seed, 0.0165% imidacloprid repelled red-winged blackbirds in captivity. Although imidacloprid appeared to have promise as a bird-repellent seed treatment (Avery et al. 1993b), no registered insecticides are currently manufactured in the United States as wildlife repellents.

Avery et al. (1998b) concluded that 0.0325% and 0.05% fipronil (CAS No. 120068-37-3) did not affect the feeding activity of red-winged blackbirds or brown-headed cowbirds (Table 8.1). Linz et al. (2006) evaluated the repellency of six insecticides with red-winged blackbirds. Compared to untreated reference groups, the consumption of sunflower was moderately reduced when it was treated with the manufacturer's label rate of Asana® XL (a.i., DuPont Chemical Company, Wilmington, DE), Endosulfan® 3EC (Gowan Company), Scout X-Tra® (Aventis Group, Bayer CropScience, Research Triangle Park, NC) and Warrior T® (Syngenta Crop Protection, Greensboro, NC). Good blackbird repellency was observed for sunflower treated with the manufacturer's label rate of Lorsban-4E® (Dow AgroSciences LLC, Indianapolis, IN), and poor repellency was observed for Baythroid 2® (Bayer CropScience) and Mustang® Maxx (FMC Corporation, Philadelphia, PA) (Linz et al. 2006).

Werner et al. (2008a) evaluated a neem oil insecticide as a blackbird repellent for rice production. No concentration–response relationship was observed among red-winged blackbirds offered 18%–100% of the manufacturer's label rate of Aza-Direct® (Gowan Company). Thus, the blackbird repellency of rice treated with Aza-Direct was unrelated to tested concentrations

(Werner et al. 2008a). In replicate feeding experiments with experimentally naïve red-winged blackbirds, Werner et al. (2008b) observed only 55% repellency for rice treated with 200% of the manufacturer's label rate of Karate® with Zeon Technology (Syngenta Crop Protection). Similarly, Werner et al. (2010) evaluated the repellent efficacy of Cobalt® insecticide (Dow AgroSciences) with red-winged blackbirds in captivity. Repellency was positively related to tested concentrations of Cobalt (25%–200% the manufacturer's label rate) and >80% repellency was observed for sunflower treated with Cobalt at ≥50% of the label rate (Werner et al. 2010).

8.6.2 Registered Fungicides

Thiram is a sulfur-based fungicide used to prevent seeds and crops (e.g., apples, wine grapes, soybean), an ectoparasiticide, and an animal repellent to protect fruit trees and ornamentals from damage by rabbits, rodents, and deer. Neff et al. (1957) observed good repellency among red-winged blackbirds offered rice seeds treated with 10% Arasan (a.i., tetramethylthiuram disulfide; DuPont Chemical Company). Wright (1962) also evaluated thiram as an avian repellent for the protection of germinating corn (Table 8.1).

Avery and Decker (1991) observed poor blackbird repellency for rice seeds treated with 0.01%–1% thiram (wt/wt). Although blackbird repellency was positively related to 25%–200% of the manufacturer's label rate of Thiram 42-S (Bayer CropScience) and Vitavax® 200 (a.i., thiram and carboxin; Bayer CropScience), maximum repellency was <50% during the concentration–response testing of these seed treatments (Werner et al. 2010). Several thiram-based animal repellents are currently manufactured in the United States. For example, DeerPro™ Winter Animal Repellent (Great Oak Inc, Redding, CT), Defiant (rabbit, deer, and rodent repellent; Taminco, Inc., Smyrna, GA), Spotrete™ F (rabbit, deer, and rodent repellent; Cleary Chemicals LLC, Dayton, NJ), and Thiram Granuflo® (rabbit, deer, and rodent repellent; Taminco, Inc., Allentown, PA) are all registered as animal repellents.

Neff et al. (1957) observed good repellency among red-winged blackbirds offered rice seeds treated with 10% phygon (CAS No. 117-80-6; Hopkins Agricultural Chemicals Company, Madison, WI) and 10% spergon (CAS No. 142655-99-0; BASF, Cambridgeshire, UK). Avery and Decker (1991) observed poor blackbird repellency for rice treated with 0.1%–1% dithane (CAS No. 12656-69-8; Dow AgroSciences) and 0.1%–1% panoctine (CAS No. 57520-17-9; Nufarm Australia Ltd, Laverton North, VIC, Australia). Moderate blackbird repellency was observed for rice treated with 0.1%–1% Kocide SD (DuPont Chemical Company; Avery and Decker 1991).

Linz et al. (2006) observed poor blackbird repellency for sunflower treated with the manufacturer's label rate of Endura® (BASF Corporation, Research Triangle Park, NC). Maximum blackbird repellency was only 37% for 100% of the manufacturer's label rate of Quadris® (Syngenta Crop Protection; Werner et al. 2010). Red-winged blackbirds exhibited 34% and 77% feeding repellency for rice treated with 100% and 200% of the manufacturer's label rate of GWN-4770 (Gowan Company), respectively (Werner et al. 2008a). Blackbirds consumed 50% fewer rice seeds treated with 91% of the manufacturer's label rate of GWN-4770 during a subsequent field efficacy experiment (Werner et al. 2008a). Although two patent applications were subsequently filed for the use of flutolanil (CAS No. 66332-96-5) as an Agrochemical Bird Repellent and Method (U.S. Patent Application No. 20,070,178,127, International Patent Application No. PCT/US2007/061231; Table 8.3), no flutolanil-based repellents are currently registered for agricultural crop protection.

A positive concentration–response relationship was observed for 25%–200% of the manufacturer's label rate of Dividend Extreme® and Tilt® fungicides (Syngenta Crop Protection), though maximum blackbird repellency was only 55% for rice treated with 200% of the Dividend Extreme label rate (Werner et al. 2008b). Blackbirds consumed 32% and 69% less rice treated with 100% and 200% of the Tilt label rate, respectively. No repellency was observed for a combination of Apron XL® LS (a.i., (R)-[(2,6-dimethylphenyl)-methoxyacetylamino]-proprionic acid methyl ester; Syngenta Crop Protection) and Maxim® 4 FS fungicides (Syngenta) during the concentration–response experiment with red-winged blackbirds. No differences were observed between untreated rice plots and those treated with Tilt during a subsequent field efficacy study. Thus, the label application of Tilt fungicide did not reduce blackbird consumption within a maturing rice field, and chemical residues of the active ingredient were insufficient for repellent efficacy (i.e., <0.1 µg/g propiconazole, CAS No. 60207-90-1; Werner et al. 2008b).

No difference was observed in the consumption of untreated rice and that treated with the manufacturer's label rate of Allegiance® FL fungicide (Bayer CropScience). Blackbirds actually preferred rice treated with Trilex® fungicide (Bayer) relative to untreated rice. Similarly, no concentration–response relationship was observed among red-winged blackbirds offered 25%–200% of the manufacturer's label rate of Allegiance FL (Werner et al. 2010).

8.6.3 Insect Repellents

Schafer and Jacobson (1983) investigated the potential avian repellency and toxicity of 55 insect repellents originating from or related to naturally occurring chemicals. Seven of the chemicals or extracts tested exhibited avian repellency and two of these were considered moderately repellent, with predicted R_{50} values of 0.237% (trans-asarone) and 0.240% (safarole, CAS No. 94-59-7; Table 8.1). None of the 55 chemicals or extracts exhibited acute oral toxicity at ≤100 mg/kg in red-winged blackbirds (Schafer and Jacobson 1983).

8.7 OTHER PLANT DERIVATIVES

Similar to the cost savings of pursuing registered pesticides as avian repellents, plant derivatives and other naturally occurring compounds can provide promising candidate repellents for registration and agricultural applications. Bullard et al. (1980) investigated the repellency and polyphenol composition of 15 varieties of bird-resistant sorghums in red-winged blackbirds (Table 8.1).

The most important observation of this study was recognition of the diversity of polyphenolic properties among bird-resistant sorghums. With one exception (WGF variety), the seven sorghum varieties that were least preferred were uniform in polyphenol properties, whereas substantial variation occurred among the eight most-preferred varieties (Bullard et al. 1980).

Mason et al. (1989b) discovered that sunflower oil concentrations of 15% (wt/wt) were reliably discriminated by red-winged blackbirds in captivity; higher oil concentrations were preferred. Conversely, all anthocyanin concentrations (0.5%–5%, wt/wt) were avoided. Thus, bird-resistant sunflower is likely affected by its relatively low oil concentration and relatively high anthocyanin concentration. Of these two characteristics, oil concentration may be relatively more important for determining the resistance of sunflower varieties to blackbird damage (Mason et al. 1989b).

Mason and Maruniak (1983) injected red-winged blackbirds subcutaneously with capsaicin (CAS No. 404-86-4) and assessed 1) changes in basal body temperature, 2) ability to discriminate warm from cool drinking water, and 3) sensitivity to oral and topical applications of capsaicin, a trigeminal irritant. As predicted from studies of mammals, the injections seemed to disrupt thermoregulation when ambient temperature increased, eliminating discrimination between warm and cool drinking water. In contrast to the effects on mammals, injections of blackbirds failed to observably diminish oral or topical sensitivity to capsaicin and apparently induced a capsaicin preference in choice drinking experiments between capsaicin and its vehicle. Thus, capsaicin may have different behavioral and physiological effects on different classes of animals (Mason and Maruniak 1983). Mason et al. (1991b) hypothesized that structural modifications of the basic capsaicin molecule, which is itself not aversive to birds, might produce aversive analogues.

To this end, European starlings and Norway rats (*Rattus norvegicus*) were given varied concentrations of synthetic capsaicin and four analogues (methyl capsaicin, veratryl amine, veratryl acetamide, vanillyl acetamide) in feeding and drinking tests. Synthetic capsaicin and vanillyl acetamide were not repellent to birds, owing to the presence of an acidic phenolic OH group. Conversely, veratryl acetamide was aversive, due to the basic nature of this compound. For rats, repellent effectiveness among compounds was reversed: synthetic capsaicin was the best repellent, while veratryl acetamide was the worst. This taxonomic reversal may reflect basic differences in trigeminal chemoreception and that chemical correlates of mammalian repellents are opposite to those that predict avian repellency (Mason et al. 1991b).

Mason (1990) evaluated the repellency of d-pulegone (CAS No. 90449-51-7) in European starlings. D-pulegone is the active flavor of pennyroyal and this compound is used as a mint additive in human foods. Concentrations as low as 0.01% (wt/wt) reduced food consumption under laboratory conditions (Mason 1990). Avery et al. (1996) discovered that 0.1%–1% pulegone suppressed rice consumption in red-winged blackbirds more effectively than 0.5% methyl anthranilate, and brown-headed cowbirds were more sensitive to pulegone than red-winged blackbirds. Belant et al. (1997a) comparatively evaluated the repellency of d-pulegone and mangone in brown-headed cowbirds. Concentrations of 0.1% d-pulegone and 0.001% mangone reduced cowbird consumption of treated feed, though consumption of mangone-treated millet was similar among no-choice tests and similar to total food consumption during choice tests. Belant et al. (1997a) concluded that mangone is less effective than d-pulegone as a blackbird repellent, and mangone would likely be ineffective as a repellent seed treatment.

Avery and Decker (1992) evaluated the repellency of cinnamic acid esters in red-winged blackbirds. Ethyl cinnamate (CAS No. 103-36-6) was moderately deterrent at 0.05%–1% concentrations. Consumption of rice treated with 1% methyl cinnamate (CAS No. 103-26-4) was virtually eliminated (Avery and Decker 1992). Jakubas et al. (1992) tested the avian repellency of coniferol (CAS No. 32811-40-8) and cinnamyl derivatives. Jakubas et al. (1992) concluded that 1) benzoate esters were more repellent than their corresponding alcohols, 2) repellency was increased by electron-donating groups, and 3) acidic functions decrease repellency. Gill et al. (1994) discovered that 0.8% cinnamamide (i.e., synthetic derivative of cinnamic acid; CAS No. 22031-64-7) prevented chestnut-capped blackbirds (*Agelaius ruficapillus*; also known as *Chrysomus ruficapillus*) from eating rice seeds.

Mason and Bonwell (1993) evaluated turpentine (CAS No. 8006-64-2) as a repellent seed treatment in brown-headed cowbirds, common grackles, and red-winged blackbirds. Although turpentine concentrations as low as 0.13% (wt/wt) were repellent to cowbirds, grackles and red-winged blackbirds demonstrated no avoidance of turpentine concentrations as high as 5%. Although turpentine was not phytotoxic, turpentine has limited value as a bird-repellent seed treatment (Mason and Bonwell 1993).

Avery et al. (2005) evaluated caffeine (CAS No. 58-08-2) as a repellent seed treatment for rice (Table 8.3). Rice seed treatments of 0.25% caffeine reduced rice consumption as much as 76% in female red-winged blackbirds and male brown-headed cowbirds. In a subsequent field study, >90% of rice seeds treated with 1% caffeine were uneaten on Day 3 of the study, whereas >80% of untreated rice was consumed by blackbirds (Avery et al. 2005). Werner et al. (2007) included sodium benzoate (CAS No. 532-32-1) in their blackbird-repellent formulations of caffeine. A positive concentration–response relationship was observed among red-winged blackbirds offered 0.025%–2% caffeine and sodium benzoate (1:1). Upon seed germination experiments, the optimal formulation enhanced the solubility of tank mixtures and ameliorated the negative impacts of caffeine seed treatments to the germination of rice seed (Werner et al. 2007). However, no caffeine-based repellents are currently available for agricultural applications.

Werner et al. (2008a) evaluated a terpenoid formulation as a blackbird repellent for rice. Gander Gone (Natural Earth Products, Winter Springs, FL) contains citrus terpenes, or plant hydrocarbons that repel arthropod and mammalian herbivores. No concentration–response relationship was observed among red-winged blackbirds offered 24%–194% of the manufacturer's recommended label rate of Gander Gone, and maximum repellency was only 25% among red-winged blackbirds offered rice treated with 1.25% Gander Gone (vol/wt; Werner et al. 2008a). Werner et al. (2010) evaluated Flock Buster® (Skeet-R-Gone, Grand Forks, ND) as a blackbird repellent for sunflower. The active ingredients of Flock Buster are lemon grass oil, garlic oil, clove oil, peppermint oil, rosemary oil, thyme oil, and white pepper. Red-winged blackbirds preferred untreated sunflower relative to sunflower treated with the manufacturer's recommended label rate of Flock Buster only on Day 1 of the 4-day preference test. Although no concentration–response relationship was observed among blackbirds offered 25%–200% of the manufacturer's recommended label rate of Flock Buster, −2.2% to −37.2% repellency (i.e., attraction) was observed for these seed treatments (Werner et al. 2010).

8.8 OTHER CANDIDATE REPELLENTS

Schafer et al. (1983) evaluated the acute oral toxicity, repellency, and hazard potential of 998 chemicals in one or more of 68 species of wild and domestic birds in captivity. Red-winged blackbirds were the most sensitive of the birds tested for a large number of chemicals. Of these chemicals, aprocarb (CAS No. 127779-20-8), Bay 22408, Bay 32651, Bay 38920, coumaphos (CAS No. 56-72-4), diazinon (CAS No. 333-41-5), DID 95, fensulfothion (CAS No. 115-90-2), Hercules AC-5727 (CAS No. 64-00-6), mexacarbate (CAS No. 315-18-4), naftalofos (CAS No. 1491-41-4), narlene, RE 5305 (CAS No. 673-19-8), and sulfotepp (CAS No. 3689-24-5) each had 1) estimated R_{50} values <5 mg/kg and 2) LD_{50} values that were greater than their R_{50} values (Table 8.1). Overall, avian repellency and toxicity were not positively correlated (i.e., toxicity varied independently with repellency) among the 998 evaluated chemicals (Schafer et al. 1983).

Schafer et al. (1986) evaluated the repellency and toxicity of 2-acetylpyridinethio-semicarbazones and related chemicals to wild birds. Two chemicals, 1-azetidinecarbothioic acid (CAS No. 71555-25-4) and 1-pyrrolidinecarbothioic acid (CAS No. 71555-26-5), were about twice as repellent to red-winged blackbirds and from 33% to 50% as toxic as methiocarb (Table 8.1). A third chemical, methyl (1-(2-pyridinyl)ethylidene) hydrazinecarbodithioate was similarly repellent to methiocarb, but almost 100 times less toxic to red-winged blackbirds than methiocarb (Schafer et al. 1986).

Many insects contain chemical defenses against avian predators. Mason et al. (1991c) evaluated the repellency of secretions produced by nymphs of the azalea lace bug (*Stephanitis pyrioides*). In the first of three experiments, adult lace bugs, which lack chemical secretions, were more palatable than nymphs. In the second experiment, nymphs that had been immersed in methylene chloride (CAS No. 75-09-2) to remove their secretions were consumed more than untreated nymphs. To test the corollary hypothesis that adults are palatable because they lack secretions, adult lace bugs and green peach aphids (*Myzus persicae*) were treated with nymph secretions in the third experiment. Treated insects of both species were avoided, while untreated insects were consumed. Mason et al. (1991c) therefore concluded that chemicals present in the secretions of lace bugs (and the defensive secretions of other insects) may represent a source of new and effective tools for wildlife management and animal damage control.

Avery et al. (1997) comparatively evaluated the repellency of rice seeds treated with anthracene (CAS No. 120-12-7) and anthrone (CAS No. 90-44-8). The repellency of rice treated with 0.5%–0.25% anthrone was comparable to that of anthraquinone at the same concentrations. Rice treated with 0.5% anthracene was the least repellent among the tested chemicals (Avery et al. 1997). Belant et al. (1997b) comparatively evaluated the repellency of dolomitic lime (CAS No. 16389-88-1), activated charcoal (CAS No. 7440-44-0), Nutra-lite (a silica-based compound), and white quartz sand (CAS No. 14808-60-7) as feeding repellents in brown-headed cowbirds. With the exception of Nutra-lite, the consumption of millet treated with 1%–4% of each of the particulate substances was less than the consumption of untreated millet. The greatest repellency was observed for lime-treated millet (Table 8.3), followed by charcoal, Nutra-lite, and sand (Belant et al. 1997b).

Belant et al. (1997c) further evaluated the repellency of dolomitic lime as a feeding repellent in brown-headed cowbirds. Lime mixed with millet or whole corn at 6.25%–25% (wt/wt) reduced cowbird feeding in captivity (Belant et al. 1997c). Clark and Belant (1998) suggested that the primary mechanism for mediating the avian repellency of agricultural lime is its pH. Cowbirds avoided millet treated with 5% agricultural lime when its pH exceeded 12.3. Moreover, if the particulate seed coating consisted of particles sized 63–150 µm and had a pH of 11.4 or less, the repellent potency was about half of that observed for raw unprocessed lime (Clark and Belant 1998).

8.9 SUGGESTED FUTURE RESEARCH

The future of blackbird-repellent research should apply the understanding provided by more than 160 published studies to date. Supplemental investigations regarding the covariance of chemical structure and avian repellency will likely foster the discovery and development of effective avian repellents for agricultural applications (Shah et al. 1992; Clark and Shah 1994). For example, steric effects and extreme delocalization of lone pairs of electrons (e.g., meta isomers and aromatic structures with multiple-substituted electron-donating groups) tend to interfere with the repellency of irritants in birds (Mason et al. 1989a; Clark and Shah 1991; Clark et al. 1991; Shah et al. 1991). Naturally occurring chemical signals, including the defensive secretions of insects (Mason et al. 1991c), should also be further investigated as avian repellents for the protection of agricultural crops (Mason et al. 1991c).

Supplemental to discovering effective active ingredients under both laboratory and field conditions, research on repellents should also be focused on developing effective application strategies and best management practices for repellent applications in the context of integrated pest management. For example, the heads of commercial sunflowers are inverted from aerial pesticide applications throughout the period of needed protection from blackbird depredation. Assuming that effective repellents will be registered and available for agricultural application, novel application strategies are needed to direct foliar applications of avian repellents to sunflower achenes prior to harvest. Because the chemical senses are fundamental to the feeding ecology of wild birds (Clark 1988; Clark and Avery 2013), additional research and development of repellent application strategies can be focused by pairing pre-ingestive sensory cues (e.g., taste, visual cues) with physiologically related, postingestive consequences (Werner and Clark 2003; Clark et al. 2014).

Additional research is also recommended for the continuation of comparative investigations among candidate repellents, pest birds, and agricultural crops. Indeed, the efficacy of some chemical repellents may be species-specific. Our inquiry and understanding of the mechanisms of interspecific differences in repellent efficacy (e.g., mammalian repellents in birds) will also advance the sciences relevant to the research and development of wildlife repellents, including blackbird repellents, for the nonlethal management of agricultural depredation.

REFERENCES

Abbott, H. G. 1958. Application of avian repellents to Eastern white pine seed. *Journal of Wildlife Management* 22:304–306.

Avery, M. L. 1989. Experimental evaluation of partial repellent treatment for reducing bird damage to crops. *Journal of Applied Ecology* 26:433–439.

Avery, M. L., J. L. Cummings, D. G. Decker, J. W. Johnson, J. C. Wise, and J. I. Howard. 1993a. Field and aviary evaluation of low-level application rates of methiocarb for reducing bird damage to blueberries. *Crop Protection* 12:95–100.

Avery, M. L., and D. G. Decker. 1991. Repellency of fungicidal rice seed treatments to red-winged blackbirds. *Journal of Wildlife Management* 55:327–334.

Avery, M. L., and D. G. Decker. 1992. Repellency of cinnamic acid esters to captive red-winged blackbirds. *Journal of Wildlife Management* 56:800–805.

Avery, M. L., D. G. Decker, and D. L. Fischer. 1994. Cage and flight pen evaluation of avian repellency and hazard associated with imidacloprid-treated rice seed. *Crop Protection* 13:535–540.

Avery, M. L., D. G. Decker, D. L. Fischer, and T.R. Stafford. 1993b. Response of captive blackbirds to a new insecticidal seed treatment. *Journal of Wildlife Management* 57:652–656.

Avery, M. L., D. G. Decker, J. S. Humphrey, E. Aronov, S. E. Linscombe, and M. O. Way. 1995. Methyl anthranilate as a rice seed treatment to deter birds. *Journal of Wildlife Management* 59:50–56.

Avery, M. L., D. G. Decker, J. S. Humphrey, and C. C. Laukert. 1996. Mint plant derivatives as blackbird feeding deterrents. *Crop Protection* 15:461–464.

Avery, M. L., J. S. Humphrey, and D. G. Decker. 1997. Feeding deterrence of anthraquinone, anthracene, and anthrone to rice-eating birds. *Journal of Wildlife Management* 61:1359–1365.

Avery, M. L., J. S. Humphrey, T. M. Primus, D. G. Decker, and A. P. McGrane. 1998a. Anthraquinone protects rice seed from birds. *Crop Protection* 17:225–230.

Avery, M. L., T. M. Primus, E. M. Mihaich, D. G. Decker, and J. S. Humphrey. 1998b. Consumption of fipronil-treated rice seed does not affect captive blackbirds. *Pesticide Science* 52:91–96.

Avery, M. L., S. J. Werner, J. L. Cummings, J. S. Humphrey, M. P. Milleson, J. C. Carlson, T. M. Primus, et al. 2005. Caffeine for reducing bird damage to newly seeded rice. *Crop Protection* 24:651–657.

Belant, J. L., S. K. Ickes, L. A. Tyson, and T. W. Seamans. 1997a. Comparison of d-pulegone and mangone as cowbird feeding repellents. *International Journal of Pest Management* 43:303–305.

Belant, J. L., S. K. Ickes, L. A. Tyson, and T. W. Seamans. 1997b. Comparison of four particulate substances as wildlife feeding repellents. *Crop Protection* 16:439–447.

Belant, J. L., L. A. Tyson, T. W. Seamans, and S. K. Ickes. 1997c. Evaluation of lime as an avian feeding repellent. *Journal of Wildlife Management* 61:917–924.

Benson, A. B., editor. 1996. *Peter Kalm's Travels in North America*. Volume 1. Dover, New York.

Bullard, R. W., M. V. Garrison, S. R. Kilburn, and J. O. York. 1980. Laboratory comparisons of polyphenols and their repellent characteristics in bird-resistant sorghum grains. *Journal of Agriculture and Food Chemistry* 28:1006–1011.

Clark, L. 1998. Physiological, ecological, and evolutionary bases for the avoidance of chemical irritants by birds. *Current Ornithology* 14:1–37.

Clark, L., and M. L. Avery. 2013. Effectiveness of chemical repellents in managing birds at airports. In *Wildlife in Airport Environments: Preventing Animal-Aircraft Collisions Through Science-Based Management,* ed. T. L. DeVault, B. F. Blackwell, and J. L. Belant, pp. 25–35. Johns Hopkins University Press, Baltimore, MD.

Clark, L., and J. L. Belant. 1998. Contribution of particulates and pH on cowbirds' (*Molothrusater*) avoidance of grain treated with agricultural lime. *Applied Animal Behaviour Science* 57:133–144.

Clark, L., J. Hagelin, and S. Werner. 2014. The chemical senses in birds. In *Sturkie's Avian Physiology*, 6th edition, ed. G. C. Whittow, pp. 89–111. Academic Press, San Diego, CA.

Clark, L., and P. S. Shah. 1991. Nonlethal bird repellents: In search of a general model relating repellency and chemical structure. *Journal of Wildlife Management* 55:538–545.

Clark, L., and P. S. Shah. 1994. Tests and refinements of a general structure-activity model for avian repellents. *Journal of Chemical Ecology* 20:321–339.

Clark, L., P. Shah, and J. R. Mason. 1991. Chemical repellency in birds: Relationship between chemical structure and avoidance response. *Journal of Experimental; Zoology* 260:310–322.

DeGrazio, J. W., J. F. Besser, T. J. DeCino, J. L. Guarino, and R. I. Starr. 1971. Use of 4-aminopyridine to protect ripening corn from blackbirds. *Journal of Wildlife Management* 35:565–569.

DeLiberto, S.T., and S. J. Werner. 2016. Review of anthraquinone applications for pest management and agricultural crop protection. *Pest Management Science* 72:1813–1825.

Dolbeer, R. A., M. L. Avery, and M. E. Tobin. 1994. Assessment of hazards to birds from methiocarb applications to fruit crops. *Pesticide Science* 40:147–161.

Dolbeer, R. A., C. R. Ingram, J. L. Seubert, A. R. Stickley, and R. T. Mitchell. 1976. 4-aminopyridine effectiveness in sweet corn related to blackbird population density. *Journal of Wildlife Management* 40:564–570.

Eisemann, J. D., S. J. Werner, and J. R. O'Hare. 2011. Registration considerations for chemical repellents in fruit crops. *Outlooks on Pest Management* 22:87–91.

Gill, E. L., M. B. Serra, S. B. Canavelli, C. J. Feare, M. E. Zaccagnini, A. K. Nadian, M. L Heffernan, et al. 1994. Cinnamamide prevents captive chestnut-capped blackbirds (*Agelaius ruficapillus*) from eating rice. *International Journal of Pest Management* 40:195–198.

Glahn, J. F., J. R. Mason, and D. R. Woods. 1989. Dimethyl anthranilate as a bird repellent in livestock feed. *Wildlife Society Bulletin* 17:313–320.

Godman, J. D. 1833. *Rambles of a Naturalist*. Thomas T. Ash – Key and Biddle, Philadelphia, PN. https://archive.org/details/ramblesnaturalist00godmrich (accessed 3 September 2016).

Guarino, J. L., W. F. Shake, and E. W. Schafer. 1974. Reducing bird damage to ripening cherries with methiocarb. *Journal of Wildlife Management* 38:338–342.

Hermann, G., and W. Kolbe. 1971. Effect of seed coating with Mesurol for protection of seed and sprouting maize against bird damage, with consideration to varietal tolerance and side-effects. *Pflanzenschutz. Nachrichten Bayer* 24:279–320.

Hilker, M., and A. Köpf. 1994. Evaluation of the palatability of chrysomelid larvae containing anthraquinones to birds. *Oecologia* 100:421–429.

Holler, N. R., H. P. Naquin, P. W. Lefebvre, D. L Otis, and D. J. Cunningham. 1982. Mesurol® for protecting sprouting rice from blackbird damage in Louisiana. *Wildlife Society Bulletin* 10:165–170.

Ingram, C. R., R.T. Mitchell, and A. R. Stickley, Jr. 1973. Hopper box treatment of corn seed with methiocarb for protecting sprouts from birds. *Bird Control Seminar* 6:206–215.

Jakubas, W. J., P. S. Shah, J. R. Mason, and D. M. Norman. 1992. Avian repellency of coniferyl and cinnamyl derivatives. *Ecological Applications* 2:147–156.

Linz, G. M., H. J. Homan, A. A. Slowik, and L. B. Penry. 2006. Evaluation of registered pesticides as repellents for reducing blackbird (Icteridae) damage to sunflower. *Crop Protection* 25:842–847.

Mason, J. R. 1989. Avoidance of methiocarb-poisoned apples by red-winged blackbirds. *Journal of Wildlife Management* 53:836–840.

Mason, J. R. 1990. Evaluation of D-Pulegone as an avian repellent. *Journal of Wildlife Management* 54:130–135.

Mason, J. R., M. A. Adams, and L. Clark. 1989a. Anthranilate repellency to starlings: Chemical correlates and sensory perception. *Journal of Wildlife Management* 53:55–64.

Mason, J. R., M. L. Avery, J. F. Glahn, D. L. Otis, R. E. Matteson, and C. O. Nelms. 1991a. Evaluation of methyl anthranilate and starch-plated dimethyl anthranilate as bird repellent feed additives. *Journal of Wildlife Management* 55:182–187.

Mason, J. R., N. J. Bean, P. S. Shah, and L. Clark. 1991b. Taxon-specific differences in responsiveness to capsaicin and several analogues: Correlates between chemical structure and behavioral averseness. *Journal of Chemical Ecology* 17:2539–2551.

Mason, J. R., and W. R. Bonwell. 1993. Evaluation of turpentine as a bird-repellent seed treatment. *Crop Protection* 12:453–457.

Mason, J. R., R. W. Bullard, R. A. Dolbeer, and P. P. Woronecki. 1989b. Red-winged blackbird (*Agelaius phoeniceus* L.) feeding response to oil and anthocyanin levels in sunflower meal. *Crop Protection* 8:455–460.

Mason, J. R., L. Clark, and T. P. Miller. 1993. Evaluation of a pelleted bait containing methyl anthranilate as a bird repellent. *Pesticide Science* 39:299–304.

Mason, J. R., J. F. Glahn, R. A. Dolbeer, and R. F. Reidinger, Jr. 1985. Field evaluation of dimethyl anthranilate as a bird repellent livestock feed additive. *Journal of Wildlife Management* 49:636–642.

Mason, J. R., and J. R. Maruniak. 1983. Behavioral and physiological effects of capsaicin in red-winged blackbirds. *Pharmacology Biochemistry & Behavior* 19:857–862.

Mason, J. R., J. Neal, J. E. Oliver, and W. R. Lusby. 1991c. Bird-repellent properties of secretions from nymphs of the azalea lace bug. *Ecological Applications* 1:226–230.

Neff, J. A., and B. Meanley. 1956. *A review of studies on bird repellents*. Progress Report 1. Wildlife Research Laboratory, Denver, CO.

Neff, J. A., and B. Meanley. 1957. *Bird repellent studies in the eastern Arkansas rice fields*. Progress Report 2. Wildlife Research Laboratory, Denver, CO.

Neff, J. A., B. Meanley, and R. B. Brunton. 1957. *Basic screening tests with caged birds and other related studies with candidate repellent formulations 1955–1957*. Progress Report 3. Wildlife Research Laboratory, Denver, CO.

Nelms, C. O., and M. L. Avery. 1997. Reducing bird repellent application rates by the addition of sensory stimuli. *International Journal of Pest Management* 43:187–190.

Schafer, E. W., and R. B. Brunton. 1971. Chemicals as bird repellents: Two promising agents. *Journal of Wildlife Management* 35:569–572.

Schafer, E. W., W. A. Bowles, Jr., and J. Hurlbut. 1983. The acute oral toxicity, repellency, and hazard potential of 998 chemicals to one or more species of wild and domestic birds. *Archives of Environmental Contamination and Toxicology* 12:355–382.

Schafer, E. W., M. L. Eschen, and A. B. DeMilo. 1986. Repellency and toxicity of 2-acetylpyridinethiosemicarbazones and related chemicals to wild birds. *Journal of Environmental Science and Health. Part A: Environmental Science and Engineering* 21:281–288.

Schafer, E. W., and M. Jacobson. 1983. Repellency and toxicity of 55 insect repellents to red-winged blackbirds (*Agelaius phoeniceus*). *Journal of Environmental Science and Health. Part A: Environmental Science and Engineering* 18:493–502.

Shah, P., L. Clark, and J. R. Mason. 1991. Prediction of avian repellency from chemical structure: The aversiveness of vanillin, vanillyl alcohol and veratryl alcohol. *Pesticide Biochemistry and Physiology* 40:169–175.

Shah, P., J. R. Mason, and L. Clark. 1992. Avian chemical repellency: A structure-activity approach and implications. In *Chemical Signals in Vertebrates VI*, eds. R. Doty, and H. Muller-Schwarze, pp. 291–296. Plenum Press, New York.

Sherburne, J. A. 1972. Effects of seasonal changes in the abundance and chemistry of the fleshy fruits of northeastern woody shrubs on patterns of exploitation by frugivorous birds. PhD Dissertation, Cornell University, Ithaca, NY.

Somers, J. D., F. F. Gilbert, D. E. Joyner, R. J. Brooks, and R. G. Gartshore. 1981. Use of 4-aminopyridine in cornfields under high foraging stress. *Journal of Wildlife Management* 45:702–709.

Starr, R. I., J. F. Besser, and R. B. Brunton. 1964. A laboratory method for evaluating chemicals as bird repellents. *Agricultural and Food Chemistry* 12:342–344.

Stickley, A. R., R. T. Mitchell, J. L. Seubert, C. R. Ingram, and M. I. Dyer. 1976. Large-scale evaluation of blackbird frightening agent 4-aminopyridine in corn. *Journal of Wildlife Management* 40:126–131.

Stone, C. P., W. F. Shake, and D. J. Langowski. 1974. Reducing bird damage to highbush blueberries with a carbamate repellent. *Wildlife Society Bulletin* 2:135–139.

Thomson, R. H. 1987. *Naturally occurring quinones. III. Recent advances.* Chapman and Hall, London, UK.

Werner, S. J., and L. Clark, L. 2003. Understanding blackbird sensory systems and how repellent applications work. In *Management of North American Blackbirds: Special Symposium of The Wildlife Society Ninth Annual Conference,* ed. G. M. Linz, pp. 31–40. National Wildlife Research Center, Fort Collins, CO. https://www.aphis.usda.gov/wildlife_damage/nwrc/symposia/blackbirds_symposium/ (accessed November 12, 2016).

Werner, S. J., J. L. Cummings, P. A. Pipas, S. K. Tupper, and R. W. Byrd. 2008a. Registered pesticides and citrus terpenes as blackbird repellents for rice. *Journal of Wildlife Management* 72:1863–1868.

Werner, S. J., J. L. Cummings, S. K. Tupper, D. A. Goldade, and D. Beighley. 2008b. Blackbird repellency of selected registered pesticides. *Journal of Wildlife Management* 72:1007–1011.

Werner, S. J., J. L. Cummings, S. K. Tupper, J. C. Hurley, R. S. Stahl, and T. M. Primus. 2007. Caffeine formulation for avian repellency. *Journal of Wildlife Management* 71:1676–1681.

Werner, S. J., H. J. Homan, M. L. Avery, G. M. Linz, E. A. Tillman, A. A. Slowik, R. W. Byrd, et al. 2005. Evaluation of Bird Shield™ as a blackbird repellent in ripening rice and sunflower fields. *Wildlife Society Bulletin* 33:251–257.

Werner, S. J., G. M. Linz, S. K. Tupper, and J. C. Carlson. 2010. Laboratory efficacy of chemical repellents for reducing blackbird damage in rice and sunflower crops. *Journal of Wildlife Management* 74:1400–1404.

Woronecki, P. P., R. A. Dolbeer, and R. A. Stehn. 1981. Responses of blackbirds to mesurol and sevin applications on sweet corn. *Journal of Wildlife Management* 45:693–701.

Wright, E. N. 1962. Experiments with anthraquinone and thiram to protect germinating maize against damage by birds. *Annales de Epiphyties* 13:27–31.

CHAPTER 9

Frightening Devices

Michael L. Avery
National Wildlife Research Center
Gainesville, Florida

Scott J. Werner
National Wildlife Research Center
Fort Collins, Colorado

CONTENTS

9.1	Auditory Frightening Devices	160
	9.1.1 Bioacoustics	160
	9.1.2 Artificial Aural Deterrents	161
	9.1.3 Combatting Habituation	162
9.2	Visual Frightening Devices	163
	9.2.1 Balloons	164
	9.2.2 Hawk Kites	164
	9.2.3 Reflective Tape	165
	9.2.4 Flags and Streamers	166
	9.2.5 Effigies and Models	166
	9.2.6 Hazing with Aircraft	166
	9.2.7 Remote Controlled Models and Drones	167
	9.2.8 Falconry	167
9.3	Light and Color	168
	9.3.1 Lasers	168
	9.3.2 Ultraviolet Wavelengths	169
	9.3.3 Aposematic Colors	169
9.4	Summary	170
References		171

By their nature, avian frightening devices are intended to provide temporary (days, weeks) relief from a specific depredation or conflict situation. Ideally, the method applied will produce an immediate fright response, causing depredating birds to leave and to stay away as long as the method is in place. Longer-term (months, years) resource protection would involve methods such as crop varietal improvement, blackbird population management, or habitat manipulation. Frightening devices primarily affect the avian auditory and visual senses. With few exceptions (e.g., avian distress or alarm calls), frightening devices are not species-specific.

Very few frightening devices have been subjected to adequate scientific evaluation, so their efficacy under field conditions is often unknown. When field tests have been conducted, flaws in experimental design and analysis have rendered most trials inconclusive as to their effectiveness (Bomford and O'Brien 1990). Anderson et al. (2013) surveyed fruit crop producers in five states and reported that >50% of respondents considered "auditory scare devices" to be "slightly effective" or "not effective" in reducing bird damage. The specific types of auditory deterrents were not indicated. Relatively few published reports of frightening devices include testing against blackbirds. Therefore, the usefulness of many aural and visual devices for managing blackbirds can only be judged by extrapolating from studies that have focused on species other than icterids, such as corvids and gulls, in settings such as landfills and orchards, which are not usually associated with blackbirds.

9.1 AUDITORY FRIGHTENING DEVICES

Red-winged blackbirds (*Agelaius phoeniceus*) live in a rich auditory environment and they employ a wide array of vocalizations affecting almost all aspects of their natural history, including mate selection, territory maintenance, brood rearing, flock foraging, predator avoidance, and migration (Yasukawa and Searcy 1995). In general, birds display less aural sensitivity and a more narrow frequency range than humans (Beason 2004). The range of sensitivity to sound frequencies has been determined for relatively few species. Redwings and brown-headed cowbirds (*Molothrus ater*) have upper limits of 9.6–9.7 kHz (Heinz et al. 1977). Similar determinations have not been made for the common grackle (*Quiscalus quiscula*) or yellow-headed blackbird (*Xanthocephalus xanthocephalus*). In discussing auditory deterrents, we distinguish naturally produced sounds from those created by humans for the purpose of scaring birds.

9.1.1 Bioacoustics

Bioacoustics is "concerned with the production of sound by and its effects on living organisms" (Merriam-Webster 2016). An alternative term, *biosonics*, is "the use of an animal's natural vocalizations to influence the behavior of that species" (Gorenzel and Salmon 2008). In the context of resource protection, researchers have investigated the potential use of avian alarm and distress calls to disperse nuisance roosts or to protect crops from depredations for several decades (e.g., Frings et al. 1955; Boudreau 1968; Brough 1969; Schmidt and Johnson 1983). Birds give alarm calls to warn of imminent danger, such as when a predator is near, and distress calls are emitted when a bird is captured or in pain (Jaremovic 1990). Actual alarm or distress calls are considered superior to artificial noises for bird control because they (1) are less prone to habituation, (2) can be broadcast at lower intensities, (3) are species-specific (although Gorenzel and Salmon 2008 list this as a disadvantage), and (4) are less annoying to humans (Jaremovic 1990). Species-specific, biologically meaningful sounds (alarm and distress calls) have greater effects on bird behavior than do devices that produce noises not biologically relevant, which at most provide short-term relief but no lasting effect on food intake and use of space (Bomford and O'Brien 1990).

Broadcast recordings of corvid distress calls have been used successfully, alone and in combination with other tactics, for roost dispersal and crop protection with various crow species (Naef-Daenzer 1983; Gorenzel and Salmon 1993; Delwiche et al. 2005; Avery et al. 2008). Distress calls of European starlings (*Sturnus vulgaris*) have long been applied successfully for roost dispersal (Frings and Jumber 1954; Zajanc 1963; Pearson et al. 1967). Berge et al. (2007) incorporated alarm and distress calls with conventional methods (reflective tape, propane cannons, pyrotechnics) to reduce grape damage by European starlings, American robins (*Turdus migratorius*), and house finches (*Carpodacus mexicanus*). Their results showed that supplementing with the calls reduced bird damage more effectively than applying the conventional methods without the calls. Heidenreich (2007) noted that "New York studies have shown distress call devices to be effective for 7–10 days in plantings with high bird pressure. Use of predator models in conjunction with distress call units gave further reduction in feeding. Best results were obtained when units were moved regularly and used in conjunction with visual scare devices." Cook et al. (2008) reached similar conclusions regarding the effectiveness of several bird control methods in dispersing gulls (*Larus fuscus*, *Larus argentatus*, and *Chroicocephalus ridibundus*) from landfills in the United Kingdom. Distress calls, falconry, and lethal and nonlethal shooting were the most effective methods tested, but Cook et al. (2008) recommended rotating the control techniques used and applying them in combination for the most effective results.

The result of years of research and development is that numerous types of avian alarm/distress call units are now marketed to discourage bird use of crop fields, airports, roosts, and so on. Some units incorporate predator calls as well as avian alarm or distress calls. Most are programmable as to the interval between calls, species of bird, and randomization of calls. Units can be battery, solar, or electrically powered. Smaller units cover up to 1.5 ha; larger, more extensive systems reportedly can cover up to 12 ha (Heidenreich 2007). Prices can range up to several thousand dollars depending on the size of the area to be protected, power supply, cables, and number of speakers needed. Some auditory units even come packaged in the form of predators such as owls and eagles (Heidenreich 2007). There is minimal information available on the reliability and effectiveness of specific brands or models, other than what is found on the websites of manufacturers and suppliers.

9.1.2 Artificial Aural Deterrents

Although many new devices have been developed and marketed during the past 30 years, there is little evidence of any marked improvement in the efficacy of auditory deterrents for crop protection. Propane cannons remain the most popular of numerous auditory methods available for scaring birds from crop fields (Bomford and O'Brien 1990). Newer models are automatic, multi- or single-shot, ground-mounted, and rotate 360° for wide coverage. The intervals between detonations are adjustable. These units cost up to several hundred dollars each. Cummings et al. (1986) tested a combined propane exploder and pop-up scarecrow in sunflower fields and found that it was effective, particularly if it was used before a habitual feeding pattern had developed. The effectiveness of propane cannons, however, was shown to be limited to relatively small areas. Cummings et al. (1986) suggested that to be effective, at least one cannon should be used for each 2–3 ha area of sunflower crop. In the upper Midwest, sunflower field sizes are often 65 ha or larger; therefore, for propane cannons to be economically effective, the expected field damage should exceed 18%, which is a high level of bird damage for sunflower in the upper Midwest (Cummings et al. 1986). For best results, cannons should be moved often, installed so that the direction and timing of the explosions vary, and be augmented with pyrotechnics or live ammunition (Linz et al. 2011).

Pyrotechnics are standard tools for dispersing problem birds (Garner 1978). The cartridges are launched from a shotgun or pistol, usually in close proximity to the target flock of depredating birds. Bird bombs or bangers create a loud bang when they detonate in the air. Screamers emit a high-pitched whistling sound and smoke trail as they fly through the air after launch. These are versatile tools, easy to obtain and use, but the effect is usually short-lived. Thus, they perform best when used in conjunction with other tactics (Cook et al. 2008; Gorenzel and Salmon 2008).

Swaddle et al. (2016) have developed and tested a method called a *sonic net* in which artificial noises that overlap the frequency range perceived by target birds are broadcast to deter the birds' use of an airport, crop field, or other site to be protected. Presumably, the broadcast noise inhibits or masks auditory communications among the birds, including predator detection, which causes the affected birds to abandon the area. The results to date have been promising, but efficacy and cost-effectiveness of this approach under various field situations remain to be determined.

Another recent development in sonic deterrents is the Long Range Acoustic Device (LRAD). LRADs can direct loud, painful sound signals (up to 160 dB) in a fairly narrow beam (30°) that can be heard over 3,000 m away (TONI 2016). The LRAD was developed for antipersonnel use by military and law enforcement, but other applications, including bird management, have surfaced. Several companies market LRADs for bird dispersal at airports, landfills, mine waste ponds, and wind farms (American Technology Corporation 2009). The units can broadcast alarm or distress calls of the target species, raptor calls, or other loud noises disturbing to birds. We are not aware of any studies that have examined the cost–benefit of employing an LRAD system for any application including management of blackbird crop depredations or roost dispersal.

Ultrasonic devices (frequency of 20 kHz or greater) are prominent on avian pest control websites. For example, Bird-X (2016) advertises a unit "tuned at 20kHz," which is recommended for use in indoor and semi-enclosed areas. Grackles, starlings, and "blackbirds" are among the target species listed. However, the upper limit of sensitivity for red-winged blackbirds, brown-headed cowbirds, and starlings is <10 kHz (Heinz et al. 1977; Dooling 1982). There are also recent claims of effectiveness of ultrasonic devices in agricultural settings in Africa (e.g., Ezeonu et al. 2012). Bird species differ in their sensitivity to frequencies of sound (Beason 2004), so conceivably some species might be responsive to ultrasound. To our knowledge, however, this has yet to be demonstrated in blackbirds (Bishop et al. 2003; Beason 2004).

Fitzgerald (2013) suggests best results with auditory deterrents are obtained when:

- Scaring devices are implemented at the early stages of crop ripening, before birds have established a habit of visiting the site.
- Sounds are presented at random intervals.
- A variety of different sounds and frequencies are used.
- The sound source is moved frequently and only used for the minimum time needed to get a response.
- Sounds are supported by other methods, such as visual deterrents.
- Sounds are reinforced by actual danger (e.g., shooting).

Fitzgerald's (2013) prescription serves as useful guidance for almost any avian crop protection situation. Adoption of an integrated avian management approach offers the highest probability for significant damage reduction.

9.1.3 Combatting Habituation

Habituation occurs with prolonged exposure to auditory devices and is the bane of effective avian crop pest control (Bomford and O'Brien 1990). Most animals will habituate to noises, even actual predator calls, that occur repeatedly but result in no real threat to them. Habituation is an adaptive response because if animals did not habituate to nonthreatening stimuli, they would be constantly expending time and energy taking flight, seeking refuge, and producing alarm calls. Most of the suggestions by Fitzgerald (2013), Heidenreich (2007), and others for increasing the effectiveness of auditory deterrents represent tactics to combat habituation.

Some authors advocate limited, judicious lethal control, usually shooting, to reinforce and enhance the effectiveness of nonlethal methods (e.g., Cleary and Dolbeer 2005; Linz et al. 2011; Fitzgerald 2013). Despite offering no supporting evidence, Beason (2004, 95) states: "The most effective use of acoustic signals is when they are reinforced with activities that produce death or a painful experience to some members of the population." This concept is intuitively appealing as a means to retard habituation to nonlethal deterrents and to minimize the application of lethal control. The acceptance of this tactic as a recommended crop protection practice, however, is based on scant quantitative information. Instead, anecdotal reports and observations by field personnel comprise the basis for its inclusion in avian crop protection planning.

One exception is the study by Baxter and Allan (2008), who explicitly assessed the effectiveness of blank rounds alone and blank rounds in combination with live rounds for dispersing gulls and corvids at a landfill in England. The numbers of gulls and corvids each declined with the onset of harassment with blank rounds, but within 5 weeks the gulls and corvids had habituated to the treatment. After a 4-wk break, control resumed using a combination of blank rounds and live ammunition (number 5 shot). Live rounds were only fired when birds attempted to land at the site. The combination treatment caused gull numbers to be reduced drastically, but corvid numbers were not affected. Baxter and Allan (2008) concluded that the departure of the gulls meant reduced competition for food, so corvids might have been induced to remain even with the onset of enhanced harassment measures.

Investigation of the concept of lethal reinforcement of nonlethal crop damage control methods could be a very fruitful area of study. Although it has become entrenched in the lore of avian pest management, to date there is no definitive information to support its general use in crop protection or other dispersal strategies. The careful study by Baxter and Allan (2008) revealed disparity in the responses of gulls and corvids to the "reinforced" harassment method. Thus, interspecific variation should be expected, and more information obtained through carefully controlled studies is needed to identify and determine the range of such differences.

Glahn (2000) employed a different study design to compare the effectiveness of pyrotechnics and lethal shooting in dispersing roosts of double-crested cormorants (*Phalacrocorax auritus*). He selected five pairs of cormorant roosts based on numbers of birds and the areas occupied. He randomly assigned the members of each roost pair to receive either pyrotechnic or shotgun dispersal. Glahn (2000) found no differences between treatments in the effort required to disperse the roosts or in the length of time before the roosts were reoccupied. He concluded that shooting and pyrotechnics were equally effective as cormorant roost dispersal tools.

9.2 VISUAL FRIGHTENING DEVICES

Scarecrows have existed for millennia. The first scarecrows in recorded history were used by Egyptians to protect wheat fields from depredating flocks of quail in the Nile valley (Warnes 2016). In North America, scarecrows used by Native Americans and European settlers resembled the human form, made to look frightening, at least in the eyes of the persons making them, so that birds would stay away (Warnes 2016). Despite centuries of experience addressing problems of bird damage to crops and other resources, human ingenuity continues to struggle to develop a consistently effective scarecrow.

Many commercially available bird scare devices today incorporate motion (e.g., Stickley and King 1995; Loria 2014). Intuitively, an animated device is more likely to draw attention and create uncertainty among a flock of birds than a static device. Nevertheless, without an added element of surprise or unpredictability even a moving scarecrow will lose effectiveness over time and with repeated exposure (Marsh et al. 1992; Stickley and King 1995). Unpredictability is achieved by using triggering systems programmed to activate at random intervals or by linking the scare device to a motion detector so that the animal itself activates the unit (Gilsdorf et al. 2002).

Motion can be imparted by wind to various types of balloons, kites, tapes, and flags. Such units sometimes provide short-term relief because of neophobic responses by the birds, but the area thus protected is limited and effectiveness wanes quickly as birds acclimate to the presence of the nonthreatening device (Bishop et al. 2003).

Figure 9.1 Eyespot balloons are sold as bird-scaring devices and might provide short-term, localized relief. (Courtesy of John S. Humphrey.)

9.2.1 Balloons

Balloons can be suspended from poles so that they swing freely in the wind, or they can be inflated with helium and tethered to float above the area to be protected (Figure 9.1).

Mott (1985) deployed helium-filled balloons of various colors in five mixed-species blackbird roosts across three nights and recorded an 82% reduction in roosting birds. When winds exceeded 16 km/hr, however, the balloons became entangled in roost vegetation. In Japan, Shirota et al. (1983) floated a 2.6-m-diameter helium-filled balloon upon which large eyespots were painted above 3.5 ha of experimental grape, cherry, and peach plantings and successfully protected the fruit from white-cheeked starling (*Spodiopsar cineraceus*) depredation. Avery et al. (1988) evaluated a smaller version of the eyespot balloon in a large flight pen and determined that it did not affect pecking of oranges by boat-tailed grackles (*Quiscalus major*). Tipton et al. (1989) painted red and black eyespots on white beach balls (50 cm diameter), which they suspended from poles about 1 m above trees in three citrus groves. Damage to the fruit by great-tailed grackles (*Quiscalus mexicanus*) was virtually the same as in groves without beach balls.

9.2.2 Hawk Kites

Results from field trials indicate that the responses vary among species and some birds habituate more rapidly than others to the presence of hawk kites (Hothem and DeHaven 1982; Conover 1983, 1984; Seamans et al. 2002). The zone of best protection is directly below the kite, so these devices have relatively small areas of effectiveness (Figure 9.2).

Densities of approximately 1 unit/ha are indicated for effective protection (Marsh et al. 1991; Seamans et al. 2002). They also require frequent monitoring to avoid deflation and entanglement with vegetation.

9.2.3 Reflective Tape

An Internet search quickly reveals numerous types of reflective tape marketed as "bird-scaring." The reflecting tape that has been evaluated most often in field tests is approximately 1 cm wide and 0.25 mm thick. This tape is usually twisted and suspended between erect poles in parallel lines above the crop. Its Mylar coating (silver on one side, red on the other) reflects sunlight, which produces a flashing effect. Twisting the tape enhances the reflecting effect and creates an illusion of motion (Figure 9.3).

In windy conditions, vibrations by the tape produce a humming or roaring noise, which might contribute to its deterrent effect (Bruggers et al. 1986; Dolbeer et al. 1986; Tobin et al. 1988). Applications of reflective tape have had mixed success in protecting crops (Conover and Dolbeer 1989). Improper installation or strong wind can cause the tape to become tangled in vegetation.

Figure 9.2 Replicas of hawks or other raptors can be suspended from tall poles or helium balloons as a means to deter depredating birds. (Courtesy of Michael L. Avery.)

Figure 9.3 The flashing appearance of thin Mylar tape suspended between support poles could contribute to successful crop protection strategies. (Courtesy of Richard A. Dolbeer.)

9.2.4 Flags and Streamers

Manikowski and Billiet (1984) installed flags of colored cloth on 2-m poles in 30 rice plots (0.25 ha each) to combat damage by red-billed quelea (*Quelea quelea*). They observed reduced numbers of quelea in flagged plots compared to adjacent untreated plots. White and red flags were the most effective. Mason et al. (1993) found that white plastic flags made from garbage bags effectively deterred wintering snow geese (*Chen caerulescens*) from 10-ha fields of rye grass and winter wheat in New Jersey. Belant and Ickes (1997) successfully disrupted herring (*L. argentatus*) and ring-billed gulls (*Larus delawarensis*) at loafing areas by deploying 1-m long streamers of reflecting tape attached to stakes and wires. This approach was ineffective, however, when applied to herring gull nesting colonies.

9.2.5 Effigies and Models

For roost dispersal, crow and vulture effigies are effective components of integrated management strategies (Avery et al. 2002a, 2008; Tillman et al. 2002; Seamans 2004). These devices are replicas of crows or vultures, sometimes even carcasses or taxidermic preparations (Figures 9.4 and 9.5). Use of actual feathers in the effigies seems to be important in their success.

Plastic models of owls or other images meant to scare birds generally are innocuous. Monk parakeets did not respond to a taxidermic parakeet effigy or a flying owl predator model (Avery et al. 2002b). Canada geese did not respond to effigies of dead geese during the nesting season and repellent effects lasted for only 5 days during the late summer postbreeding season (Seamans and Bernhardt 2004). To our knowledge, there have been no tests of effigies in blackbird roosts.

9.2.6 Hazing with Aircraft

Flying fixed-wing aircraft over sunflower fields beset with flocks of depredating blackbirds proved marginally useful in North Dakota (Handegard 1988). Some rice producers also employ aircraft to haze blackbirds from their fields, but the extent to which this is used and the efficacy of the technique is not known (Cummings et al. 2005). High operating costs and safety have been major impediments to widespread use of this crop protection method (DeHaven 1971; Linz et al. 2011). Mott (1983) observed that low-level helicopter flights over mixed blackbird–starling winter roosts

Figure 9.4 Commercially available crow replicas, or effigies, can be useful in integrated roost dispersal efforts. (Courtesy of John S. Humphrey.)

Figure 9.5 Taxidermic vulture effigies, suspended upside down, are effective tools for vulture roost dispersal. (Courtesy of Eric A. Tillman.)

caused birds to flush on clear nights but not when it was overcast. Birds did not abandon the roosts but instead resettled soon after the aircraft left.

9.2.7 Remote Controlled Models and Drones

Managers have operated model aircraft to haze birds at airports, landfills, and aquaculture facilities (e.g., Solman 1981; Coniff 1991). Constraints include that it requires a skilled operator, only relatively small areas can be covered, inclement weather inhibits operation, and it is labor-intensive. With new technological advances, drones have replaced model aircraft in this method of pest bird management (Figure 9.6). Many are advertised online for bird control applications (BBC 2014).

Automated drone technology incorporating GPS-guided, programmed flight paths offers promise for new, improved effective hazing options in the near future (Ampatzidis et al. 2015).

9.2.8 Falconry

The sight of a live raptor, especially one in flight, evokes alarm calls and, if the perceived threat is sufficiently close, evasive action by feeding birds. The presence of an airborne raptor might cause the feeding flock to leave the crop area. Such a deterrent effect will persist as long as the raptor is present. When the raptor is gone, so is the threat, and the feeding flock is free to resume depredations.

Figure 9.6 Unmanned aerial vehicles (drones) offer new opportunities for scaring depredating birds from large crop fields. (Courtesy of Michael L. Avery.)

Erickson et al. (1990) considered falconry to be too costly as a tool for dispersing birds in agricultural settings. Dolbeer (1998) found no evidence that a falconry program reduced bird strikes at JFK International Airport beyond levels already achieved with a conventional program of bird frightening, shooting, and habitat management. Nevertheless, falconry has great human-interest appeal and in some cases might prove useful in an integrated management context.

9.3 LIGHT AND COLOR

9.3.1 Lasers

The potential for unexpected or especially intense lights to be aversive to birds has been posited for many years (e.g., Lustick 1973). For the most part this has been unfounded as a basis for bird management, and in some cases lights have been demonstrated to be attractive to birds rather than aversive (Gorenzel and Salmon 2008). Lasers are the principal exception.

Development of the laser dates to the late 1950s. The term *laser* is an acronym for "light amplification by stimulated emission of radiation," first articulated by Gould (1959). Once exotic and seemingly formidable, lasers are now common in many aspects of everyday life, including medical care, consumer electronics, entertainment, business and industry, law enforcement, and national defense. Modern communications through fiber optic systems rely on lasers. The unique aspect of the laser lies in the coherence of the emitted light. Coherence allows the laser beam to be very narrowly focused and for the beam to remain narrow over long distances. Since the early 1990s, laser devices have been marketed for bird dispersal uses (Glahn et al. 2001; Blackwell et al. 2002). Currently, both red and green lasers are commercially available specifically for bird management.

Properly used, the commercial bird-deterrent lasers are safe for birds and people, and they have utility as a management tool. However, there are constraints to their use and to their effectiveness. Laser pointers and similar devices are readily available to the public, and their misuse has generated legitimate concerns for human health and safety as related to aircraft piloting. There have been numerous instances of persons on the ground shining laser pointers or other laser devices at low flying aircraft and causing pilot disorientation. Such dangerous actions can cast suspicion on the legitimate use of lasers by trained wildlife personnel engaged in bird control. Thus, before applying a laser to roost dispersal or other bird management efforts, always consult local laws, ordinances, and regulations governing their use.

In terms of bird management, there are two basic limitations to laser use:

- Lasers consist of light, so in order to affect bird behavior, the laser light needs to be visible to the birds. In sunny, daylight conditions, lasers are at best barely visible at short range. Operators cannot see the beam to aim and use it effectively. And birds cannot detect the beam and thus are unaffected. So, lasers are not useful in well-lit situations. Because blackbirds feed during the day, lasers are not appropriate as a tool to prevent feeding in crops. Homan et al. (2010) did note that the green laser they tested appeared much brighter in daylight than the red one.
- Lasers affect birds by startling them. Birds are unfamiliar with the red or green beam and therefore respond as they would to a sudden unexpected loud noise, such as from a propane cannon. Thus, as with pyrotechnics or other loud noises, birds are initially startled by laser lights but are not driven off permanently. Roosting blackbirds appear to be particularly recalcitrant to laser dispersal, although the number of properly controlled trials is minimal (Homan et al. 2010).

The advantage to laser dispersal compared to pyrotechnics is that lasers are silent. Lasers can disperse roosting birds from towers or other structures without aggravating nearby residents. However, as with pyrotechnics, lasers do not provide permanent solutions. When the laser stimulus is removed, so is the deterrent effect and birds readily repopulate the roost unless other actions are implemented (e.g., Avery et al. 2002b; Gorenzel et al. 2002). For dispersal of crow or vulture roosts, lasers are potentially very useful components of integrated management strategies (Avery et al. 2008).

9.3.2 Ultraviolet Wavelengths

Among diurnal birds, there are two distinct classes of color vision, violet sensitive (VS) and ultraviolet sensitive (UVS). Retinal cones in VS species have maximum absorption in the 402–426 nm range, whereas UVS species have cones maximally sensitive in the 355–380 nm range (Odeen and Hasted 2013). The red-winged blackbird is among the avian species having retinal cones that are maximally sensitive in the near ultraviolet (UV), at 370 nm (Chen et al. 1984). The ability of blackbirds and many other species to see in the UV portion of the spectrum is a trait they share with insects and numerous other taxa but one that is distinct from humans and most other mammals. UV perception in birds is known to function in mate selection (e.g., Bennett et al. 1997), foraging (e.g., Burkhardt 1982; Schaefer et al. 2006), and flight orientation (e.g., Kreithen and Eisner 1978).

Management applications of avian UV sensitivity have started to emerge. Pole-mounted windmills with spinning blades painted with UV-reflecting paint are sold as "scare windmills" to frighten birds including wild turkeys (*Meleagris gallopavo*) and geese (Anonymous 2002; JWB Marketing 2014). We are aware of no objective controlled evaluation of these units, but they are frequently included in discussions of management alternatives for protecting crops from bird damage (e.g., Eaton 2010).

9.3.3 Aposematic Colors

Avian response to color has management applications in areas of crop protection (Greig-Smith and Rowney 1987), pesticide avoidance (Kalmbach and Welch 1946; Gionfriddo and Best 1996), and collision avoidance (Blackwell et al. 2012). Birds react differently to different colors. Avery et al. (1999) exposed red-winged blackbirds and boat-tailed grackles to rice seeds of several colors in a series of cage and pen tests; blue was consistently the least-preferred color.

Color alone will not prevent birds from feeding on a crop or occupying a roost site, but color can serve as a sign or warning signal. Such warning, or aposematic, colors are found throughout the natural world and advertise to predators the presence of toxic or debilitating chemical compounds in a potential prey item (e.g., Berenbaum 1995). Although blackbirds are frequently used as test subjects in feeding trials and they can be conditioned to avoid food associated with specific colors (e.g., Mason and Reidinger 1983; Werner et al. 2008), it is unclear that an aposematic color approach can be successfully applied in crop protection scenarios. Limited attempts to improve efficacy of crop protection through application of color cues in field applications of chemical repellents have produced mixed results. Elmahdi et al. (1985) reported enhancement of methiocarb repellency against red-billed quelea with calcium carbonate added to the treatment on sorghum. They concluded that the presence of the white paint residue from calcium carbonate signaled to depredating birds that the crop was inedible even after the methiocarb was no longer present. Conversely, Dolbeer et al. (1992) reported that the white markings left on sorghum plants by calcium carbonate spray did not reduce blackbird damage beyond applications of methiocarb alone.

Avery (2002) hypothesized that the UV reflectance of anthraquinone enhances the birds' ability to associate the appearance of treated food with the adverse postingestional consequences and thereby learn more rapidly to avoid the treated food. Werner et al. (2012) exposed red-winged blackbirds to sunflower treated with 0.25% anthraquinone (wt/wt) during 2 days of repellent conditioning. Relative to unconditioned blackbirds, three test groups previously exposed to the UV-absorbent, postingestive repellent subsequently avoided sunflower treated with 0.2% of an UV-absorbent cue and 0%, 0.025%, or 0.05% anthraquinone throughout a 14-day preference test. Similarly, an independent group of red-winged blackbirds exposed to the UV-absorbent, postingestive repellent subsequently avoided UV-reflective food (Werner et al. 2012). In the absence of negative postingestive consequences, however, UV cues alone are unlikely to elicit food avoidance among wild birds (Lyytinen et al. 2001).

Relative to the repellency of food treated only with an anthraquinone-based repellent, synergistic repellency (i.e., a 45%–115% increase) was observed when 0.2% of the UV feeding cue was combined with 0.02% or 0.035% anthraquinone (wt/wt; Werner et al. 2014). In contrast, <10% repellency was observed for 0.2% of a non-UV feeding cue (red number 40 aluminum lake dispersion) paired with 0.02% anthraquinone. Aversion performance was therefore not attributed to characteristics of either conditioned or unconditioned stimuli but their combinations, and enhanced repellency of anthraquinone plus the UV-absorbent cue was attributed to UV wavelengths. Thus, the addition of an UV feeding cue can enhance avian repellency at repellent concentrations realized from previous field applications on agricultural crops (e.g., <0.1% anthraquinone; Werner et al. 2014).

9.4 SUMMARY

The number and variety of auditory and visual frightening devices available for short-term bird management are greater than ever, as is the need for such tools. Human ingenuity coupled with technological advances and economic incentive ensure that new, improved options will continue to be developed. Much less available, however, are objective evaluations of the efficacy, or cost-effectiveness, of frightening devices, especially as related to blackbird management. For many devices, performance measures are little more than testimonials on a vendor's website. Even when a field trial of a frightening device is conducted, appropriate scale, replication, and controls are seldom part of the study design, perhaps because of constraints imposed by cost and resource limitations (Bomford and O'Brien 1990). Producers thus continue to rely on familiar tools such as propane cannons and shooting as short-term blackbird management options.

REFERENCES

American Technology Corporation. 2009. *American Technology announces LRAD® order for bird and waterfowl control and protection.* http://www.marketwired.com/press-release/american-technology-announces-lradr-order-bird-waterfowl-control-protection-nasdaq-atco-1246276.htm (accessed August 30, 2016).

Ampatzidis, Y., J. Ward, and O. Samara. 2015. Autonomous system for pest bird control in specialty crops using unmanned aerial vehicles. *2015 ASABE Annual International Meeting.* New Orleans, LA.

Anderson, A., C. A. Lindell, K. M. Moxcey, W. F. Siemer, G. M. Linz, P. D. Curtis, J. E. Carroll, C. L., et al. 2013. Bird damage to select fruit crops: The cost of damage and the benefits of control in five states. *Crop Protection* 52:103–109.

Anonymous. 2002. "Scare Windmill" repels pest birds. *Farm Magazine* 26:27.

Avery M. L. 2002. Avian repellents. In: Plimmer J. R., editor. *Encyclopedia of agrochemicals.* Hoboken, NJ: Wiley. p. 122–128.

Avery, M. L., D. E. Daneke, D. G. Decker, P. W. Lefebvre, R. E. Matteson, and C. O. Nelms. 1988. Flight pen evaluation of eyespot balloons to protect citrus from bird depredations. *Vertebrate Pest Conference* 13:277–280.

Avery, M. L., E. C. Greiner, J. R. Lindsay, J. R. Newman, and S. Pruett-Jones. 2002b. Monk parakeet management at electric utility facilities in south Florida. *Vertebrate Pest Conference* 20:140–145.

Avery, M. L., J. S. Humphrey, D. G. Decker, and A. P. McGrane. 1999. Seed color avoidance by captive red-winged blackbirds and boat-tailed grackles. *Journal of Wildlife Management* 63:1003–1008.

Avery, M. L., J. S. Humphrey, E. A. Tillman, K. O. Phares, and J. E. Hatcher. 2002a. Dispersing vulture roosts on communication towers. *Journal of Raptor Research* 36:45–50.

Avery, M. L., E. A. Tillman, and J. S. Humphrey. 2008. Effigies for dispersing urban crow roosts. *Vertebrate Pest Conference* 23:84–87.

Baxter, A. T., and J. R. Allan. 2008. Use of lethal control to reduce habituation to blank rounds by scavenging birds. *Journal of Wildlife Management* 72:1653–1657.

BBC. 2014. Suffolk farm uses drone to scare pigeons off rape crop. http://www.bbc.com/news/uk-england-suffolk-27069825 (accessed August 30, 2016).

Beason, R. C. 2004. What can birds hear? *Vertebrate Pest Conference* 21:92–96.

Belant, J. L., and S. K. Ickes. 1997. Mylar flags as gull deterrents. *Great Plains Wildlife Damage Control Workshop* 13:73–79.

Bennett, A. T. D., I. C. Cuthill, J. C. Partridge, and K. Lunau. 1997. Ultraviolet plumage colors predict mate preference in starlings. *Proceedings of the National Academy of Science U S A* 94:8618–8621.

Berenbaum, M. R. 1995. Aposematism and mimicry in caterpillars. *Journal of the Lepidopterists' Society* 49:386–396.

Berge, A., M. Delwiche, W. P. Gorenzel, and T. Salmon. 2007. Bird control in vineyards using alarm and distress calls. *American Journal of Enology and Viticulture* 58:135–143.

Bird-X. 2016. QuadBlaster QB-4. http://www.bird-x.com/quadblaster-qb-4-products-82.php?page_id=103 (accessed September 15, 2016).

Bishop, J., H. McKay, D. Parrott, and J. Allan. 2003. Review of international research literature regarding the effectiveness of auditory bird scaring techniques and potential alternatives. Unpublished report prepared by Central Science Laboratory for the Department for Environment, Food & Rural Affairs, London.

Blackwell, B. F., G. E. Bernhardt, and R. A. Dolbeer. 2002. Lasers as nonlethal avian repellents. *Journal of Wildlife Management* 66:250–258.

Blackwell, B. F., T. L. DeVault, T. W. Seamans, S. L. Lima, P. Baumhardt, and E. Fernandez-Juricic. 2012. Exploiting avian vision with aircraft lighting to reduce bird strikes. *Journal of Applied Ecology* 49:758–766.

Bomford, M., and P. H. O'Brien. 1990. Sonic deterrents in animal damage control: A review of device tests and effectiveness. *Wildlife Society Bulletin* 18:411–422.

Boudreau, G. W. 1968. Alarm sounds and responses of birds and their application in controlling problem species. *Living Bird* 7:27–46.

Brough, T. 1969. The dispersal of starlings from woodland roosts and the use of bioacoustics. *Journal of Applied Ecology* 6:403–410.

Bruggers, R. L., J. E. Brooks, R. A. Dolbeer, P. P. Woronecki, R. K. Pandit, T. Tarimo, All-India Co-ordinated Research Project on Economic Ornithology, and M. Hoque. 1986. Responses of pest birds to reflecting tape in agriculture. *Wildlife Society Bulletin* 14:161–170.

Burkhardt, D. 1982. Birds, berries and UV. *Naturwissenschaften* 69:153–157.

Chen, D-M., J. S. Collins, and T. H. Goldsmith. 1984. The ultraviolet receptor of bird retinas. *Science* 225:337–340.

Cleary, E. C., and R. A. Dolbeer. 2005. *Wildlife hazard management at airports: A manual for airport personnel.* Second edition. Washington: Federal Aviation Administration, Office of Safety and Standards.

Coniff, R. 1991. Why catfish farmers want to throttle the crow of the sea. *Smithsonian* 22:44–55.

Conover, M. R. 1983. Pole-bound hawk-kites failed to protect maturing cornfields from blackbird damage. *Bird Control Seminar* 9:85–90.

Conover, M. R. 1984. Comparative effectiveness of avitrol, exploders, and hawk-kites in reducing blackbird damage to corn. *Journal of Wildlife Management* 48:109–116.

Conover, M. R., and R. A. Dolbeer. 1989. Reflecting tapes fail to reduce blackbird damage to ripening cornfields. *Wildlife Society Bulletin* 17:441–443.

Cook, A., S. Rushton, J. Allan, and A. Baxter. 2008. An evaluation of techniques to control problem bird species on landfill sites. *Environmental Management* 41:834–843.

Cummings, J. L., C. E. Knittle, and J. L. Guarino. 1986. Evaluating a pop-up scarecrow coupled with a propane exploder for reducing blackbird damage to ripening sunflower. *Vertebrate Pest Conference* 12:286–291.

Cummings, J. L., S. A. Shwiff, and S. K. Tupper. 2005. Economic impacts of blackbird damage to the rice industry. *Wildlife Damage Management Conference* 11:317–322.

DeHaven, R. W. 1971. Blackbirds and the California rice crop. *Rice Journal* 74:7–8.

Delwiche, M. J., A. P. Houk, W. P. Gorenzel, and T. P. Salmon. 2005. Electronic broadcast call unit for bird control in orchards. *Applied Engineering in Agriculture* 21:721–727.

Dolbeer, R. A. 1998. Evaluation of shooting and falconry to reduce bird strikes with aircraft at John F. Kennedy International Airport. *International Bird Strike Committee Meeting* 24:145–158.

Dolbeer, R. A., P. P. Woronecki, and R. L. Bruggers. 1986. Reflecting tapes repel blackbirds from millet, sunflowers, and sweet corn. *Wildlife Society Bulletin* 14:418–425.

Dolbeer, R. A., P. P. Woronecki, and R. W. Bullard. 1992. Visual cue fails to enhance bird repellency of methiocarb in ripening sorghum. In *Chemical signals in vertebrates VI*. eds. R. L. Doty and D. Müller-Schwarze, pp. 323–330. New York: Plenum Press.

Dooling, R. 1982. Auditory perception in birds. In *Acoustic communication in birds* (Vol. 1), eds. D. Kroodsma and E. Miller, pp. 95–130. New York: Academic Press.

Eaton, A. 2010. *Bird damage prevention for northern New England fruit growers.* University of New Hampshire Cooperative Extension Service. https://extension.unh.edu/resources/files/Resource001797_Rep2514.pdf (accessed July 11, 2016).

Elmahdi, E. M., R. W. Bullard, and W. B. Jackson. 1985. Calcium carbonate enhancement of methiocarb repellency for quelea. *Tropical Pest Management* 31:67–72.

Erickson, W. A., R. E. Marsh, and T. P. Salmon. 1990. A review of falconry as a bird-hazing technique. *Vertebrate Pest Conference* 14:314–316.

Ezeonu, S. O., D. O. Amaefule, and G. N. Okonkwo. 2012. Construction and testing of ultrasonic bird repeller. *Journal of Natural Sciences Research* 2:8–17.

Fitzgerald, S. 2013. *Managing bird damage in crops.* Ontario Fruit and Vegetable Growers Association. https://onvegetables.files.wordpress.com/2013/06/managing-bird-damage-in-crops-factsheet-final.pdf (accessed July 11, 2016).

Frings, H., M. Frings, B. Cox, and L. Peissner. 1955. Recorded calls of herring gulls (*Larus argentatus*) as repellents and attractants. *Science* 121:340–341.

Frings, H., and J. Jumber. 1954. Preliminary studies on the use of a specific sound to repel starlings (*Sturnus vulgaris*) from objectionable roosts. *Science* 119:318–319.

Garner, K. M. 1978. Management of blackbird and starling winter roost problems in Kentucky and Tennessee. *Vertebrate Pest Conference* 8:54–59.

Gilsdorf, J. M., S. E. Hygnstrom, and K. C. Vercauteren. 2002. Use of frightening devices in wildlife damage management. *Integrated Pest Management Review* 7:29–45.

Gionfriddo, J. P., and L. B. Best. 1996. Grit color selection by house sparrows and northern bobwhites. *Journal of Wildlife Management* 60:836–842.

Glahn, J. F. 2000. Comparison of pyrotechnics versus shooting for dispersing double-crested cormorants from their night roosts. *Vertebrate Pest Conference* 19:44–48.

Glahn, J. F., G. Ellis, P. Fioranelli, and B. S. Dorr. 2001. Evaluation of moderate and low-powered lasers for dispersing double-crested cormorants from their night roosts. *Wildlife Damage Management Conference* 9:34–48.

Gorenzel, W. P., B. F. Blackwell, G. D. Simmons, T. P. Salmon, and R. A. Dolbeer. 2002. Evaluation of lasers to disperse American crows, *Corvus brachyrhynchos*, from urban night roosts. *International Journal of Pest Management* 48:327–331.

Gorenzel, W. P. and T. P. Salmon. 1993. Tape-recorded calls disperse American crows from urban roosts. *Wildlife Society Bulletin* 21:334–338.

Gorenzel, W. P. and T. P. Salmon. 2008. *Bird hazing manual; techniques and strategies for dispersing birds from spill sites.* Publication 21638. University of California Agriculture and Natural Resources Communication Services, Oakland, CA.

Gould, G. R. 1959. The LASER, Light Amplification by Stimulated Emission of Radiation. In *The Ann Arbor Conference on Optical Pumping*, eds. P. A. Franken and R. H. Sands, p. 128, University of Michigan, Abstract. June 15–18, 1959. Ann Arbor, MI.

Greig-Smith, P. W., and C. M. Rowney. 1987. Effects of colour on the aversions of starlings and house sparrows to five chemical repellents. *Crop Protection* 6:402–409.

Handegard, L. L. 1988. Using aircraft for controlling blackbird/sunflower depredations. *Vertebrate Pest Conference* 13:293–294.

Heidenreich, C. 2007. *Bye birdie – Bird management strategies for small fruit.* Cornell Cooperative Extension. http://www.fruit.cornell.edu/berry/ipm/ipmpdfs/byebyebirdiesmallfruit.pdf (accessed July 11, 2016).

Heinz, R., J. M. Sinnott, and M. B. Sachs. 1977. Auditory sensitivity of the redwing blackbird (*Agelaius phoeniceus*) and brown-headed cowbird (*Molothrus ater*). *Journal of Comparative and Physiological Psychology* 91:1365–1376.

Homan, H. J., A. A. Slowik, B. F. Blackwell, and G. Linz. 2010. Field testing class IIIb handheld lasers to disperse roosting blackbirds. *National Sunflower Association Sunflower Research Forum*, January 13–14, 2010. Fargo, ND. http://u95010.eos-intl.net/U95010/OPAC/Common/Pages/GetDoc.aspx?ClientID=MU95010&MediaCode=4818003

Hothem, R. L., and R.W. DeHaven. 1982. Raptor-mimicking kites for reducing bird damage to wine grapes. *Vertebrate Pest Conference* 10:171–178.

Jaremovic, R. 1990. 1990. Bioacoustical scaring trials. In *National Bird Pest Workshop Proceedings*, ed. P. Fleming, I. Temby and J. Thompson, pp. 98–110. University of New England, Armidale, New South Wales, Australia.

JWB Marketing. 2014. *Scare Windmill: $89*. http://www.birdcontrolsupplies.com/windTurkey.htm (accessed August 30, 2016).

Kalmbach, E. R., and Welch, J. F. 1946. Colored rodent baits and their value in safeguarding birds. *Journal of Wildlife Management* 10:353–360.

Kreithen, M. L. and T. Eisner. 1978. Ultraviolet light detection by homing pigeon. *Nature* 273:347–348.

Linz, G. M., H. J. Homan, S. W. Werner, H. M. Hagy, and W. J. Bleier. 2011. Assessment of bird management strategies to protect sunflower. *BioScience* 61:960–970.

Loria, K. 2014. Good gyrations; inflatable dancers scare off birds. *Fruit Growers News* 53(3):1, 8.

Lustick, S. 1973. The effect of intense light on bird behavior and physiology. *Bird Control Seminar* 6:171–186.

Lyytinen, A., R. V. Alatalo, L. Lindström, and J. Mappes. 2001. Can ultraviolet cues function as aposematic signals? *Behavioral Ecology* 12:65–70.

Manikowski, S., and F. Billiet. 1984. Coloured flags protect ripening rice against *Quelea guelea*. *Tropical Pest Management* 30:148–150.

Marsh, R. E., W. A. Erickson, and, T. P. Salmon. 1991. Bird hazing and frightening methods and techniques (with emphasis on containment ponds). *Other Publications in Wildlife Management.* Paper 51. University of California, Davis, CA.

Marsh, R. E., W. A. Erickson, and, T. P. Salmon. 1992. Scarecrows and predator models for frightening birds from specific areas. *Vertebrate Pest Conference* 15:112–114.

Mason, J. R., L. Clark, and N. J. Bean. 1993. White plastic flags repel snow geese (*Chen caerulescens*). *Crop Protection* 12:497–500.

Mason, J. R., and R. F. Reidinger, Jr. 1983. Importance of color for methiocarb-induced food aversions in red-winged blackbirds. *Journal of Wildlife Management* 47:383–393.

Merriam-Webster. 2016. *Definition of bioacoustics*. http://www.merriam-webster.com/dictionary/bioacoustics (accessed September 15, 2016).

Mott, D. F. 1983. Influence of low-flying helicopters on the roosting behavior of blackbirds and starlings. *Bird Control Seminar* 9:81–84.

Mott, D. F. 1985. Dispersing blackbird-starling roosts with helium-filled balloons. *Eastern Wildlife Damage Control Conference* 2:156–162.

Naef-Daenzer, L. 1983. Scaring of carrion crows (*Corvus corone corone*) by species-specific distress calls and suspended bodies of dead crows. *Bird Control Seminar* 9:91–95.

Odeen, A., and O. Hastad. 2013. The phylogenetic distribution of ultraviolet sensitivity in birds. *BMC Evolutionary Biology* 13:36. DOI: 10.1186/1471-2148-13-36.

Pearson, E. W., P. R. Skon, G. W. Corner. 1967. Dispersal of urban roosts with records of starling distress calls. *Journal of Wildlife Management* 31:502–506.

Schaefer, H. M., D. J. Levey, V. Schaefer, and M. L. Avery. 2006. The role of chromatic and achromatic signals for fruit detection by birds. *Behavioral Ecology* 17:784–789.

Seamans, T. W. 2004. Response of roosting turkey vultures to a vulture effigy. *Ohio Journal of Science* 104:136–138.

Seamans, T. W., and G. E. Bernhardt. 2004. Response of Canada geese to a dead goose effigy. *Vertebrate Pest Conference* 21:104–106.

Seamans, T. W., B. F. Blackwell, and J. T. Gansowski. 2002. Evaluation of the Allsopp Helikite® as a bird scaring device. *Vertebrate Pest Conference* 20:123–128.

Schmidt, R. H., and R. J. Johnson. 1983. Bird dispersal recordings: An overview. In *Vertebrate pest control and management materials: Fourth symposium ASTM STP 817*, ed. D. E. Kaukeinen, pp. 43–65. American Society for Testing and Materials, Philadelphia, PA.

Shirota, Y., M. Sanada, and S. Masaki. 1983. Eyespotted balloons as a device to scare grey starlings. *Applied Entomology and Zoology* 18:545–549.

Solman, V.E.F. 1981. Birds and aviation. *Environmental Conservation* 8:45–52.

Stickley, Jr., A. R., and J. O. King. 1995. Long-term trial of an inflatable effigy scare device for repelling cormorants from catfish ponds. *Eastern Wildlife Damage Control Conference* 6:89–92.

Swaddle, J. P., D. L. Moseley, M. K. Hinders, and E. P. Smith. 2016. A sonic net excludes birds from an airfield: Implications for reducing bird strike and crop losses. *Ecological Applications* 26:339–345.

Tillman, E. A., J. S. Humphrey, and M. L. Avery. 2002. Use of vulture carcasses and effigies to reduce vulture damage to property and agriculture. *Vertebrate Pest Conference* 20:123–128.

Tipton, A. R., J. H. Rappole, A. H. Kane, R. H. Flores, D. B. Johnson, J. M. Hobbs, P. Schulz, S. L. Beasom, and J. Palacios. 1989. Use of monofilament line, reflective tape, beach balls, and pyrotechnics for controlling grackle damage to citrus. *Great Plains Wildlife Damage Control Workshop* 9:126–128.

Tobin, M. E., P. P. Woronecki, R. A. Dolbeer, and R. L. Bruggers. 1988. Reflecting tape fails to protect ripening blueberries from bird damage. *Wildlife Society Bulletin* 16:300–303.

TONI Bird Control Solutions. 2016. *Acoustic solution – Directional beam*. http://www.birdcontrolsolutions.net/en/solutions/lrad.php (accessed September 15, 2016).

Warnes, K. 2016. *Scarecrows historically speaking*. http://historybecauseitshere.weebly.com/scarecrows-historically-speaking.html (accessed June 25, 2016).

Werner, S. J., S. T. DeLiberto, S. E. Pettit, and A. M. Mangan. 2014. Synergistic effect of an ultraviolet feeding cue for an avian repellent and protection of agricultural crops. *Applied Animal Behavior Science* 159:107–113.

Werner, S. J., B. A. Kimball, and F. D. Provenza. 2008. Food color, flavor, and conditioned avoidance among red-winged blackbirds. *Physiology and Behavior* 93:110–117.

Werner, S. J., S. K. Tupper, J. C. Carlson, S. E. Pettit, J. W. Ellis, and G. M. Linz. 2012. The role of a generalized ultraviolet cue for blackbird food selection. *Physiology and Behavior* 106:597–601.

Yasukawa, K., and W. A. Searcy. 1995. Red-winged blackbird (*Agelaius phoeniceus*). No. 184. *The Birds of North America*, ed. P. G. Rodewald. Cornell Lab of Ornithology, Ithaca, New York. https://birdsna.org/Species-Account/bna/species/rewbla (accessed September 25, 2016).

Zajanc, A. 1963. Methods of controlling starlings and blackbirds. *Vertebrate Pest Conference* 1:190–212.

CHAPTER 10

Strategies for Evading Blackbird Damage

George M. Linz and Page E. Klug
National Wildlife Research Center
Bismarck, North Dakota

CONTENTS

10.1	Cultural Practices	176
10.2	Advancing Harvest Date	177
10.3	Wildlife Conservation Food Plots	177
10.4	Management of Wetland Roost Sites	180
	10.4.1 Glyphosate Herbicide	181
	10.4.2 Ecological Effects	181
	10.4.3 Economics	182
	10.4.4 Operational Program	182
10.5	Management of Upland Roost Sites	183
10.6	Bird-Resistant Crops	183
	10.6.1 Corn	184
	10.6.2 Sunflower	184
10.7	Summary	185
References		185

Foraging blackbird flocks have great mobility as they search for food that is plentiful, is easily accessed, and has a high nutritional value. Ripening corn, rice, and sunflower fit those criteria, as does seeded rice. The birds will move from field to field to find the ideal combination of energy spent to discover food versus the energy value of the food (Pyke et al. 1977). An extraordinary effort is often needed to actively move the birds from a foraging location with a high positive value (e.g., close to roost and early ripening sunflower) versus a location with a low value (e.g., far from roost and mature corn). Indeed, Handegard (1988) relayed that despite the intense use of low-flying aircraft and live shot-shell ammunition, field specialists were not able to move blackbirds from sunflower fields located near cattail-dominated (*Typha* spp.) wetland roosts. In this case, birds were also undergoing their annual feather molt, which hampered flight and increased the energetic cost of moving to a new foraging site (Linz et al. 1983).

For this reason, we believe that harvest advancement through desiccation (i.e., crop phenology), wildlife conservation food plots (WCFP), and habitat management should form the foundation of any blackbird management scheme that might include a suite of potential damage control options. These methods help reduce damage by manipulating the environment within and surrounding crop fields.

10.1 CULTURAL PRACTICES

An obvious bird management strategy is to abandon a bird-susceptible crop and substitute other crops (e.g., soybeans, flax) that are not damaged by blackbirds. Dedicated growers, however, recognize the value of bird-susceptible crops (e.g., sunflower, rice, and corn) in their crop rotation and have opted to use time-tested and new cultural practices to keep damage at economically acceptable levels (generally ≤5%; Linz and Homan 2011; Dolbeer and Linz 2016). Savvy growers plant less vulnerable crops near known wetland roost sites, coordinate planting time with neighbors to eliminate early and late ripening fields, plant large fields to spread damage over more heads so remaining seeds can undergo compensatory growth, leave harvested fields untilled to provide alternate food sources, create vehicle pathways to facilitate bird hazing efforts with pyrotechnics and shotguns, and control weeds and insects that attract birds (Sedgwick et al. 1986; Wilson et al. 1989; Dolbeer 1990; Linz et al. 2011; Dolbeer and Linz 2016).

Finally, although an untested concept, short-stature (SS) sunflower that provide less height than standard hybrids for birds to scan their surroundings for predators, especially birds of prey, might reduce the suitability of the foraging site. Short-stature sunflower also might allow for more effective use of scare devices (e.g., propane cannons, pyrotechnics, unmanned aircraft systems, shotgun shells), because the sound is not muffled. Additionally, high-clearance ground spray equipment can more easily apply bird repellents and other pesticides (Figure 10.1). Currently, improved SS sunflower varieties are not available for widespread planting in northern sunflower growing areas in North America. However, sunflower breeders are rapidly developing SS sunflower that is comparable to standard-height sunflower for such agroeconomic characteristics as days to maturity, yield, oil content, and disease tolerance (Mullally 2013; Linz and Hanzel 2015).

Figure 10.1 Short-stature sunflower allows high-clearance ground sprayers to easily apply bird repellents and desiccants as part of a bird damage management program. (Courtesy of USDA Wildlife Services.)

10.2 ADVANCING HARVEST DATE

Advancing harvest date can reduce bird damage and even out the grower's harvest schedule. This strategy works particularly well for ripening sunflowers, which require about 38 days from last anther to physiological maturity (R9) when the seeds contain about 35% moisture (Putnam et al. 1990). Sunflowers are mature long before they are dry enough for combining and long-term storage. Thus, the grower must wait for natural desiccation, which usually occurs after freezing temperatures, or dry the seeds before storing in bins, which can be costly. Natural desiccation can be slow and uneven, and inclement weather can reduce quality and yield through stem breakage and shattering. Waiting for natural desiccation also increases the time that sunflowers are susceptible to predation by blackbirds. Artificially advancing the harvest date reduces the amount of time that the crop is susceptible to blackbird predation, especially by late-migrating blackbirds.

Growers can use paraquat, sodium chloride, saflufenacil, or a tank mix of glyphosate and saflufenacil herbicides to desiccate physiologically mature non–genetically modified organism (GMO) crops and advance harvest as much as 20 days over nondesiccated sunflowers (Dow AgroSciences 2016). The latter is a particularly popular option because glyphosate kills grass, and saflufenacil, which is a broad-leafed herbicide, dries down sunflower faster than a glyphosate application alone. Paraquat is an effective desiccant but changes the cell structure of the head, allowing moisture to enter during a precipitation event, which could result in a delayed harvest. Use of sodium chloride declined with the introduction of glyphosate as a harvest aid because it is relatively expensive and must be applied at high volumes with a ground sprayer (Linz et al. 2011).

10.3 WILDLIFE CONSERVATION FOOD PLOTS

Linz et al. (2004) coined the moniker *wildlife conservation sunflower plots* to emphasize that sunflower plots can be planted with the needs of all wildlife in mind, in particular providing a refuge for local and migrating birds, including blackbirds dispersed from ripening crops such as sunflower. Here, we generalize the term *wildlife conservation sunflower plots* to *wildlife conservation food plots* (WCFP) to include all food varieties provided to attract wildlife. WCFP (also known as lure, decoy, food, trap, supplemental feeding, and diversionary plots) typically are small acreages (0.8–1.6 ha) strategically placed to provide food for wildlife (Cummings et al. 1987; Hagy et al. 2008, 2010; Tranel et al. 2008; U.S. Department of Agriculture 2013; Kubasiewicz et al. 2016). Entire fields are sometimes planted to a bird-susceptible crop (e.g., wheat, sunflower, corn, rice) or planted to attract wildlife that might otherwise forage in commercial crops (Gustad 1979; Cummings et al. 1987; Knittle and Porter 1988). Aside from reducing damage in a commercial field where damage >5% is economically important, food remaining in WCFP is available for both migrating birds and resident animals (Tranel et al. 2008; Galle et al. 2009; Hagy et al. 2010). Additionally, WCFP might be considered to support a population of an endangered species (Ewen et al. 2015).

The U.S. Department of Agriculture Wildlife Services program supports the use of lure crops (i.e., WCFP) to divert wildlife from damaging agricultural resources, especially where crop damage is recurrent (e.g., near historical roosts) and other methods are deemed ineffective (U.S. Department of Agriculture 2003).

Avery (2003) advised that blackbirds with no alternative quality food will endure otherwise effective repellents (i.e., blackbirds need to eat to survive). That notion can be expanded to include mechanical scare devices, pyrotechnics, and bird-resistant crops (Dolbeer et al. 1984; Linz et al. 2011). The keys to successful use of WCFP are location close to roosts, protection from predators, and caloric content (Linz et al. 2008). Here, we present a case history on the use of WCFP to reduce blackbird damage to sunflower and advocate the continued development of perennial sunflower as a potential cost-effective approach to developing a long-term management scheme to reduce damage to grain crops (Glover et al. 2010; Kantar et al. 2012, 2014; Linz et al. 2014; Linz and Hanzel 2015). This approach could attract the support of private conservation groups, state and federal resource agencies, and agriculturalists.

In the early 1980s, Cummings et al. (1987) offered alternative feeding locations to reduce blackbird damage to ripening sunflower. Cooperating growers planted nine 10-ha oilseed sunflower plots and one 14-ha field planted with both corn and sunflower near commercial fields (Cummings et al. 1987). Blackbirds used the lure fields heavily, and the economic analysis indicated that commercial fields had attained an average positive cost–benefit ratio of 1.0:4.0 (i.e., one unit of cost provided four units of benefit). Although the results were promising, no government entities were willing to formally implement a WCFP program.

Hagy et al. (2007, 2008, 2010) revisited the use of WCFP as a bird management tool in 2004 and 2005. Scientists offered candidate sunflower producers US$375.00/ha to plant 35 8-ha WCFP near cattail-dominated wetlands with histories of elevated blackbird damage (Figure 10.2). Blackbird damage in the WCFP plots was highly variable, ranging from 0% to 100%. Across both years of the study, WCFP produced an average of 1,290 kg/ha and birds removed 435 kg/ha, valued at US$160.95/ha (US$0.37/kg). Hagy et al. (2007, 2008, 2010) assumed, as did Cummings et al. (1987), that birds feeding in the fields would have caused the same amount of damage to commercial sunflower fields. In comparison to the research by Cummings et al. (1987), Hagy et al. (2008) concluded that the cost–benefit ratio was 3.4:1.0, indicating a negative economic return. The cost–benefit ratio did not include the intrinsic values of WCFP, such as use of the plots by 34 nonblackbird species (Hagy et al. 2010).

Given the expense of annually planting WCFP plots, Hagy et al. (2008) concluded that WCFP are best used to protect high-value oil and confectionery sunflower varieties planted either near roosts or under flight lines of blackbirds emanating from roosts. Planting of oilseed sunflower WCFP near commercial confectionery sunflower, the latter being much more valuable, could offset field planting costs if blackbird damage in the WCFP was ≥12%, a level of damage found in 74% of the WCFP (Hagy et al. 2008). Additionally, planting less valuable crops (e.g., corn, perennial sunflower, millet) near wetlands also might serve as alternative feeding sites and lower sunflower damage. This strategy might be especially effective if sunflower growers are actively dispersing (e.g., via pyrotechnics, shotgun shells, airplanes) the birds from their crop (Avery 2003).

Figure 10.2 Wildlife conservation food plots provides alternative food for blackbirds and other wildlife. (Courtesy of Heath Hagy, North Dakota State University.)

However, the behavior of blackbirds foraging under stressful conditions associated with harassment has not been assessed.

Since Hagy et al. (2008, 2010) conducted their study, collaborating scientists at the University of Minnesota and USDA Agricultural Research Service (USDA-ARS) have advanced the development of a perennial sunflower that could reduce planting costs for WCFP and serve as a potential tool to alleviate blackbird damage in commercial (i.e., annual) sunflower (Linz et al. 2011; Kantar et al. 2012, 2014). Perennial sunflower fits the notion of WCFP because this cultivar would provide a pesticide-free food source for beneficial insects, such as honeybees (*Apis mellifera*), help stabilize highly erodible lands with low production potential near wetlands, and offer year-round habitat for wildlife (Cox et al. 2010; Glover et al. 2010; U.S. Department of Agriculture 2015). Initial plantings on a working farm in North Dakota showed that the perennial habit was retained through two years, but additional development is needed to improve agronomic qualities such as oil quality and head and achene size (Linz et al. 2014; Figure 10.3). This research is ongoing with further field testing likely as the hybrid is improved (R. Stupar, personal communication).

Sunflowers growers often contend that, in addition to planting costs, WCFP take valuable agricultural land out of production. On the other hand, the grower might suffer high damage if steps are not taken to ameliorate damage—a catch-22. We suggest that growers can harvest the WCFP and recover planting costs if blackbirds do not use the plot. In some situations planting WCFP on federal and state wildlife lands is a viable alternative to planting on private lands. Landowners also can plant small food plots on a portion of Conservation Reserve Program lands to benefit all wildlife, including blackbirds dispersed from nearby fields (U.S. Department of Agriculture 2013).

Current WCFP planting recommendations differ little from previous research (Cummings et al. 1987; Hagy et al. 2008), including planting plots near cattail-dominated wetlands, which are favored as night roosts, and near, but not immediately adjacent to, commercial fields (Figure 10.4; Linz et al. 2008). Planting varieties that ripen earlier than nearby commercial fields may habituate birds to the plots, thereby further increasing the efficacy of WCFP. The use of other less valuable but desirable crops to further buffer sunflower should also be explored. Finally, additional research is needed on the best planting practices, including economic evaluation, selection of plot locations, planting times, field size, and variety selection (Cummings et al. 1987; Hagy et al. 2008, Ewen et al. 2015).

Figure 10.3 Perennial sunflower developed by University of Minnesota scientists planted in North Dakota. Perennial grain crops have the potential of providing cost-effective alternative food sources for wildlife. (Courtesy of USDA Wildlife Services.)

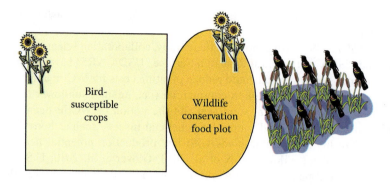

Figure 10.4 Wildlife conservation food plots preferred by blackbirds (e.g., sunflower, corn, millet, rice) should be planted between roost sites and bird-susceptible crops to provide alternative food sources. Areal coverage of plot is dependent on anticipated seed production and number of blackbirds. (Linz et al. 2008.)

10.4 MANAGEMENT OF WETLAND ROOST SITES

In late summer, after the nesting season, blackbirds begin roosting in wetlands dominated by cattails (*Typha* × *glauca*) and phragmites (*Phragmites australis*) when available because they offer protection from predators and inclement weather (Meanley 1965; Yasukawa and Searcy 1995; Linz and Homan 2011). Although native common cattail (*Typha latifolia*) can be found in the United States, almost all of the cattails in the Prairie Pothole Region are a hybrid between the invasive exotic narrow-leaved cattail (*Typha angustifolia* L.) and common cattail (Kantrud 1990). Hybrid cattails are a fast-growing and robust cattail that forms dense homogenous stands that tolerate seasonal water draw-downs and inundation (Weller 1975; van der Valk and Davis 1978). Likewise, the native *Phragmites* subspecies (*Phragmites australis americanus*) can be found in isolated locations in North America, but again almost all *Phragmites* is a nonnative subspecies (*Phragmites australis australis*) originating in Europe (Tulbure et al. 2007). The aggressive habits of both the cattail hybrid and the nonnative *Phragmites* allow them to outcompete and displace native plants important to animals dependent on wetlands for survival and reproduction.

Otis and Kilburn (1988) found that the main predictor of the severity of blackbird damage to sunflower is the presence or absence of nearby wetlands, with fields located near wetlands receiving two to four times more damage. Choosing the closest available high-quality food source (e.g., rice, sunflower, corn) fits the optimal foraging theory, suggesting that birds seek food to enhance their probability of surviving and reproducing (Pyke et al. 1977). Commercial sunflower provides blackbirds with the energy to replace feathers during their annual molt and helps them accumulate energy reserves for migration (Linz et al. 1983).

Growers planting within the foraging range (≤8 km) of wetland blackbird roosts absorb most of the losses (Dolbeer 1990; Klosterman et al. 2013). In southern winter roosts, Meanley (1965) reported that blackbirds readily use cattails and phragmites, with some roosts harboring a million individuals. Many of these roosts were located near rice-growing areas, providing access to planted rice in the spring and ripening rice in late summer.

Linz et al. (1992a, 1992b) proposed that breaking the link between wetland roosts and ripening crops was a reasonable approach for reducing damage. Attempts at controlling cattail and phragmites in wetlands with mechanical methods such as grazing, mowing, burning, and disking produces poor results because both of these species have a large rhizome root system that allows the plant to regenerate quickly (Tu et al. 2001). Additionally, these methods are nearly

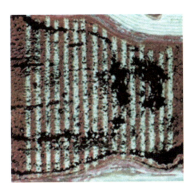

Figure 10.5 Color infrared photo showing strips of live cattail (red) and open water (white) created by an aerial application of glyphosate herbicide to reduce blackbird roost substrate and improve habitat for waterfowl. (Courtesy of USDA Wildlife Services.)

useless where soils are water-saturated or there is standing water. To overcome these drawbacks, scientists from the USDA-APHIS-WS National Wildlife Research Center (NWRC) and North Dakota State University (NDSU) proposed the use of glyphosate, a systemic herbicide, to fragment cattail-dominated wetlands and reduce blackbird roosting habitat (Linz et al. 1992a; Linz and Homan 2011; Figure 10.5).

This idea followed research by Solberg and Higgins (1993) of South Dakota State University (SDSU) showing that glyphosate could be used to manage cattails to enhance waterfowl production. Over the next decade, the NWRC, NDSU, SDSU, and USFWS cooperated on a multifaceted series of studies to assess the efficacy, cost-benefits, and environmental effects of using an aquatic herbicide to reduce blackbird roosting habitat by fragmenting cattail-dominated wetlands (Linz et al. 1992b; Henry et al. 1994; Leitch et al. 1997; Linz and Homan 2011). Here, we briefly summarize key findings of these studies.

10.4.1 Glyphosate Herbicide

Glyphosate (*N*-(phosphonomethyl)glycine) is a systemic, broad-spectrum, post-emergence herbicide that was discovered by John E. Franz in the early 1970s, while working for Monsanto (St. Louis, MO). Today, numerous companies formulate and distribute glyphosate herbicides approved for use in aquatic environments. Glyphosate blocks the shikimic acid pathway, which is essential for protein synthesis, and as a result kills the plant (Cole 1985; Linz and Homan 2011). Applications are most effective in late summer when cattails are actively metabolizing and transporting carbohydrates to their rhizomes. Glyphosate is rapidly adsorbed by soil particles and sediment (Bronstad and Friestad 1985).

10.4.2 Ecological Effects

Glyphosate, formulated for aquatic use and applied at labelled rates, does not adversely affect aquatic invertebrates, which are a critical part of waterfowl diets during the reproductive season (Henry et al. 1994). We caution that glyphosate formulations containing polyoxyethylene tallow amine surfactant are not registered for use in wetlands because of toxic effects on aquatic organisms (Henry et al. 1994; Relyea 2005). Glyphosate applied to dense cattail stands results in massive amounts of decaying vegetation that could result in low dissolved oxygen (DO) levels, decreasing invertebrate populations and affecting invertebrate survival. However, Linz et al. (1999) showed that DO levels were similar between glyphosate-treated and reference wetlands, thus corroborating with the conclusion by Cole (1985) that wind-driven waves and spray in open areas of wetlands increase the absorptive surface at the air–water interface, offsetting any reduction in DO from decomposition.

Linz et al. (1996a, 1996b) also investigated the effects of eliminating cattails (and presumably *Phragmites*) on the density of birds requiring emergent vegetation as nest substrate and cover from predators. They found that herbicide applications in cattail marshes resulted in fewer nesting marsh wrens (*Cistothorus palustris*), red-winged blackbirds, and yellow-headed blackbirds (*Xanthocephalus xanthocephalus*) but more waterfowl, agreeing with the conclusion of Solberg and Higgins (1993) that creating openings in wetlands was beneficial for waterfowl.

10.4.3 Economics

Linz and Homan (2011) reported treatment costs, including glyphosate, surfactant, and helicopter application, of about US$95/ha in North Dakota. The use of helicopters essentially eliminates complaints about chemical drift onto shoreline vegetation. Linz and Homan (2011) recommended that cattails be treated with an aqueous solution containing 2.2 kg/ha glyphosate and 1% v/v surfactant. For our discussion on the economics of glyphosate applications, we assume that current application costs are similar across the United States for both cattails and phragmites. Growers are encouraged to consult with state and federal wildlife and agriculture officials to obtain information on wetland regulations prior to engaging in a wetland management program.

Here, we present an example for sunflower (US$0.55 kg), which is a high-value crop compared to corn at US$0.21/kg and rice at US$0.32/kg. We assume that daily sunflower consumption by one blackbird is 0.009 kg/day and each bird will damage 0.27 kg over a 30-day damage period (Peer et al. 2003). With sunflower's 5-year (2011–2015) market price valued at US$0.55 kg (National Agricultural Statistics Service 2016), a single blackbird (combining sexes and species) damages about US$0.15 of sunflower/year. Thus, growers must anticipate an average of 633 blackbirds/ha (US$95/ha) of cattail to justify treatment costs. Regrowth of cattail following treatment is contingent on water levels. If water depths remain stable at >30 cm, there should be few living cattails for at least 4 years and perhaps up to 6 years post-treatment (Linz and Homan 2011). A treatment that is effective for at least 4 years requires only 158 blackbirds/day/ha of cattail to justify costs, provided that sunflower is planted every year on lands somewhere near the treated wetland. Wetlands and grain crops planted in juxtaposition are scattered throughout the United States, with roosts containing a few hundred to several million blackbirds (Meanley 1965; Dolbeer 1990; Linz et al. 2003).

10.4.4 Operational Program

From 1991 to 2010, the USDA-APHIS-WS conducted a cattail management demonstration program in North Dakota and South Dakota. During that time, the USDA-APHIS-WS annually sprayed <1% (1,500 ha) of cattail-dominated wetlands in the Dakotas using aerial applications of glyphosate herbicide (Ralston et al. 2007; Linz and Homan 2011). WS sprayed about 70% of the emergent vegetation (largely cattails), which was sufficient to minimize or reduce roosting blackbirds but provide cover and nesting substrate for other birds. This limited spray coverage, combined with the findings of numerous field studies on ecological and environmental effects, led Linz and Homan (2011) to conclude that glyphosate has minimal impact on wetland fauna. Indeed, numerous wetland species benefited from the treatment, including waterfowl, an economically important species in North Dakota. Although statistical evidence is lacking, managing cattails appears to help sunflower producers disperse blackbirds and thus reduce the severity of damage sustained in fields located near cattail-dominated wetlands (Linz et al. 1995).

The USDA-APHIS-WS cattail management program appears to meet the requirements of wildlife interests and agriculture (McEnroe 1992; Stromstad 1992). Fragmenting dense cattail stands returns wetlands to their original configuration, which promotes avian diversity while preventing the formation of large roosting aggregations of blackbirds. The federally funded USDA-APHIS-WS cattail management program ended in 2010. However, individual growers can use the techniques developed over 25 years of research and operational experience.

10.5 MANAGEMENT OF UPLAND ROOST SITES

In areas of the United States where wetlands dominated by emergent vegetation are relatively uncommon, blackbirds establish large roosts that can number in the millions in dense tree stands in urban and rural environments (Mott 1984; Glahn et al. 1994). The U.S. Fish and Wildlife Service conducted national winter roost surveys in 1974–1975 and 1976–1977 and found 825 roosts, with 54% harboring over 1 million birds each. These large roosts were commonly found in conifers (33%), hardwoods (23%), wetlands (12%), cane (12%), and 21% in other habitats (Glahn et al. 1994). The most important attributes of a roost site are high tree densities and dense canopy that afford the birds protection from predators and adverse weather (Meanley 1965, 1971; Lyon and Caccamise 1981).

In these circumstances, the public sometimes seeks help from wildlife agencies because of agricultural damage (e.g., rice), health risks (e.g., histoplasmosis), and general aesthetic problems (e.g., fecal matter on sidewalks and backyards; Heisterberg et al. 1987) caused by large roosts. Options for resolving conflicts with blackbirds include no action, moving the roost by modifying habitat, dispersing birds with mechanical repellents (e.g., pyrotechnics, distress calls, shooting, lasers, and propane cannons), and population reduction (Garner 1978; Booth 1994; Conover 2002).

In urban environments, thinning tree canopies might be an effective first step in moving roosting populations to more desirable rural locations. Loud sound resulting from pyrotechnics (e.g., shell crackers) and firing shotgun shells can be effective if used when the birds first start arriving at the roost in the late afternoon and continuing until dark. A persistent effort occurring over a week or more might be needed to force the birds to move from a well-established roost.

In agricultural areas, local population management might be achieved by broadcasting avicide-laced baits (Linz et al. 2015). Glahn and Wilson (1992) broadcast rice baits treated with DRC-1339 avicide (a.i., 3-chloro-p-toluidine hydrochloride, also 3-chloro-4-methylbenzenamine hydrochloride) near a large spring blackbird roost in Louisiana and killed an estimated 4 million blackbirds. A subsequent survey of rice producers estimated that damage to sprouting rice was reduced 83% compared to previous years. This suggested that lethal control of local blackbird populations might reduce local crop damage, but Glahn and Wilson (1992) concluded that such toxic baiting programs should only be used after other methods have failed. Finally, Linz et al. (2015) conducted a comprehensive review of attempts to reduce crop damage using various population management strategies and found no situation in North America or South America where lethal control, as a stand-alone tactic, met the criteria of practicality, environmental safety, cost-effectiveness, and wildlife stewardship.

10.6 BIRD-RESISTANT CROPS

The search for bird-resistant crops peaked in the 1980s, when a concerted effort was made to discover and develop bird-resistant corn and sunflower varieties (Dolbeer et al. 1982, 1984, 1986a, 1986b; Seiler and Rogers 1987; Mah et al. 1990; Mah and Nuechterlein 1991). The efficacy of bird-resistant crops and many other bird management techniques (e.g., mechanical and chemical repellents) is largely contingent on the availability of quality alternative food sources, a practice encouraged by Dolbeer et al. (1984) and Avery (2003). Here, we present research aimed at the discovery of bird-resistant corn and sunflower.

10.6.1 Corn

It takes about 60 days for corn kernels to become physiologically mature after pollination (Nielsen 2013). Blackbirds can severely damage corn during the milk (R3) and dough (R4) developmental stages, a period of 3–4 weeks when kernel moisture content is about 80% and 70%, respectively (Dolbeer 1990). After that time, the corn kernel hardens, reducing its attractiveness to blackbirds, particularly female red-winged blackbirds and yellow-headed blackbirds (Yasukawa and Searcy 1995). Male blackbirds can damage field corn for several more weeks as the kernels mature and dry down to 25% at harvest maturity (Linz et al. 1983; Nielsen 2013). Further, birds open the protective husks that allows mold, fungus, and insects to inflict additional damage and cause economic losses (Nielsen 2009).

Linehan (1967) showed that corn hybrids vary widely in their susceptibility to bird damage. Dolbeer et al. (1982, 1986a) quantified the importance of numerous corn traits in aviary tests and concluded that long and heavy husks were the most difficult for birds to access the kernels. In field tests, however, these characteristics were not important predictors of bird damage, leading Dolbeer et al. (1984) to conclude that the yield and timing of maturity were the most important factors influencing damage levels.

Dolbeer et al. (1986b) reasoned that there is little incentive for corn seed companies to develop new lines solely for bird resistance because the number of fields with economically important bird damage was small. Their fallback position was to suggest that a bird-resistance rating system should be developed for commercially marketed seed corn hybrids. Farmers in high damage areas would have the option of considering bird resistance along with other important agronomic characteristics such as yield, maturation time, stem durability, and dry-down in selecting a hybrid. With profit margins historically narrow, present-day corn growers are unlikely to give up yield in exchange for bird resistance unless anticipated bird damage is overwhelming (>15%), in which case planting a less susceptible crop such as soybean near known roost sites is another alternative.

10.6.2 Sunflower

In 1979, sunflower breeders from NDSU, with technical assistance from the USDA-ARS and private industry, were funded by the Denver Wildlife Research Center (now the National Wildlife Research Center) to assess various chemical and morphological traits that might thwart blackbird feeding on sunflower achenes (Guarino 1984). The goal was to develop a bird-resistant sunflower while maintaining palatability, yield, and oil content. Scientists surmised the features needed to inhibit perch-feeding and seed access included a flat or concave head shape, tightly held achenes, thick fibrous hulls, hulls with high levels of anthocyanins, long chaffs, long wrap around bracts, a head-to-stem distance of more than 15 cm, and ground-facing flowers (Parfitt 1984; Seiler and Rogers 1987; Gross and Hanzel 1991). Additionally, the percentage of oil, which is correlated with hull thickness, was thought to be a key reason for birds selecting particular varieties of sunflower (Mason et al. 1991). Mason et al. (1991) conducted a series of experiments to determine whether red-winged blackbirds were in fact capable of discerning the difference in oil content among sunflower varieties. Indeed, they found that low oil content associated with heavy hulled achenes reduced red-winged blackbird feeding when given a choice of achenes with higher oil content. Thus, the birds rejected some varieties of sunflower over others not because of bird-resistant features but because of the oil content. Resistant genotypes developed at NDSU were similar to confectionery varieties that feature heavy hulls with large seeds and low oil content. Compared to oil seed varieties, confectionery varieties are difficult for small birds (e.g., female yellow-headed blackbirds and red-winged blackbirds) to extract and dehull, especially after the achenes hardens at physiological maturity (Linz et al. 1983; Twedt et al. 1991). Thus, the oil content must be a controlled variable to test for true bird resistance.

In the early 1990s, the bird-resistant sunflower-breeding program was abandoned because of the prohibitive technical challenges involved in developing a commercially competitive hybrid that would have the combination of bird-resistant traits and high oil content and yield. Since then, the development of doubled-haploid technology might revive the quest for bird-resistant varieties by allowing plant breeders to rapidly develop and evaluate new cultivars for specific plant traits (Maluszynski 2003; Jan et al. 2011; Lilliboe 2011). Conventional plant inbreeding procedures might take multiple generations to evaluate a particular plant trait, whereas doubled-haploid technology achieves the same aim in one generation at presumably reduced costs. Doubled-haploid technology is used to develop plant characteristics, such as disease resistance, drought tolerance, and yield, in many crops (e.g., corn, small grains, fruits, and vegetables; Maluszynski 2003). Future commercial sunflower and corn varieties might also be developed that have some bird-resistant qualities while maintaining oil content (sunflower) and yield. In the meantime, we suggest that the new double-haploid technology could help with the development of a perennial sunflower and improvement of the agronomics of SS sunflower. Both of these products will likely meet better success for managing bird damage to sunflower.

10.7 SUMMARY

Blackbirds are highly mobile and undergo local movements in search of suitable roosting habitat and food sources. Blackbirds also can delay migration in northern areas if weather conditions are benign and ample food is available (T. Turner, National Sunflower Association of Canada, personal communication). Whereas simply planting crops not susceptible to blackbird damage (e.g., soybeans, flax) is an obvious bird management strategy, it may not be the best economic decision when damage levels are relatively low (<5%). Advancing crop harvest by artificially drying the crop ahead of normal is another plausible tactic. Growers might consider managing cattail-dominated wetlands and thinning tree canopies to reduce their attractiveness to roosting blackbirds. Another option is to plant WCFP near favored roost sites to serve as a buffer around an economically valuable crop. Finally, researchers need to investigate blackbird feeding behavior when presented with SS and perennial sunflower as well as response to hazing devices, especially unmanned aircraft systems (Ampatzidis et al. 2015). Regardless, using these evading strategies in combination with other bird management tactics could result in an effective integrated pest management plan.

REFERENCES

Ampatzidis, Y., J. Ward, and O. Samara. 2015. Autonomous system for pest bird control in specialty crops using unmanned aerial vehicles. *American Society of Agricultural and Biological Engineers (ASABE) 2015 Annual International Meeting*, Paper No. 152181748. http://elibrary.asabe.org/azdez.asp?JID=5&AID=45793&CID=norl2015&v=&i=&T=1&refer=7&access=&dabs=Y (accessed March 19, 2016).

Avery, M. L. 2003. Avian repellents. In *Encyclopedia of Agrochemicals, Vol. 1*, ed. J. R. Plimmer, pp. 122–128, Wiley, New York.

Booth, T. W. 1994. Bird dispersal techniques dispersal techniques. In *Prevention and Control of Wildlife Damage*, eds. S. E. Hygnstrom, R. M. Timm, and G. E. Larson, pp. E19–E24. University of Nebraska Cooperative Extension Service, Lincoln, NE.

Bronstad, H. O., and J. D. Friestad. 1985. Behavior of glyphosate in the aquatic environment. In *The Herbicide Glyphosate*, eds. E. Grossbard and D. Atkinson, pp. 200–205. Butterworth, London, UK.

Cole, D. J. 1985. Mode of action of glyphosate – A literature analysis. In *The Herbicide Glyphosate*, eds. E. Grossbard and D. Atkinson, pp. 48–74. Butterworth, London, UK.

Conover, M. R. 2002. *Resolving Human-Wildlife Conflicts: The Science of Wildlife Damage Management*. CRC Press, Boca Raton, FL.

Cox, T. S., D. L. Van Tassel, C. M. Cox, and L. R. DeHaan. 2010. Progress in breeding perennial grains. *Crop Pasture Science* 61:513–521.

Cummings, J. L., J. L. Guarino, C. E. Knittle, and W. C. Royal, Jr. 1987. Decoy plantings for reducing blackbird damage to nearby commercial sunflower fields. *Crop Protection* 6:56–60.

Dolbeer, R. A. 1981. Cost-benefit determination of blackbird damage control for cornfields. *Wildlife Society Bulletin* 9:44–51.

Dolbeer, R. A. 1990. Ornithology and integrated pest management: Red-winged blackbirds *Agelaius phoeniceus* and corn. *Ibis* 132:309–322.

Dolbeer, R. A., and G. M. Linz. 2016. *Blackbirds*. Wildlife Damage Management Technical Series, U.S. Department of Agriculture, Animal & Plant Health Inspection Service, Wildlife Services, Washington, DC.

Dolbeer, R. A., P. P. Woronecki, and R. A. Stehn. 1982. Effect of husk and ear characteristics on resistance of maize to blackbird (*Agelaius phoeniceus*) damage in Ohio, USA. *Crop Protection* 4:127–139.

Dolbeer, R. A., P. P. Woronecki, and R. A. Stehn. 1984. Blackbird (*Agelaius phoeniceus*) damage to maize: Crop phenology and hybrid resistance. *Protection Ecology* 7:43–63.

Dolbeer, R. A., P. P. Woronecki, and R. A. Stehn. 1986b. Blackbird-resistant hybrid corn reduces damage but does not increase yield. *Wildlife Society Bulletin* 14:298–301.

Dolbeer, R. A., P. P. Woronecki, R. A. Stehn, G. J. Fox, J. J. Hanzel, and G. M. Linz. 1986a. Field trials of sunflower resistant to bird depredation. *North Dakota Farm Research Journal* 43:21–24, 28.

Dow AgroSciences. 2016. *Sunflower Harvesting: Maturity is Critical for Desiccation and Harvest*. Agronomy Bulletin 50, Indianapolis, IN. http://agronomy.mycogen.com/bulletin/maturity-is-critical-for-desiccation-and-harvest (accessed November 12, 2015).

Ewen, J. G., L.Walker, S.Canessa, and J. J. Groombridge. 2015. Improving supplementary feeding in species conservation. *Conservation Biology* 29:341–349.

Galle, A. M., G. M. Linz, H. J. Homan, and W. J. Bleier. 2009. Avian use of harvested crop fields in North Dakota during spring migration. *Western North America Naturalist* 69:491–500.

Garner, K. M. 1978. Management of blackbird and starling winter roost problems in Kentucky and Tennessee. *Vertebrate Pest Conference* 8:54–59.

Glahn, J. F., R. D. Flynt, and E. P. Hill. 1994. Historical use of bamboo/cane as blackbird and starling roosting habitat: Implications for roost management. *Journal of Field Ornithology* 65:237–246.

Glahn, J. F., and E. A. Wilson. 1992. Effectiveness of DRC-1339 baiting for reducing blackbird damage to sprouting rice. *Eastern Wildlife Damage Control Conference* 5:117–123.

Glover, J. D., J. P. Reganold, L. W. Bell, et al. 2010. Increased food and ecosystem security via perennial grains. *Science* 328:1638–1639.

Gross, P. L., and J. J. Hanzel. 1991. Stability of morphological traits conferring bird resistance to sunflower across different environments. *Crop Science* 31:997–1000.

Guarino, J. L. 1984. Current status of research on the blackbird-sunflower problem in North Dakota. *Vertebrate Pest Conference* 11:211–216.

Gustad, O. C. 1979. New approaches to alleviating migratory bird damage. *Great Plains Wildlife Damage Control Workshop Proceedings* 4:166–175.

Hagy, H. M., G. M. Linz, and W. J. Bleier. 2007. Are sunflower fields for the birds? *Wildlife Damage Management Conference* 12:61–71.

Hagy, H. M., G. M. Linz, and W. J. Bleier. 2008. Optimizing the use of decoy plots for blackbird control in commercial sunflower. *Crop Protection* 27:1442–1447.

Hagy, H. M., G. M. Linz, and W. J. Bleier. 2010. Wildlife conservation sunflower plots and croplands as fall habitat for migratory birds. *American Midland Naturalist* 164:119–135.

Handegard, L. L. 1988. Using aircraft for controlling blackbird/sunflower depredations. *Vertebrate Pest Conference* 13:293–294.

Heisterberg, J. F., A. R. Stickley, Jr., K. M. Garner, and P. D. Foster, Jr. 1987. Controlling blackbirds and starlings at winter roosts using PA-14. *Eastern Wildlife Damage Control Conference* 3:350–356.

Henry, C. J., K. F. Higgins, and K. J. Buhl. 1994. Acute toxicity and hazard assessment of Rodeo®, X-77 Spreader®, and Chem-Trol® to aquatic invertebrates. *Archives of Environmental Contamination and Toxicology* 27:392–399.

Jan, C. C., L. Qi, B. Hulke, and X. Fu. 2011. Present and future plans of the sunflower "doubled haploid" project. *Sunflower Magazine*, Mandan, ND. www.sunflowernsa.com/uploads/resources/561/jan_present.futureplansdoubledhaploid.pdf (accessed February 5, 2016).

Kantar, M., K. Betts, B. S. Hulke, R. M. Stupar, and D. Wyse. 2012. Breaking tuber dormancy in *Helianthus tuberosus* L. and interspecific hybrids of *Helianthus annuus* L. x *Helianthus tuberosus*. *Horticultural Science* 47:1342–1346.

Kantar, M. B., K. Betts, J. Michno, et al. 2014. Evaluating an interspecific *Helianthus annuus* x *Helianthus tuberosus* population for use in a perennial sunflower breeding program. *Field Crops Research* 155:254–264.

Kantrud, H. A. 1990. *Effects of Vegetation Manipulation on Breeding Waterfowl in Prairie Wetlands—A Literature Review*. U.S. Fish and Wildlife Technical Report 3, U.S. Department of Interior, Fish and Wildlife Service, Washington, DC.

Klosterman, M. E., G. M. Linz, A. A. Slowik, and H. J. Homan. 2013. Comparisons between blackbird damage to corn and sunflower in North Dakota. *Crop Protection* 53:1–5.

Knittle, C. E., and R. D. Porter. 1988. *Waterfowl Damage and Control Methods in Ripening Grain: An Overview*. U. S. Fish and Wildlife Service Technical Report 14, U.S. Department of Interior, Fish and Wildlife Service, Washington, DC.

Kubasiewicz, L. M., N. Bunnefeld, A. I. T. Tulloch, C. P. Quine, and K. J. Park. 2016. Diversionary feeding: An effective management strategy for conservation conflict? *Biodiversity and Conservation* 25:1–22.

Leitch, J. A., G. M. Linz, and J. F. Baltezore. 1997. Economics of cattail (*Typha* spp.) control to reduce blackbird damage to sunflower. *Agriculture, Ecosystems & Environment* 65:141–149.

Lilliboe, D. 2011. Doubled Haploids. *Sunflower Magazine*, Mandan, ND. http://www.sunflowernsa.com/magazine/articles/default.aspx?articleID=3368 (accessed June 16, 2016).

Linehan, J. T. 1967. Measuring bird damage to corn. *Vertebrate Pest Conference* 3:50–53.

Linz, G. M., D. L. Bergman, and W. J. Bleier. 1992a. Evaluating Rodeo herbicide for managing cattail-choked marshes: Objectives and methods. In *Proceedings of a Cattail Management Symposium*, ed. G. M. Linz, pp. 21–27. U.S Department of Agriculture and U.S. Department of Interior, Fargo, ND. https://www.aphis.usda.gov/wildlife_damage/nwrc/publications/92pubs/92-72.pdf (accessed June 28, 2016).

Linz, G. M., D. L. Bergman, and W. J. Bleier. 1992b. Progress on managing cattail marshes with Rodeo® herbicide to disperse roosting blackbirds. *Vertebrate Pest Conference* 15:56–61.

Linz, G. M., D. L. Bergman, H. J. Homan, and W. J. Bleier. 1995. Effects of herbicide-induced habitat alterations on blackbird damage to sunflower. *Crop Protection* 14:625–629.

Linz, G. M., W. J. Bleier, J. D. Overland, and H. J. Homan. 1999. Response of invertebrates to glyphosate-induced habitat alterations in wetlands. *Wetlands* 19:220–227.

Linz, G. M., S. B. Bolin, and J. F. Cassel. 1983. Postnuptial and postjuvenal molts of red-winged blackbirds in Cass County, North Dakota. *Auk* 100:206–209.

Linz, G. M., D. C. Blixt, D. L. Bergman, and W. J. Bleier. 1996a. Responses of red-winged blackbirds, yellow-headed blackbirds and marsh wrens to glyphosate-induced alterations in cattail density. *Journal of Field Ornithology* 67:167–176.

Linz, G. M., D. C. Blixt, D. L. Bergman, and W. J. Bleier. 1996b. Response of ducks to glyphosate-induced habitat alterations in wetlands. *Wetlands* 16:38–44.

Linz, G. M., E. H. Bucher, S. B. Canavelli, E. Rodriguez, and M. L. Avery. 2015. Limitations of population suppression for protecting crops from bird depredation: A review. *Crop Protection* 76:46–52.

Linz, G. M., and J. J. Hanzel. 2015. Sunflower and bird pests. In *Sunflower: Chemistry, Production, Processing, and Utilization*, eds. E. M. Force, N. T. Dunford, and J. J. Salas, pp. 175–186. AOCS Press, Urbana, IL.

Linz, G. M., and H. J. Homan. 2011. Use of glyphosate for managing invasive cattail (*Typha* spp.) to protect crops near blackbird (Icteridae) roosts. *Crop Protection* 30:98–104.

Linz, G. M., H. J. Homan, H. M. Hagy, J. M. Raetzman, L. B. Penry, and W. J. Bleier. 2008. *A Grower's Guide for Planting Wildlife Conservation Food Plots*. National Sunflower Association, Mandan, ND. https://www.sunflowernsa.com/Research/searchable-database-of-forum-papers/ (accessed October 23, 2016)

Linz, G. M., H. J. Homan, S. J. Werner, H. M. Hagy, and W. J. Bleier. 2011. Assessment of blackbird management strategies to protect sunflower. *BioScience* 61:960–970.

Linz, G., B. S. Hulke, M. B. Kantar, J. Homan, R. M. Stupar, and D. L. Wyse. 2014. Potential use of perennial sunflower to reduce blackbird damage to sunflower. *Vertebrate Pest Conference* 26:356–359.

Linz, G. M., R. A. Sawin, M. W. Lutman, H. J. Homan, L. B. Penry, and W. J. Bleier. 2003. Characteristics of spring and fall blackbird roosts in the northern Great Plains. *Wildlife Damage Management Conference* 10:220–228.

Linz, G. M., D. A. Schaaf, P. Mastrangelo, H. J. Homan, L. B. Penry, and W. J. Bleier. 2004. Wildlife conservation sunflower plots as a dual-purpose wildlife management strategy. *Vertebrate Pest Conference* 21:294–294.

Lyon, L. A., and D. F. Caccamise. 1981. Habitat selection by roosting blackbirds and starlings: Management implications. *Journal of Wildlife Management* 45:435–443.

Mah, J., G. M. Linz, and J. J Hanzel. 1990. Relative effectiveness of individual sunflower traits for reducing red-winged blackbird depredation. *Crop Protection* 9:359–362.

Mah, J., and G. L. Nuechterlein. 1991. Feeding behavior of red-winged blackbirds on bird-resistant sunflowers. *Wildlife Society Bulletin* 19:39–46.

Maluszynski, M., K. J. Kasha, B. P. Forster, and I. Szarejko. 2003. *Doubled haploid production in crop plants: A manual.* Kluwer Academic, Dordrecht, The Netherlands.

Mason, J. R., G. Nuechterlein, G. Linz, R. A. Dolbeer, and D. L. Otis. 1991. Oil concentration differences among sunflower achenes and feeding preferences of red-winged blackbirds. *Crop Protection* 10:299–304.

McEnroe, M. 1992. Cattail management: Views of the U.S. Fish and Wildlife Service. In *Proceedings of a Cattail Management Symposium*, ed. G. M. Linz, pp. 42–44. U.S Department of Agriculture and U.S. Department of Interior, Fargo, ND. https://www.aphis.usda.gov/wildlife_damage/nwrc/publications/92pubs/92-72.pdf (accessed August 2, 2016).

Meanley, B. 1965. The roosting behavior of the red-winged blackbird in the southern United States. *Wilson Bulletin* 77:217–228.

Meanley, B. 1971. *Blackbirds and the Southern Rice Crop.* U.S. Department of Interior, U.S. Fish and Wildlife Service, Resource Publication 100. U.S. Government Print Office, Washington, DC. http://pubs.usgs.gov/rp/100/report.pdf (accessed February 3, 2016).

Mott, D. F. 1984. Research on winter roosting blackbirds and starlings in the southeastern United States. *Vertebrate Pest Conference* 11:183–187.

Mullally, S. 2013. Options growing with short-stature hybrids. *Sunflower Magazine*, Mandan, ND. http://www.sunflowernsa.com/magazine/articles/default.aspx?articleID=3509 (Accessed September 25, 2016).

Nielsen, (B.) R. L. 2009. Corn ear damage caused by bird feeding. *Corny News Network*. Cooperative Extension Service, Purdue University, West Lafayette, IN. https://extension.entm.purdue.edu/pestcrop/2009/issue24 (accessed February 5, 2016).

Nielsen, R. L. 2013. Grain fill stages in corn. *Corny News Network.* Agronomy Department, Purdue University, West Lafayette, IN. https://www.agry.purdue.edu/ext/corn/news/timeless/grainfill.html (accessed March 12, 2016).

Otis, D. L., and C. M. Kilburn. 1988. *Influence of Environmental Factors on Blackbird Damage to Sunflower.* U.S. Fish and Wildlife Service Technical Report 16. Washington, DC.

Parfitt, D. E. 1984. Relationship of morphological plant characteristics of sunflower to bird feeding. *Canadian Journal of Plant Science* 64:37–42.

Peer, B. D., H. J. Homan, G. M. Linz, and W. J. Bleier. 2003. Impact of blackbird damage to sunflower: Bioenergetic and economic models. *Ecological Applications* 13:248–256.

Putman, D. H., E. S. Oplinger, D. R. Hicks, et al. 1990. *Sunflower. Alternative Field Crops Manual*, University of Wisconsin, Madison, WI. http://corn.agronomy.wisc.edu/Crops/Sunflower.aspx (accessed June 23, 2016).

Pyke, G. H., H. R. Pulliam, and E. L. Charnov. 1977. Optimal foraging: A selective review of theory and tests. *The Quarterly Review of Biology* 52:137–154.

Ralston, S. T., G. M. Linz., W. J. Bleier, and H. J. Homan. 2007. Cattail distribution and abundance in North Dakota. *Journal of Aquatic Plant Management* 45:21–24.

Relyea, R. A. 2005. The lethal impact of roundup on aquatic and terrestrial amphibians. *Ecological Applications* 15:1118–1124.

Sedgwick, J. A., J. L. Oldemeyer, and E. L. Swenson. 1986. Shrinkage and growth compensation in common sunflowers: Refining estimates of damage. *Journal of Wildlife Management* 50:513–520.

Seiler, G. J., and C. E. Rogers. 1987. Influence of sunflower morphological characteristics on achene depredation by birds. *Agricultural, Ecosystems, and Environment* 20:59–70.

Solberg, K. L., and K. F. Higgins. 1993. Effects of glyphosate herbicide on cattails, invertebrates, and waterfowl in South Dakota wetlands. *Wildlife Society Bulletin* 21:299–307.

Stromstad, R. 1992. Cattail management: The North Dakota Game and Fish department perspective. In *Proceedings of a Cattail Management Symposium*, ed. G. M. Linz, pp. 38–41. U.S Department of Agriculture and U.S. Department of Interior, Fargo, ND. https://www.aphis.usda.gov/wildlife_damage/nwrc/publications/92pubs/92-72.pdf (accessed June 28, 2016).

Tranel, M. A., W. Bailey, and K. Haroldson. 2008. Functions of food plots for wildlife management on Minnesota's wildlife management areas. In *Summaries of wildlife research findings 2007*, eds. M. W. DonCarlos, R. O. Kimmel J. S. Lawrence, M. S. Lenarz, pp. 70–79. Minnesota Department of Natural Resources, St. Paul, MN. https://www.leg.state.mn.us/docs/2008/mandated/080748.pdf (accessed June 28, 2016).

Tu, M., C. Hurd, and J. M. Randall. 2001. *Weed control methods handbook: Tools and techniques for use in natural areas*. The Nature Conservancy, Arlington, VA. http://www.invasive.org/gist/products/handbook/methods-handbook.pdf (accessed February 5, 2016).

Tulbure, M. G., C. A. Johnston, and D. L. Auger. 2007. Rapid invasive of Great Lakes coastal wetland by nonnative *Phragmites australis* and *Typha*. *Journal of Great Lakes Research* 33:269–279.

Twedt, D. J., W. J. Bleier, and G. M. Linz. 1991. Geographic and temporal variation in the diet of yellow-headed blackbirds. *Condor* 93:975–986.

U. S. Department of Agriculture. 2003. WS Directive 2.460 06/11/03 *LURE CROPS*, Washington, DC. https://www.aphis.usda.gov/wildlife_damage/directives/2.460_lure_crops.pdf (accessed June 23, 2016).

U. S. Department of Agriculture. 2013. *Fact Sheet: Conservation Reserve Program Sign-Up 45 Environmental Benefits Index (EBI)*, 7 pp. www.fsa.usda.gov/Internet/FSA_File/su45ebifactsheet.pdf (accessed February 5, 2016).

U. S. Department of Agriculture. 2015. *Colony Collapse Disorder and Honey Bee Health Action Plan*. CCD and Honey Bee Health Steering Committee, Washington, DC. http://www.ree.usda.gov/ree/news/CCD-HBH_Action_Plan_05-19-2015-Dated-FINAL.pdf (accessed August 2, 2016).

U.S. Department of Agriculture. 2016. *National Agricultural Statistics Service*, Washington, DC. http://www.nass.usda.gov (accessed February 5, 2016).

Van der Valk, A. G., and C. B. Davis. 1978. The role seed banks in the vegetation dynamics of prairie glacial marshes. *Ecology* 59:322–335.

Weller, M. W. 1975. Studies of cattail in relation to management for marsh wildlife. *Iowa State Journal of Research* 49:383–412.

Wilson, E. A., E. A. LeBoeuf, K. M. Weaver, and D. J. LeBlanc. 1989. Delayed seeding for reducing blackbird damage to sprouting rice in southwestern Louisiana. *Wildlife Society Bulletin* 17:165–171.

Yasukawa, K., and W. A. Searcy. 1995. Red-winged blackbird (*Agelaius phoeniceus*). No. 184. *The Birds of North America*, ed. P. G. Rodewald. Cornell Lab of Ornithology, Ithaca, NY. https://birdsna.org/Species-Account/bna/species/rewbla (accessed September 25, 2016).

CHAPTER 11

Allowable Take of Red-Winged Blackbirds in the Northern Great Plains

Michael C. Runge and John R. Sauer
Patuxent Wildlife Research Center
Laurel, Maryland

CONTENTS

11.1 Introduction	192
11.1.1 The Legal Context for Allowable Take	192
11.1.2 A Quantitative Framework for Allowable Take	192
11.2 Intrinsic Growth Rate	194
11.2.1 First and Last Ages of Reproduction	195
11.2.2 Adult and Juvenile Survival Rates	195
11.2.3 Reproductive Rate	196
11.2.4 Estimate of Intrinsic Growth Rate	196
11.3 Population Size in the Northern Great Plains	198
11.3.1 Methods	198
11.3.1.1 Analysis of BBS Trends	198
11.3.1.2 Ground-Based Surveys	199
11.3.1.3 Computing Numbers of Females from Breeding Pairs	200
11.3.1.4 Computing Adjustment Factors and Extrapolating to Estimate Population Size	200
11.3.2 Results	200
11.4 Allowable Take	201
11.4.1 Territoriality and The Units of Take	202
11.4.2 Allowable Take of Breeding Females	202
11.4.3 Allowable Take of Other Sex and Age Classes	203
11.5 Discussion	203
References	204

Red-winged blackbirds (*Agelaius phoeniceus*) are protected under the Migratory Bird Treaty Act (MBTA), which has provisions against take. Blackbirds may be taken legally without a federal permit, however, under an existing depredation order (50 CFR 21.43), which allows for take of blackbirds that are in the process of doing, or about to do, agricultural damage. Modeling the effect of take on blackbird population allows us to balance the conservation protections of the MBTA with the protection of agricultural interests. A quantitative framework based on harvest theory, demography, and population status has been used to assess the allowable take of a number of species of birds under the MBTA. In this chapter, we calculate allowable levels of take for two populations of red-winged blackbirds in the northern Great Plains from estimates of intrinsic growth rate and population size.

11.1 INTRODUCTION

As one of the most abundant granivorous bird species in North America, red-winged blackbirds can create conflict with agricultural interests, especially at times of the year when they gather in very large flocks (Chapter 2, this book). Nonlethal methods to resolve these conflicts, including habitat manipulation (Bergman et al. 1997; Leitch et al. 1997), mechanical frightening devices (Bergman et al. 1997; Linz et al. 2012; Chapter 9, this book), chemical repellants (Mason and Clark 1992; Werner et al. 2010), chemical frightening agents (Woronecki et al. 1967), and decoy crops (Hagy et al. 2008; Linz et al. 2012) are employed preferentially, but lethal methods are sometimes called upon (Blackwell et al. 2003). However, lethal take is generally prohibited under the MBTA unless a specific exception is granted by the U.S. Fish and Wildlife Service (USFWS). A central question that must be addressed before granting exceptions is what constitutes an allowable level of lethal take; this question has both legal and scientific elements (Runge et al. 2009).

11.1.1 The Legal Context for Allowable Take

The MBTA (16 USC §§703–712) prohibits take of migratory birds in the United States, except as allowed by the Secretary of the Interior "having due regard ... to the distribution, abundance, economic value, breeding habits, and times and lines of migratory flight of such birds" (16 USC §704). Such take is typically authorized under a federal permit issued on a case-by-case basis by the USFWS, but there are occasions when blanket permission can be granted. An existing depredation order (50 CFR 21.43) allows nonlethal and lethal control of blackbirds, cowbirds, crows, grackles, and magpies under certain circumstances, including when they "are causing serious injuries to agricultural or horticultural crops or to livestock feed." Take under this Depredation Order needs to be reported annually to the USFWS.

In issuing a permit for take or in establishing a depredation order, the USFWS needs to evaluate whether the proposed take is compatible with the MBTA, as well as any other applicable laws. In previous work (Runge et al. 2009), we have argued that an important standard for consideration is whether the proposed take is sustainable, in the sense that if it recurred at regular intervals indefinitely, the population in question would persist. This standard allows the use of harvest theory as an assessment framework, and has since been employed to evaluate lethal take of black vultures (*Coragyps atratus*) (Runge et al. 2009), double-crested cormorants (*Phalacrocorax auritus*), songbirds taken from the wild as part of the caged-bird trade (Johnson et al. 2012), scaly-naped pigeons (*Patagioenas squamosa*) (Rivera-Milán et al. 2014), and several other species.

11.1.2 A Quantitative Framework for Allowable Take

Sustainable removal from a population, whether harvest, incidental take, or deliberate take, is possible because of density-dependent feedback mechanisms. Through a variety of processes, a reduction in the population size through removal frees up resources for use by the remaining individuals, leading to increased demographic rates (survival or reproduction). The increase in the demographic rates creates a surplus of individuals relative to what is needed to maintain the population size, thus compensating for the loss through take. The simplest form of a population model that can capture these dynamics is the discrete logistic model (Runge et al. 2004, 2009); this model is also appropriate for species like blackbirds without a complex age structure that affects the population dynamics. The properties of the discrete logistic model illustrate the general properties that arise from harvest theory. A population that begins at its carrying capacity (K) and is subjected to a fixed rate of removal (h) will initially decline until the density-dependent responses compensate for the removal rate and the population reaches a lower equilibrium size. This equilibrium population size is a function of the removal rate (as well as the underlying demographic parameters of the population). When the removal rate is applied to the equilibrium population size, a sustainable annual take is produced (Figure 11.1). As the removal rate increases, the equilibrium population size decreases; this tension creates a parabolic curve (a *yield curve*) relating absolute annual take to equilibrium population size. The yield curve shows that there is a maximum annual take that can be sustainably removed; for a logistic model (with linear density-dependence), this occurs when the removal rate is one-half the intrinsic rate of growth (r_{max}) and the equilibrium population size is one-half the carrying capacity. Sustainable harvest is possible at removal rates higher than $r_{max}/2$ (but less than r_{max}), but results in equilibrium population sizes less than $K/2$. A fixed removal rate is a fairly robust strategy: if the population size drops below the corresponding equilibrium point, density-dependent feedback allows the population size to rebound; likewise if the population size drifts above the corresponding equilibrium point, the number of animals removed will exceed the sustainable amount and reduce the population size, maintaining the population around its equilibrium level.

These features of harvest theory were used to develop a method called the *prescribed take level* (PTL) approach (Runge et al. 2009) that makes the distinction between the policy and scientific

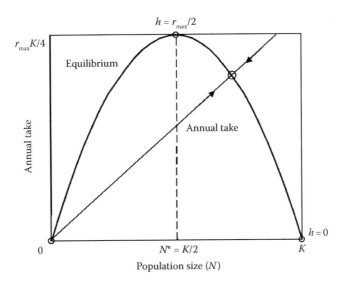

Figure 11.1 Yield curve for a logistic population model subject to take, with a fixed-rate strategy superimposed. In the absence of take (removal rate $h = 0$), the equilibrium population size is at the carrying capacity (K), and there is no sustainable take. At the maximum sustainable yield ($h = r_{max}/2$), the equilibrium population size is at $K/2$ and the sustainable take is $r_{max}K/4$. Under a strategy that has a fixed rate of annual take (annual take proportional to population size), the population size will reach an equilibrium level (open circle): if the annual take is less than the sustainable level for the corresponding population size, the population will increase; if the annual take is more than the sustainable level, the population will decrease.

tasks transparent. The PTL approach is a fixed-rate strategy for take that specifies the allowable level of annual take as

$$PTL_t = F_O \frac{\tilde{r}_{max}}{2} \tilde{N}_t \qquad (11.1)$$

where r_{max} is the intrinsic rate of growth of the population, N_t is the population size at time t, and F_O is a management factor ($0 < F_O < 2$) that reflects the policy objectives of the agency responsible for managing take. The tildes (~) acknowledge that r_{max} and N_t are estimated with uncertainty, and the decision-maker must choose points on those uncertainty distributions that reflect the agency's tolerance for risk. Thus, to determine the level of allowable take, the scientific tasks are to estimate distributions for r_{max} and N_t that accurately reflect the uncertainty in those quantities. The policy tasks are to specify the management factor, F_O, and the quantiles on the uncertainty distributions that capture the appropriate tolerance for risk.

The intrinsic rate of growth, r_{max}, is the growth rate the population would experience in the absence of anthropogenic mortality and at low density (when density-dependent processes are not depressing the demographic rates). This growth rate is rarely observed because most populations are not at such low densities relative to their carrying capacity, and there is often some form of anthropogenic removal. Thus, the intrinsic rate of growth usually needs to be inferred from other measures. There are a variety of methods that have been used to estimate the intrinsic rate of growth. Slade et al. (1998) described an approach for estimating the asymptotic growth rate (λ) that requires knowledge of a few basic life-history parameters: the age at first reproduction (α), the age at last reproduction (ω), the adult survival rate (p), the reproductive rate (b), and the survival rate from birth to the age of first reproduction (l_α). The asymptotic growth rate is found by solving the equation

$$1 = p\lambda^{-1} + l_\alpha b \lambda^{-\alpha} - l_\alpha b p^{(\omega-\alpha+1)} \lambda^{-(\omega+1)}. \qquad (11.2)$$

The intrinsic rate of growth is found from the asymptotic growth rate, $r_{max} = \lambda - 1$. We use this method to estimate the intrinsic rate of growth for red-winged blackbirds in the northern Great Plains.

The current population size, N_t, can be estimated by many methods, the appropriate method depending on the life history of the species and the practical realities of monitoring. We use a combination of trend information from the North American Breeding Bird Survey (BBS, Sauer et al. 2013) and population estimates from ground-based total-area counts in North Dakota to estimate the population size of red-winged blackbirds in Bird Conservation Regions (BCR, Sauer et al. 2003) 11 (Prairie Potholes) and 17 (Badlands and Prairies) over time.

11.2 INTRINSIC GROWTH RATE

To estimate the intrinsic growth rate for red-winged blackbirds in the northern Great Plains, we use the method of Slade et al. (1998), which calculates the growth rate from component life-history parameters. We used published estimates of life-history parameters in our analysis, with three considerations. First, to the extent possible, we relied on estimates from the northern Great Plains, recognizing that the intrinsic growth rate may vary across the wide range of red-winged blackbirds. Second, we attempted to account for density effects. If the estimate of growth rate from Slade's equation is meant to represent the intrinsic growth rate in the discrete logistic model, then the life-history parameters need to represent conditions of low density and the absence of anthropogenic take.

Third, we were explicit about representing uncertainty in the life-history parameters, so we could account for that uncertainty in our estimate of the intrinsic growth rate.

11.2.1 First and Last Ages of Reproduction

Most female passerines first attempt to reproduce as yearlings (Noon and Sauer 2001) and evidence suggests this is true for red-winged blackbirds (Orians and Beletsky 1989; Yasukawa and Searcy 1995). Thus, we set $\alpha = 1$ with certainty.

Although experimental evidence suggests that there is not reproductive senescence in passerines (Holmes et al. 2003), we did include a range of possible ages of last reproduction. For a lower bound, we considered the field observations of breeding by known individuals of red-winged blackbirds: Orians and Beletsky (1989) observed a female that reproduced for at least 10 years. For an upper bound, we considered estimates based on longevity: Wasser and Sherman (2010) listed the maximum longevity of red-winged blackbirds in the wild as 15.8 years. We represented uncertainty in the age of last reproduction as a discrete uniform distribution between the values of 10 and 16, inclusive.

11.2.2 Adult and Juvenile Survival Rates

Annual survival of red-winged blackbirds has been studied in three ways, all involving unique marking of individual birds: by tracking the return rates of territorial males (e.g., Yasukawa 1987); through recovery of dead, banded birds (Fankhauser 1967, 1971; Searcy and Yasukawa 1981); and through repeated recapture of banded birds (Fankhauser 1967, 1971; Searcy and Yasukawa 1981). In nearly all published analyses, modern capture–mark–recapture methods for analysis have not been employed; instead, a variety of *ad hoc* calculations have been used. Only Dyer et al. (1977) and Stehn (1989a) published survival estimates using likelihood-based estimators (Seber 1970; Brownie et al. 1978); the full details of the data collection are only provided by Stehn (1989a).

For the purposes of this analysis, we restricted our attention to the analysis of Stehn (1989a), which was based on all bandings and incidental recoveries in North America, from 1955 to 1975. Stehn (1989a) found no differences in survival rate by age, sex, or region. We calculated our survival estimate from his estimates of annual survival for adult males and females in four geographic areas (Table 4 in Stehn 1989a), weighting the annual survival rates by the number of incidental recoveries. Stehn (1989a) did not provide estimates of standard error in his Table 4; we assumed the standard error scales with the square root of the number of recoveries and used this relationship to infer the standard error and confidence interval for our estimate of the weighted annual adult survival rate. Based on 723 recoveries of male and female adult red-winged blackbirds, we describe uncertainty in the adult annual survival rate as normally distributed (mean 0.6146, SE 0.009678).

The Slade equation also requires an estimate of survival rate to breeding age (l_α), which, in this case, is survival to age 1 year. There can be some ambiguity in the Slade equation between which life-history stages go into the reproductive rate and which go into the juvenile survival rate; we define this breakpoint at fledging. Thus, survival from egg to chick to fledgling is included in the reproductive rate (discussion in Section 11.2.3) and survival from fledging to yearling is the juvenile survival rate. This juvenile survival rate is the least-studied life-history parameter for red-winged blackbirds. The most vulnerable period appears to be the first 4–8 weeks after fledging; after independence from their parents, juvenile blackbirds appear to survive at the same rates as adults (Stehn 1989a). The survival rate during this vulnerable period (prior to parental independence) is very difficult to measure. Stehn (1989b) calculated that a juvenile female survival rate of 0.42 was needed for replacement. Thus, the juvenile survival rate must be between 0.42 and 0.61 (the adult survival rate), but we have little other information to indicate where in that interval it might fall. Johnson et al. (2012), using a demographic invariant approach that relates life-history parameters to body mass, estimated a juvenile survival rate of 0.50 (SD 0.123), an estimate roughly centered on the range we have concluded. For the purposes of this analysis, we represent uncertainty about juvenile female survival rate with a normal distribution (mean 0.515, SD 0.0485), the 95% confidence interval for which is (0.42, 0.61).

None of the studies of survival in red-winged blackbirds considered the effects of anthropogenic mortality or density-dependence. Thus, if anything, our estimates are an underestimate of the survival rates we are seeking (i.e., survival at low density and in the absence of take). However, if anthropogenic take of blackbirds over large geographical areas is not a major component of mortality and adult survival rate is relatively insensitive to density-dependent effects, then our estimates are reasonable.

11.2.3 Reproductive Rate

Based on how we have defined the juvenile survival rate, the reproductive rate of interest is the number of female fledglings produced per adult female. We assume that all females age 1 year and older attempt to breed; there is no indication that the propensity to breed is anything other than 1.0 for females of all ages. Reproductive rates are perhaps the most studied life-history parameter for red-winged blackbirds with dozens of field studies scattered across the species' range (Dyer et al. 1977). Reproductive rates are higher in wetland habitats than upland habitats (Dyer et al. 1977; Besser et al. 1987; Stehn 1989c). We assume that red-winged blackbirds preferentially select wetland breeding habitat when it is available, only spilling into upland habitats when the wetland territories are full. The calculation of r_{max} is meant to reflect the growth rate when resources are not limiting; thus the reproductive rate in wetlands is more relevant than the reproductive rate in upland habitat. We confine our attention to two studies of reproduction in wetlands: Besser et al. (1987), because it is from the northern Great Plains; and Stehn (1989c) because it summarizes data from many studies in a thorough manner, correcting for several strong sources of sampling bias. These sources include the frequency of visitation and repeated years of visitation, both of which appear to increase the predation rate. Besser et al. (1987) estimated 1.78 fledglings of both sexes per female in marsh habitats. Stehn (1989c) estimated 1.93 fledglings of both sexes per female in marsh habitats, using data from field studies (not nest cards, Pettingill 1965). Thus, the two studies suggest a range of 0.89–0.965 for the number of female fledglings per female in marsh habitats, correcting for sampling issues, at average densities.

The reproductive rate of interest is the rate at low density. None of the reproductive studies were conducted in populations of blackbirds that were severely depleted; indeed, in many of the studies, the populations were stable and possibly at carrying capacity. The theory of life-history evolution suggests that reproductive rates are likely to show high environmental variation and to be sensitive to density (Pfister 1998). Neither of the studies we used corrected for density-dependence, although Stehn (1989b) suggested, without evidence, that the asymptotic reproductive rate could be 40% higher than the rate at equilibrium. If we entertain a multiplicative asymptotic factor between 20% and 50%, this provides a range of 1.07–1.45 for the number of female fledglings per adult female in wetland habitat at low density. We used a uniform distribution over this interval to represent uncertainty.

Red-winged blackbird nests are often parasitized by brown-headed cowbirds (*Molothrus ater*) in the northern Great Plains. We have not explicitly accounted for the effect of cowbird parasitism on blackbird reproductive rate, but the field studies we relied on calculated the number of *blackbird* young fledged per female blackbird and did not include cowbird fledglings in the reproductive rate. This is the appropriate reproductive rate to use for calculating r_{max} for blackbirds.

11.2.4 Estimate of Intrinsic Growth Rate

To estimate the intrinsic rate of growth, we sampled 10,000 replicates independently from the distributions that represented uncertainty in the demographic parameters. We solved Slade's equation (Equation 11.2) numerically for the asymptotic growth rate (λ) and subtracted 1 to calculate the intrinsic rate of growth (r_{max}). The distribution of the 10,000 replicates of r_{max} (Figure 11.2)

ALLOWABLE TAKE OF RED-WINGED BLACKBIRDS IN THE NORTHERN GREAT PLAINS

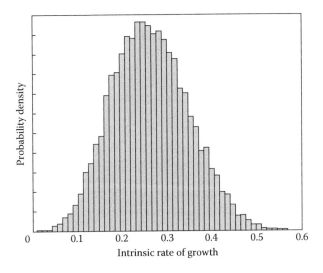

Figure 11.2 Histogram of the intrinsic rate of growth (r_{max}) for red-winged blackbirds, as estimated from Slade's equation.

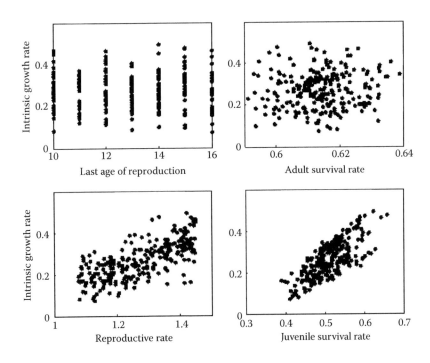

Figure 11.3 Sensitivity of the intrinsic rate of growth (r_{max}) to the component demographic parameters in Slade's equation.

describes the uncertainty in the growth rate. The median growth rate was 0.260 (95% credible interval, 0.110–0.436; 60% credible interval, 0.189–0.337).

The sensitivity of the estimate of r_{max} to the uncertainty in the component life-history parameters can be visualized by plotting r_{max} against the individual demographic rates for each replicate (Figure 11.3). The estimate of the intrinsic growth rate is most sensitive to uncertainty in the juvenile survival rate, quite sensitive to uncertainty in the reproductive rate, and not at all sensitive to the last age of reproduction or the adult survival rate, over the ranges of those variables that represent the uncertainty.

11.3 POPULATION SIZE IN THE NORTHERN GREAT PLAINS

The second piece of information needed to calculate an allowable level of take is an estimate of the size of the population, preferably a probability distribution that captures the uncertainty about the estimate. Here we estimate the population size in BCRs 11 and 17, using temporal trends from the BBS and point estimates from ground-based surveys. We focus on the U.S. portion of BCR 11 (which spans the United States and Canada) and all of BCR 17 (which falls entirely within the United States) because take is authorized under national statutes, not international treaties. In this chapter, unless explicitly noted, a reference to BCR 11 is to the U.S. portion of that region.

11.3.1 Methods

The BBS is the primary source of information on population status and trends for >420 species of North American birds, including red-winged blackbirds (Sauer et al. 2013). The BBS is a roadside survey, and the sample unit is a roadside route consisting of 50 3-minute point counts, separated by 400 m, that are conducted by a single observer during the breeding season (late May to early July). Although the BBS does not provide direct estimates of population size (due to limitations of point counts and roadside surveys), analyses of BBS data that control for observer and route effects allow for estimation of population change, producing annual indexes that are scaled to a "bird/route" index. Because population management often requires population size estimates, N_t (e.g., for setting population goals or assessing allowable take), several approaches have been devised for scaling the BBS index to a population size estimate. One approach is to estimate adjustment factors that scale the BBS for incomplete counts and on-road versus off-road population levels (Rosenberg and Blancher 2005; Runge et al. 2009). This approach can be effective if estimates of the adjustment factors are available (e.g., Runge et al. 2009); unfortunately, for many songbirds these data are not available (Thogmartin et al. 2006). An alternative strategy is to calibrate the BBS index with surveys that do provide a population size estimate for the species of interest, then use the estimated scaling factor between the BBS index and the population size estimate to scale BBS data for the region and time of interest. Typically, these population size surveys are small scale and short term but useful in that they allow us to use the BBS population change estimates to extrapolate population size estimates to regions and years outside the scope of the original population size estimates (e.g., Millsap et al. 2013; Zimmerman et al. 2015). This calibration approach to estimation of population size requires that a more intensive survey exist that overlaps with BBS data; for most species and regions these surveys do not exist. However, population estimates exist for red-winged blackbird populations in North Dakota, as quadrat surveys for breeding birds were conducted in 1990 (Nelms et al. 1994) and 1992–1993 (Igl and Johnson 1997). BCRs 11 and 17 cover North Dakota; this adjustment factor is assumed to be applicable to the entirety of BCRs 11 and 17. Here, we use these data to estimate an adjustment factor with the 1990, 1992, and 1993 BBS annual indexes from North Dakota, then apply the adjustment factor to BBS results for red-winged blackbirds in BCRs 11 and 17 to obtain estimates of N_t for modeling of allowable take.

11.3.1.1 Analysis of BBS Trends

Although started in the eastern United States in 1966, the BBS was initiated in the upper Great Plains in 1967, but the sample size of routes covered was not adequate until 1968. We analyzed data from 1968 to 2015 for our analysis. BBS results are analyzed using strata defined as the intersection of states and BCRs; a log-linear hierarchical model is used to estimate a composite time series of annual indexes for each stratum, then these composite annual indexes are area-weighted among strata to obtain overall estimated annual indices. In the log-linear model, counts are assumed to be independent overdispersed Poisson random variables. Expected values of counts from stratum i, route/observer j, and year t, $\lambda_{i,j,t}$, are modeled as follows:

$$ln(\lambda_{i,j,t}) = S_i + \beta_i(t - t^*) + \omega_j + \gamma_{i,t} + \eta I_{j,t} + \varepsilon_{i,j,t} \qquad (11.3)$$

with the following explanatory variables: slope of centered year (β, t^* is fixed year 1986); stratum intercepts (S); route/observer (ω); year (γ); start-up effects η ($I(j,t)$ takes value 1 for first year of observer's surveying a route, 0 otherwise); and overdispersion effects (ε). (Note that the use of λ here refers to the expected values of counts, whereas the use of λ in Equation 11.2 refers to the intrinsic growth rate for the population. These are different quantities.) See Sauer et al. (2013) for details of the model and model fitting.

We used Bayesian methods to fit this model to the BBS data, employing Markov chain Monte Carlo (MCMC) methods to estimate posterior distributions for the parameters of interest, from which medians and credible intervals are used for inference. For these analyses, prior distributions must be assigned to parameters. As in other BBS analyses, a normal prior distribution with mean zero and variance 1×10^6 was assigned to η, S_i, and β_i. Route/observer effects (ω) and overdispersion effects (ε) were identically distributed mean zero normal distributions with common hierarchically structured variances σ_ω^2 and σ_ε^2, respectively. Year effects (γ) had variances $\left(\sigma_{y,i}^2\right)$ that differed among strata. Variances were assigned noninformative inverse gamma prior distributions.

Annual indices are defined as functions of the model parameters. Stratum-specific annual indices were defined as follows:

$$n_{i,t} = z_i \exp\left(S_i + \beta_i(t - t^*) + \gamma_{i,t} + 0.5\sigma_\omega^2 + 0.5\sigma_\varepsilon^2\right), \qquad (11.4)$$

where z_i is the stratum-specific proportion of routes containing the species (added because routes on which the species was not encountered are not included in the analysis). Annual indices of larger areas such as states or BCRs were defined as area-weighted averages of the annual indices, (i.e., for $i = 1 \ldots I$ regions, $n_t = \left[\sum_{i=1}^{I} A_i\, n_{i,t}\right] / \sum_{i=1}^{I} A_i$, A_i is the area of stratum i). We define $N'_t = \sum_{i=1}^{I} A_i n_{i,t}$ as the BBS total for the stratum. In this analysis, we compute BBS totals for BCRs 11 and 17 in North Dakota (i.e., $N'_{11ND,t}$, $N'_{17ND,t}$), as well as overall BBS total for BCR 11 in the United States and BCR 17. The North Dakota totals are used to develop adjustment factors to population estimates from ground counts in North Dakota, and these adjustment factors are used to scale BBS totals from BCRs 11 (U.S. data) and 17 to provide the overall population estimates for BCRs 11 (U.S. data) and 17.

11.3.1.2 Ground-Based Surveys

Nelms et al. (1994) and Igl and Johnson (1997) conducted surveys on randomly-selected quarter-section plots across North Dakota in 1990 and 1992–1993. The goal of their surveys was to estimate population densities of avian species in North Dakota and compare results to surveys conducted on the same plots in 1967 (Stewart and Kantrud 1972) to assess population change. The surveys provided estimates of population density of breeding pairs of red-winged blackbirds (breeding pairs/km²) in North Dakota of 6.24 (SE = 0.977, n = 129) in 1990, 7.13 (0.793, 128) in 1992, and 8.38 (0.869, 128) in 1993. Surveys were stratified by physiographic regions within the state, but regions did not correspond to BCRs and we could not estimate population densities by BCR within the state.

11.3.1.3 Computing Numbers of Females from Breeding Pairs

Red-winged blackbirds are polygynous, and an "indicated territory," as defined by the presence of a breeding male, likely indicates the presence of >1 female. Many investigators have estimated the number of females per breeding males, and the numbers vary from 1.57 (Goddard and Board 1967) to 3.72 (Orians 1961). Using information from a variety of studies, we estimated a mean of 2.48 females per breeding male (SE = 0.156, n = 15 studies) (Nero 1956; Orians 1961; Meanley and Webb 1963; Goddard and Board 1967; Holm 1973; Weatherhead and Robertson 1977a, 1977b; Hurly and Robertson 1985; Muldal et al. 1986; Besser et al. 1987).

11.3.1.4 Computing Adjustment Factors and Extrapolating to Estimate Population Size

Adjustment factors were computed and adjusted population estimates were calculated as derived statistics within the Bayesian analysis. The MCMC fitting procedure is iterative, and results converge over iterations to form sequences of estimates from which the posterior distributions are estimated. To compute the adjustment factor and adjusted population estimates in such a way as to incorporate uncertainty in the population estimates, we treated the density estimates from 1990, 1992, and 1993 as random variables, generating normally distributed variables (d_t) with means corresponding to the estimates of density by year and precision defined by the estimated standard errors. To compute the adjustment factor, we used a combined ratio estimator:

$$v = \frac{A_{ND}(d_{1990} + d_{1992} + d_{1993})}{N'_{11ND,1990} + N'_{11ND,1992} + N'_{11ND,1993} + N'_{17ND,1990} + N'_{17ND,1992} + N'_{17ND,1993}}. \tag{11.5}$$

Here, v is the sum of the total indicated pairs from the plot surveys for the 3 years divided by the sum of the BBS indexes for the states for the 3 years. To compute the adjusted population index for BCRs 11 and 17, we estimated $N'_{BCR11+BCR17,t} = \left[\sum_{m}^{I} A_m\, n_{m,t}\right]$, for all strata m in BCRs 11 and 17 in the United States and multiplied it by v to obtain an estimate of $N_{BCR11+BCR17,t}$. We used $N_{BCR11+BCR17,2015}$ as our estimate of indicated territories for assessing allowable take.

To compute total number of females, we generated normally distributed variables with means corresponding to the mean estimate of females/males and precision defined by the standard errors estimated among the studies. We then multiplied the number of indicated pairs for regions by this sex ratio to estimate the total number of females.

We used the program JAGS (Plummer 2003) to fit the hierarchical model. JAGS uses MCMC for estimation of posterior distributions of parameters. Our experience with these models has shown that a burn-in of length 10,000 is sufficient for BBS analyses; after this burn-in, we used the next 10,000 iterations to estimate medians and 2.5%, 20%, 80%, and 97.5% percentiles of the posterior distributions. We used these quantities as our estimates and associated credible intervals of parameters.

11.3.2 Results

The number of breeding adult red-winged blackbirds remained relatively steady over the period 1968–2015, perhaps with a small dip in 1990–1995 in the U.S. portion of BCR 11, and small dips in 1990–1995 and 2005–2010 in BCR 17 (Figure 11.4). These dips might reflect population declines

ALLOWABLE TAKE OF RED-WINGED BLACKBIRDS IN THE NORTHERN GREAT PLAINS

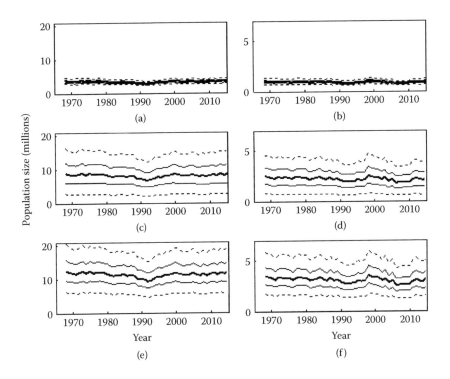

Figure 11.4 Population size of red-winged blackbirds in the northern Great Plains in Bird Conservation Region 11 (Prairie Potholes, U.S. portion only) and Bird Conservation Region 17 (Badlands and Prairies), 1968–2015. (a, b) Indicated territories (breeding males); (c, d) adult females; (e, f) total breeding adults. The bold line is the median estimate from the posterior distribution for population size; the solid lines are the 20th and 80th percentiles; the dashed lines are 2.5th and 97.5th percentiles of the same distribution (and thus encompass the 95% credible interval for population size).

during dry periods; for example, several breeding bird species in this region experienced population dips in the early 1990s that coincide with extreme drought conditions during this period (Peterjohn and Sauer 1993; Igl and Johnson 1997). With an average number of females per territory of 2.48 (standard error 0.16; 95% confidence interval [CI], 2.17–2.79), the total number of breeding adults was roughly 3.5 times the number of indicated territories. In 2015 in BCR 11 (Prairie Potholes, U.S. portion), the estimated median number of indicated territories was 3.54 million (95% credible interval, 2.82–4.40 million; 20th percentile, 3.22 million), the estimated median number of breeding females was 8.74 million (95% CI, 2.71–15.58 million; 20th percentile, 6.02 million), and the estimated number of breeding adults was 12.19 million (95% CI, 6.11–19.56 million; 20th percentile, 9.46 million). In 2015 in BCR 17 (Badlands and Prairies), the estimated median number of indicated territories was 0.95 million (95% credible interval, 0.71–1.28 million; 20th percentile, 0.84 million), the estimated median number of breeding females was 2.34 million (95% CI, 0.72–4.32 million; 20th percentile, 1.61 million), and the estimated number of breeding adults was 3.27 million (95% CI, 1.60–5.51 million; 20th percentile, 2.51 million).

11.4 ALLOWABLE TAKE

The calculation of allowable take requires scientific estimates of the intrinsic growth rate and population size, a description of the uncertainty surrounding the estimates, and policy judgments regarding the management factor (F_O) and the risk tolerance to use in the face of uncertainty about the growth rate and population size. The scientific estimates are described in Sections 11.2 and 11.3. Policy judgements regarding F_O for nuisance control of red-winged blackbirds in the United States fall under the purview of the U.S. Department of Agriculture's Animal and Plant Health Inspection Service Wildlife Services, the agency most likely to be requesting and overseeing removal of blackbirds. The management factor would reflect the amount by which population reduction is needed to reduce damage and could be influenced by the enabling legislation of the agency as well as the National Environmental Policy Act (42 USC §4321 et seq.). The policy judgments regarding the risk tolerance in the face of uncertainty about the intrinsic rate of growth and population size are probably under the purview of the U.S. Fish and Wildlife Service, the agency with authority under the MBTA to oversee the conservation of migratory birds. A brief discussion of the legal considerations is found in Runge et al. (2009). For purposes of illustration, and without suggesting that these are the correct policy judgments, we consider two values for F_O (1.0 and 0.5), use the median estimate for r_{max} (0.260), and use the 20th percentile of the 2015 estimate of population size.

11.4.1 Territoriality and The Units of Take

Red-winged blackbirds have a polygynous mating system in which adult males hold territories and breed with one or multiple females (Searcy and Yasukawa 2014). Nearly all females first breed as yearlings, but males are typically older, on average, when they are able to first establish and defend a territory. This arrangement creates a surplus of nonterritorial males, or floaters, who challenge territorial males and quickly replace them if they die (Sawin et al. 2003; Searcy and Yasukawa 2014). The ground-based estimates of population size count the number of breeding territories, reflecting the number of breeding adult males. We have used estimates of the "harem size" (the number of breeding females per territory) to calculate the number of adult females, as well as the total number of male and female breeders, but these calculations leave out floaters.

Depending on the methods and timing of take, different segments of the population may be removed. For instance, methods of removal that take place during the breeding season and focus on active territories may preferentially remove breeding adults (with an unknown fraction of floaters included), but removal methods that focus on flocking aggregations in the fall may target all adults equally, as well as juvenile (hatch-year) birds. Biologically, the most relevant unit of take would be adult females, as nearly all of them are breeders, and their absolute numbers determine the annual production. Thus, the take of females affects the population dynamics in the manner expressed by the underlying models from which the sustainable take formula (Equation 11.1) is derived. The take of males will have a very different effect. Because of the quick replacement of breeding males with floaters, mortality of a substantial portion of males will have little to no effect on population dynamics until the floater population is gone; at this point additional mortality would reduce the number of breeding territories and, we assume, the productivity of the population. If the number of floaters per breeding male is greater than 1 (Searcy and Yasukawa 2014), the removal rate of males would need to be higher than 50% before effects on the population dynamics were seen. For these reasons, we focus on take of adult females, but provide brief comments on take of other sexes and ages of blackbirds.

11.4.2 Allowable Take of Breeding Females

Using Equation 11.1, the estimates of intrinsic growth rate and population size described in Sections 11.2 and 11.3, and the policy assumptions discussed in Section 11.4, the annual allowable take of breeding female red-winged blackbirds, based on the 2015 population size in the U.S. portion of BCR 11 would be 783,000 if using $F_O = 1$ or 392,000 if using $F_O = 0.5$. The corresponding annual allowable take in BCR 17 would be 210,000 for $F_O = 1$ or 105,000 for $F_O = 0.5$. These correspond to removal rates of 0.130 (using $F_O = 1$) and 0.065 (using $F_O = 0.5$).

Measurement of take against this standard could occur in one of two ways, by assessing the absolute removal of adult females or by assessing the removal rate of adult females. To assess the absolute removal of adult females, the total number of birds removed and the fraction of those birds that were adult females would need to be estimated; the methods to make such estimates would depend on the methods of removal. The removal rate of adult females could be estimated with band recovery models (Brownie et al. 1978) if an adequate fraction of the population was banded and if the removed birds could be examined for bands.

11.4.3 Allowable Take of Other Sex and Age Classes

As noted above, the allowable take of males is complicated by the polygynous breeding structure. The allowable take of males is certain to be higher than the allowable take of females, but how much higher will depend on the fraction of adult males that are territory holders rather than floaters, a number that is not easily estimated and one that will change under a sustained program of removal. In the face of this uncertainty, and in the absence of focused monitoring of adult sex ratio, removal of adult males in the same numbers as adult females will be at least as conservative as the stated values of F_O seek.

Programs of removal in the fall and winter might include take of hatch-year birds. Like adult males, hatch-year males likely can be removed at a higher rate than adult females. Hatch-year females, on the other hand, should only be removed at the same rate as adult females, because most hatch-year females that survive the winter will become breeders. The absolute numbers of hatch-year birds that can be removed (as opposed to the rates of removal) will depend on the timing of removal; we cannot estimate those ratios from the information provided in this chapter. The simplest approach that meets the tenets of the take framework expressed in Equation 11.1 is to remove all age and sex classes in proportion to their frequency in the population, and to insure that the removal of adult females does not exceed the numbers and rates specified in Section 11.4.2. More aggressive patterns of removal are possible but would require more careful monitoring.

11.5 DISCUSSION

We have estimated the intrinsic growth rate for red-winged blackbirds in the northern Great Plains, the population size and composition in Bird Conservation Regions 11 (Prairie Potholes) and 17 (Badlands and Prairies), and allowable take of adult females in these same two BCRs, based on the 2015 population sizes. The policy judgments we made to provide estimates of allowable take are illustrative placeholders; the agencies engaged in requesting and permitting lethal removal of blackbirds may need to examine these policy judgements carefully against their statutory and regulatory guidance. As Runge et al. (2009) noted, biological thresholds may or may not be relevant in establishing policy thresholds, and it is sometimes difficult to infer the intended degree of risk tolerance from statutory language. Nevertheless, the PTL framework provides specific policy parameters that need to be determined and thus provides clarity about the policy judgments that need to be made to set levels of allowable take.

The scientific data available for red-winged blackbirds in the northern Great Plains was not collected for the express purpose of estimating the parameters needed for determining allowable take at the BCR level; rather, we examined the whole body of literature on red-winged blackbirds that has been collected for a wide variety of purposes. From this body of work, we inferred the intrinsic rate of growth and population size of blackbirds, taking pains to quantitatively express the uncertainty in these estimates. These estimates may be adequate for the purposes of authorizing take, especially if the desired level of take falls below the estimates we have provided. If greater levels of take are desired to manage conflict with human endeavors, it might be valuable to conduct research specifically designed to estimate the demographic and population parameters of interest.

It is possible that our estimates of intrinsic growth rate for red-winged blackbirds are negatively biased. Johnson et al. (2012) took a different approach to estimating adult survival rate and intrinsic growth rate, using a demographic invariants approach to develop a relationship between adult survival and body mass and to infer growth rate from that approach. From these relationships, their estimate of adult survival was 0.70 (SD 0.090) and their estimate of intrinsic growth rate was 0.45 (SD 0.51). These estimates are higher than ours; the differences could possibly be explained by negative bias in our estimates owing to effects of take, density, and emigration. However, the only species-specific information from red-winged blackbirds that Johnson et al. (2012) used to inform their estimates was mean body mass. Thus, we use our estimate for adult survival but note the possibility that our estimate is negatively biased. There is a need for a deeper understanding of the relative reliability of the two methods for estimating intrinsic growth rate.

There are a number of other hidden assumptions we have made in framing this analysis of blackbird take in the northern Great Plains. First, we have assumed that the relevant spatial scale to evaluate take is at the level of the BCR. Other scales, from township to county to state and even to areas larger than BCRs, might be more relevant. The question is: at what scale does the MBTA ask that take be sustainable? Using a large scale (like BCR) provides considerable spatial flexibility in managing take: take could be unsustainable at, say, the level of several townships yet be sustainable at the BCR level if take is lower across much of the BCR than in the focal townships. This policy judgment bears consideration.

Second, we have focused on expressing allowable take in terms of adult females. There are biological reasons that make this the easiest metric to use, and there are many removal strategies for which this metric is feasible. However, there could be removal strategies that focus on the need to remove specific subsets of the blackbird population at specific times of year for which our calculations do not provide adequate guidance. Such situations would require a more in-depth analysis and possibly the collection of focused data that currently are unavailable.

Third, we have assumed that nuisance problems would be managed by lethal removal of birds. There are, however, other means of take, including chemosterilization or nonlethal harassment, that could have demographic impacts. The methods of this chapter do not provide either scientific or policy guidance for such means. These are, nevertheless, interesting and practical problems and further research to explore them would be fruitful.

Human activities rely on, are affected by, and impact the natural world; the challenge is in finding a balance among many desired outcomes. Removal of blackbirds to address various impacts on human activities is emblematic of such challenges. Assessing the allowable level of take requires both scientific understanding and policy judgment, made in the face of uncertainty. This chapter provides a framework for tackling this challenge. Whether or not lethal take is the best method for addressing conflict with blackbirds, however, involves an even larger set of questions, including the issues of practicality, environmental safety, cost-effectiveness, and wildlife stewardship (Linz et al. 2015); these topics are outside the scope of this chapter.

REFERENCES

Bergman, D. L., T. L. Pugh, and L. E. Huffman. 1997. Nonlethal control techniques used to manage blackbird damage to sunflower. *Great Plains Wildlife Damage Control Workshop* 13:81–87.

Besser, J. F., O. E. Bray, J. W. DeGrazio, J. L. Guarino, D. L. Gilbert, R. R. Martinka, and D. A. Dysart. 1987. Productivity of red-winged blackbirds in South Dakota. *Prairie Naturalist* 19:221–232.

Blackwell, B. F., E. Huszar, G. M. Linz, and R. A. Dolbeer. 2003. Lethal control of red-winged blackbirds to manage damage to sunflower: An economic evaluation. *Journal of Wildlife Management* 67:818–828.

Brownie, C., D. R. Anderson, K. P. Burnham, and D. S. Robson. 1978. *Statistical inference from band recovery data: A handbook.* Resource Publication No. 131. U.S. Department of the Interior, U.S. Fish and Wildlife Service, Washington, DC.

Dyer, M. I., J. Pinowski, and B. Pinowska. 1977. Population dynamics. In *Granivorous birds in ecosystems: Their evolution, populations, energetics, adaptations, impact and control*, eds. J. Pinowski, and S. C. Kendeigh, 53–105. Cambridge University Press, Cambridge, UK.

Fankhauser, D. P. 1967. Survival rates in red-winged blackbirds. *Bird-Banding* 38:139–142.

Fankhauser, D. P. 1971. Annual adult survival rates of blackbirds and starlings. *Bird-Banding* 42:36–42.

Goddard, S. V., and V. V. Board. 1967. Reproductive success of red-winged blackbirds in north central Oklahoma. *Wilson Bulletin* 79:283–289.

Hagy, H. M., G. M. Linz, and W. J. Bleier. 2008. Optimizing the use of decoy plots for blackbird control in commercial sunflower. *Crop Protection* 27:1442–1447.

Holm, C. H. 1973. Breeding sex ratios, territoriality, and reproductive success in the red-winged blackbird (*Agelaius phoeniceus*). *Ecology* 54:356–365.

Holmes, D. J., S. L. Thomson, J. Wu, and M. A. Ottinger. 2003. Reproductive aging in female birds. *Experimental Gerontology* 38:751–756.

Hurly, T. A., and R. J. Robertson. 1985. Do female red-winged blackbirds limit harem size? I. A removal experiment. *Auk* 102:205–209.

Igl, L. D., and D. H. Johnson. 1997. Changes in breeding bird populations in North Dakota: 1967 to 1992–93. *Auk* 114:74–92.

Johnson, F. A., M. A. H. Walters, and G. S. Boomer. 2012. Allowable levels of take for the trade in Nearctic songbirds. *Ecological Applications* 22:1114–1130.

Leitch, J. A., G. M. Linz, and J. F. Baltezore. 1997. Economics of cattail (*Typha* spp.) control to reduce blackbird damage to sunflower. *Agriculture, Ecosytems & Environment* 65:141–149.

Linz, G. M., E. H. Bucher, S. B. Canavelli, E. Rodriguez, and M. L. Avery. 2015. Limitations of population suppression for protecting crops from bird depredation: A review. *Crop Protection* 76:46–52.

Linz, G. M., H. J. Homan, S. Werner, J. C. Carlson, and W. J. Bleier. 2012. Sunflower growers use nonlethal methods to manage blackbird damage. *Wildlife Damage Management Conference* 14:114–118.

Mason, R., and L. Clark. 1992. Nonlethal repellents: The development of cost-effective, practical solutions to agricultural and industrial problems. *Vertebrate Pest Conference* 15:115–129.

Meanley, B., and J. S. Webb. 1963. Nesting ecology and reproductive rate of the red-winged blackbird in tidal marshes of the upper Chesapeake Bay region. *Chesapeake Science* 4:90–100.

Millsap, B. A., G. S. Zimmerman, J. R. Sauer, R. M. Nielson, M. Otto, E. Bjerre, and R. Murphy. 2013. Golden eagle population trends in the western United States: 1968–2010. *Journal of Wildlife Management* 77:1436–1448.

Muldal, A. M., J. D. Moffatt, and R. J. Robertson. 1986. Parental care of nestlings by male red-winged blackbirds. *Behavioral Ecology and Sociobiology* 19:105–114.

Nelms, C. O., W. J. Bleier, D. L. Otis, and G. M. Linz. 1994. Population estimates of breeding blackbirds in North Dakota, 1967, 1981–1982 and 1990. *American Midland Naturalist* 132:256–263.

Nero, R. W. 1956. A behavior study of the red-winged blackbird: II. Territoriality. *Wilson Bulletin* 68:129–150.

Noon, B. R., and J. R. Sauer. 2001. Population models for passerine birds: Structure, parameterization, and analysis. In *Wildlife 2001: Populations*, eds. D. R. McCullough, and R. H. Barrett, 441–464. Elsevier Applied Science, New York.

Orians, G. H. 1961. The ecology of blackbird (*Agelaius*) social systems. *Ecological Monographs* 31:285–312.

Orians, G. H., and L. D. Beletsky. 1989. Red-winged blackbird. In *Lifetime reproduction in birds*, ed. I. Newton, 183–197. Academic Press, New York.

Peterjohn, B. G., and J. R. Sauer. 1993. North American Breeding Bird Survey annual summary 1990–1991. *Bird Populations* 1:52–67.

Pettingill, O. S., Jr. 1965. A North American nest-record card program. *Bird-Banding* 36:65–66.

Pfister, C. A. 1998. Patterns of variance in stage-structured populations: Evolutionary predictions and ecological implications. *Proceedings of the National Academy of Sciences of the United States of America* 95:213–218.

Plummer, M. 2003. JAGS: A program for analysis of Bayesian graphical models using Gibbs sampling. *International Workshop on Distributed Statistical Computing* 3:1–10.

Rivera-Milán, F. F., G. S. Boomer, and A. J. Martínez. 2014. Monitoring and modeling of population dynamics for the harvest management of scaly-naped pigeons in Puerto Rico. *Journal of Wildlife Management* 78:513–521.

Rosenberg, K. V., and P. J. Blancher. 2005. Setting numerical population objectives for priority landbird species. In *Bird conservation implementation and integration in the Americas, General Technical Report PSW-GTR-191*, eds. C. J. Ralph, and T. D. Rich, 57–67. U.S. Department of Agriculture Forest Service, Albany, CA.

Runge, M. C., W. L. Kendall, and J. D. Nichols. 2004. Exploitation. In *Bird ecology and conservation: A handbook of techniques*, eds. W. J. Sutherland, I. Newton, and R. E. Green, 303–328. Oxford University Press, Oxford, UK.

Runge, M. C., J. R. Sauer, M. L. Avery, B. F. Blackwell, and M. D. Koneff. 2009. Assessing allowable take of migratory birds. *Journal of Wildlife Management* 73:556–565.

Sauer, J. R., J. E. Fallon, and R. Johnson. 2003. Use of North American Breeding Bird Survey data to estimate population change for bird conservation regions. *Journal of Wildlife Management* 67:372–389.

Sauer, J. R., W. A. Link, J. E. Fallon, K. L. Pardieck, and D. J. Ziolkowski, Jr. 2013. The North American Breeding Bird Survey 1966–2011: Summary analysis and species accounts. *North American Fauna* 79:1–32.

Sawin, R. S., G. M. Linz, R. L. Wimberly, M. W. Lutman, and W. J. Bleier. 2003. Estimating the number of nonbreeding male red-winged blackbirds in central North Dakota. In *Management of North American blackbirds*, ed. G. M. Linz, 97–102. USDA National Wildlife Research Center, Fort Collins, CO.

Searcy, W. A., and K. Yasukawa. 1981. Sexual size dimorphism and survival of male and female blackbirds (Icteridae). *Auk* 98:457–465.

Searcy, W. A., and K. Yasukawa. 2014. *Polygyny and sexual selection in red-winged blackbirds*. Princeton University Press, Princeton, NJ.

Seber, G. A. F. 1970. Estimating time-specific survival and reporting rates for adult birds from band returns. *Biometrika* 57:313–318.

Slade, N. A., R. Gomulkiewicz, and H. M. Alexander. 1998. Alternatives to Robinson and Redford's method of assessing overharvest from incomplete demographic data. *Conservation Biology* 12:148–155.

Stehn, R. A. 1989a. *Adult survival rate of red-winged blackbirds*. Bird Section Research Report 431. U.S. Department of Agriculture, Denver Wildlife Research Center, Sandusky, OH.

Stehn, R. A. 1989b. *Population ecology and management strategies for red-winged blackbirds*. Bird Section Research Report 432. U.S. Department of Agriculture, Denver Wildlife Research Center, Sandusky, OH.

Stehn, R. A. 1989c. *Reproductive rate in red-winged blackbirds*. Bird Section Research Report 430. U.S. Department of Agriculture, Denver Wildlife Research Center, Sandusky, OH.

Stewart, R. E., and H. A. Kantrud. 1972. Population estimates of breeding birds in North Dakota. *Auk* 89:766–788.

Thogmartin, W. E., F. P. Howe, F. C. James, D. H. Johnson, E. T. Reed, J. R. Sauer, and F. R. Thompson, III. 2006. A review of the population estimation approach of the North American Landbird Conservation Plan. *Auk* 123:892–904.

Wasser, D. E., and P. W. Sherman. 2010. Avian longevities and their interpretation under evolutionary theories of senescence. *Journal of Zoology* 280:103–155.

Weatherhead, P. J., and R. J. Robertson. 1977a. Harem size, territory quality, and reproductive success in the redwinged blackbird (*Agelaius phoeniceus*). *Canadian Journal of Zoology* 55:1261–1267.

Weatherhead, P. J., and R. J. Robertson. 1977b. Male behavior and female recruitment in the red-winged blackbird. *Wilson Bulletin* 89:583–592.

Werner, S. J., G. M. Linz, S. K. Tupper, and J. C. Carlson. 2010. Laboratory efficacy of chemical repellents for reducing blackbird damage in rice and sunflower crops. *Journal of Wildlife Management* 74:1400–1404.

Woronecki, P. P., J. L. Guarino, and J. W. De Grazio. 1967. Blackbird damage control with chemical frightening agents. *Vertebrate Pest Conference* 3:54–56.

Yasukawa, K. 1987. Breeding and nonbreeding season mortality of territorial male red-winged blackbirds (*Agelaius phoeniceus*). *Auk* 104:56–62.

Yasukawa, K., and W. A. Searcy. 1995. Red-winged blackbirds (*Agelaius phoeniceus*). In *The birds of North America*, ed. P. G. Rodewald. No. 184. Cornell Lab of Ornithology, Ithaca, NY. doi: http://dx.doi.org/10.2173/bna.184.

Zimmerman, G. S., J. R. Sauer, K. Fleming, W. A. Link, and P. R. Garrettson. 2015. Combining waterfowl and breeding bird survey data to estimate wood duck breeding population size in the Atlantic Flyway. *Journal of Wildlife Management* 79:1051–1061.

CHAPTER 12

The Economic Impact of Blackbird Damage to Crops

Stephanie A. Shwiff, Karina L. Ernest, Samantha L. Degroot, and Aaron M. Anderson
National Wildlife Research Center
Fort Collins, Colorado

Steven S. Shwiff
Texas A&M University
College Station, Texas

CONTENTS

12.1 Examples of Direct or Primary Damage Estimates 208
    12.1.1 Rice 208
    12.1.2 Sunflower 208
    12.1.3 Corn 209
12.2 Assessing Costs and Benefits of Bird Management Tools 210
12.3 Estimates of Indirect or Secondary Damage 212
12.4 Case Study—Sunflower Damage 212
    12.4.1 Introduction 212
    12.4.2 Methods 212
    12.4.3 Results 212
    12.4.4 Summary 214
12.5 Research Needs 214
References 215

There are nearly 1,000 species of birds in North America, some of which provide obvious economic benefits like egg production, meat production, bird watching, or hunting (American Bird Association 2016). Some bird species, however, can cause a considerable amount of damage to U.S. agriculture, with estimates of annual damage caused by birds in the United States exceeding US$4.7 billion (Pimentel et al. 2005). Blackbirds (Icteridae) are one group of birds in North America that can cause significant economic damage to commercial grain crops, and to a lesser extent vine and tree crops (Wilson et al. 1989; Dolbeer 1990; Linz et al. 2011; Anderson et al. 2013).

Four species of blackbirds—red-winged blackbird (*Agelaius phoeniceus*), common grackle (*Quiscalus quiscula*), yellow-headed blackbird (*Xanthocephalus xanthocephalus*), and brown-headed cowbird (*Molothrus ater*)—are primarily responsible for damage to sprouting and ripening grain crops (Lowther 1993; Twedt and Crawford 1995; Yasukawa and Searcy 1995; Peer and Bollinger 1997). During late winter, these species commonly can be found feeding on food present in concentrated animal feedlot operations (Dolbeer et al. 1978). For much of the year, however, these birds forage on insects, waste grain, and weed seeds, thus providing valuable ecological services.

The direct economic damage created by birds typically falls into the three broad categories of destruction, depredation, and disease transmission. Total economic damage (D) of a particular bird species is the sum across these three categories and across time, as follows:

$$D = \sum_{n=1}^{t} \left(\text{Destruction}_n + \text{Depredation}_n + \text{Disease}_n \right)$$

Destruction refers to destroyed property (e.g., bird strikes to aircraft and defecation on statues, golf courses, buildings, and bridges), equipment (e.g., vehicles, farm equipment, cables, irrigation equipment), nonconsumptive damage to crops, usually associated with roosting behaviors, and contamination of water, grains, and livestock feed. A substantial portion of the overall economic impact of birds is through depredation of crops, the focus of this paper.

In this chapter, we review (1) available crop damage data collected over the last five decades, (2) economic analyses that define the level of damage, and (3) estimates of the costs and benefits of particular blackbird management methods. We then present an economic analysis example using regional economic models known as input–output (IO) models (Richardson 1972; Treyz et al. 1991). These models take into account the effects of damage on the economy as a whole, including loss of jobs as a result of reduced production. These analyses provide data for documenting economic losses and justifying the use of resources to reduce damage.

12.1 EXAMPLES OF DIRECT OR PRIMARY DAMAGE ESTIMATES

Although blackbirds are known to damage many crops, we focus on direct damage to rice, corn and sunflower because these commodities are their favored foods over large geographic areas and therefore have garnered much attention from scientists (Linz et al. 2015). Detailed food habit and damage analyses are presented elsewhere in this book.

12.1.1 Rice

Rice planted in the southeastern United States and California is available to foraging birds after seeding in the spring and while ripening prior to harvest in the summer and fall. Recent objective damage surveys are not available; however, in the 1980s blackbird damage to newly planted rice in southwestern Louisiana and east Texas amounted to US$8 million (Wilson et al. 1989; Decker et al. 1990).

Cummings et al. (2005) used a mail survey to gather information on bird damage from rice growers in the United States and found that between 1996 and 2000 the average annual blackbird damage to newly planted rice ranged from 6% to 15%, and the average percent loss to ripening rice ranged from 6% to 14%. Growers in Arkansas and Louisiana reported the highest damage, with blackbirds causing US$8.7 million in damage to rice in 2001. Cummings et al. (2005) estimated the total damage in the United States to be US$13.4 million.

12.1.2 Sunflower

Sunflower is a minor crop in North America that is typically grown in a semi-arid climate (U.S. Department of Agriculture 2016). Blackbird damage is the most common reason that sunflower producers in North Dakota stop planting sunflower (Linz et al. 2011; Hulke and Kleingartner

THE ECONOMIC IMPACT OF BLACKBIRD DAMAGE TO CROPS

Figure 12.1 Typical damage by red-winged blackbirds to a ripening oilseed sunflower head in North Dakota. (Courtesy of USDA Wildlife Services.)

2014; Figure 12.1). Ripening sunflower is particularly vulnerable to blackbirds because the crop is susceptible for 8 weeks, from early seed-set in mid-August until harvest in mid-October (Cummings et al. 1989; Linz et al. 2011).

When bird damage to sunflower became an economic issue in the 1970s, scientists sought to define the extent, magnitude, and frequency of sunflower losses across the sunflower-growing areas of North Dakota, South Dakota, and Minnesota (Guarino 1984). In 1979 and 1980, Hothem et al. (1988) conducted statewide bird damage survey in these states and found that blackbird damage averaged 1.4% across years and was valued at US$6.5 million annually. Further, nearly 21% of the fields received >1% damage, while 5% showed >10% damage.

In 2003, 20 years after Hothem and colleagues conducted their survey, Peer et al. (2003) updated the sunflower damage estimates using a bioenergetic approach combined with population data. They calculated that the combined fall population of male and female red-winged blackbirds, yellow-headed blackbirds, and common grackles was 75 million and that each bird ate an average of US$0.072 of sunflower annually. Total loss was estimated to be 1.7% and was valued at US$5.4 million.

In 2009 and 2010, Klosterman et al. (2013) used 120 3.2 × 3.2 km block sampling design established by Ralston et al. (2007) to objectively assess bird damage to randomly selected sunflower fields in North Dakota's Prairie Pothole Region, a core sunflower-growing area. They found that average annual blackbird damage was 2.7%, valued at US$3.5 million.

Finally, from 2001 to 2013 (except 2004), national surveys of blackbird damage in physiologically mature sunflower fields were conducted throughout the foremost sunflower-growing states (Kandel and Linz 2016). These surveys are expected to be carried out on a periodic basis for the foreseeable future. We detail these results in a case study presented in Section 12.4 using regional economic models (also known as *IO models*; Richardson 1972; Treyz et al. 1991).

12.1.3 Corn

While rice and sunflower are grown in specific regions of North America, corn is planted widely (U.S. Department of Agriculture 2016). Corn loss to birds in 24 states in the United States in 1970 was estimated to be about US$9 million (Stone et al. 1972). From 1977 to 1979, damage surveys conducted in Ohio, Michigan, Kentucky, Tennessee, and Ontario showed that primary damage (the actual corn removed by the birds) averaged about 0.6% (Dolbeer 1981), with <2.5% of cornfields in Ohio having losses >5%. In 1981, Besser and Brady (1986) conducted a survey of bird damage to ripening field corn in 10 major producing states and found that only 2% of the corn ears were damaged.

Weatherhead et al. (1982) used bioenergetics and population data to estimate that in 1979 blackbirds damaged 0.41% of the field corn grown in Quebec valued at about C$279,000. This number was substantially lower than a government estimate of C$16 million. Their study and others have pointed out that qualitative judgements on crop damage should not be used as a basis for determining the economic impact of wildlife damage (e.g., Dolbeer 1981; Besser 1985; Dolbeer et al. 1994).

In 1993, a field survey on corn crops in the top 10 corn-producing states showed that birds damaged 0.19% of field corn, resulting in a total of US$25 million in damages (Wywialowski 1996). Finally, Klosterman et al. (2013), using the same study design previously discussed for sunflower, determined that bird damage to cornfields in North Dakota in 2008 and 2009 averaged 0.2%, valued at US$1.3 million.

We conclude from these examples that overall blackbird damage to crops is low industry-wide. However, bird damage is economically important to a small percentage of producers that farm near favored roost sites. We are encouraged that wildlife managers and industry executives are seeking data to clearly define the problem and more effectively allocate resources to manage damage. Finally, we noted that objective surveys of damage to sunflower are relatively recent, whereas the data for corn is aged, and little objective data has been gathered to clarify the level and geographic extent of bird damage to rice.

12.2 ASSESSING COSTS AND BENEFITS OF BIRD MANAGEMENT TOOLS

Modern commodity producers evaluate the cost and benefits of using crop inputs to maximize profits. The use of various bird damage management techniques can be costly and therefore warrant close scrutiny prior to providing recommendations for their use. Here, we provide several examples where scientists evaluated the cost and benefits of various blackbird management strategies.

DeGrazio et al. (1971, 1972) and Stickley et al. (1976) provided overall cost estimates of using a chemical-frightening agent to protect corn. However, Dolbeer (1981) was an early advocate of using a cost–benefit equation to determine when the use of damage-control measures was warranted economically. He found that in most damage scenarios the cost of using a chemical-frightening agent exceeded the dollars saved by reducing damage. He further concluded that damage would need to exceed 4.6% in a field to be justified. His conclusion was based on the cost of aerially applied treated baits (US$13.71/ha) versus the value of the crop US$2.15/bushel and yields of 250 bushel/ha.

Weatherhead et al. (1980) tested the idea of using decoy traps in cornfields to capture blackbirds and therefore reduce damage. They calculated that the cost of the trap, food, and labor over 98 days amounted to about C$333.00 and the cost per captured bird was C$1.01/bird. Corn was valued at C$0.145/kg at the time of the study. Given an individual bird might eat 0.53 kg during the damage season, an individual bird was worth only C$0.08. The authors concluded that this method of population management was not economically efficient.

Cummings et al. (1986) evaluated a combined propane exploder and CO_2 pop-up scarecrow in sunflower and found that it was effective, particularly if used before an ingrained feeding pattern had developed. The effectiveness of propane cannons, however, was shown to be limited to relatively small areas and damage needed to exceed 18% to be cost beneficial (Table 12.1).

Table 12.1 Methods That Are Commercially Available to Sunflower Producers to Help Reduce Sunflower Damage Caused by Blackbirds in the Prairie Pothole Region of the United States

Method	Cost[a]	Threshold[b]	Comments
Propane cannons	$110/ha	120 birds/ha[c]	1 unit/3 ha
Repellents			
Flock Buster®	$50/ha	–	Questionable efficacy
Bird Shield™	$42/ha	–	Questionable efficacy
Decoy crops	$375/ha	800 birds/ha[d]	Situational efficacy
Desiccation	$24/ha	1000 birds/ha[e]	Saflufenacil + glyphosate
Roost-site destruction	$95/ha	238 birds/ha[f]	Aquatic glyphosate

Source: Linz et al. 2011.
Note: Costs (US$) and economic threshold-for-use are estimates.
[a] When applicable, includes estimated cost of aerial application (US$12/ha).
[b] The number of birds per hectare at the breakeven point of application cost.
[c] Amortized over a 10-year life expectancy for propane cannon.
[d] Cost is based on loss of opportunity of lands in agricultural production. Costs are less for lands not in agricultural production (e.g., CRP). Also, the threshold estimate is based on decoy crops protecting crops of confectionery sunflower (Hagy et al. 2008).
[e] Based on advancement of harvest of 7 days and 0.009 kg sunflower eaten per day per bird @ US$0.37 per kg sunflower (Peer et al. 2003). Does not include savings related to faster dry down to avoid plant lodging due to insect and disease damage.
[f] Amortized over 4-year life expectancy of treatment.

Blackwell et al. (2003) assessed the economics of proposed use of lethal control to manage an estimated population of 27 million red-winged blackbird in the sunflower-growing region of North Dakota. They modelled the potential population effects of removing a maximum of 2 million red-winged blackbirds annually over a 5-year campaign during spring migration. Assuming US$0.07 in damage per bird and variable annual culls of 1.2 million with density compensation, Blackwell et al. (2003) calculated that after 5 years the population would remain between 78.5% and 93.2% of the original 27 million individuals, with a total net benefit of US$386,103–US$776,026 over 5 years. The results from these models led to the conclusion that culling red-winged blackbirds at this level would produce only a marginal economic impact on the sunflower industry; therefore a lethal control program was not pursued.

Hagy et al. (2008) calculated the costs and benefits of using wildlife conservation food plots (WCFP) as a blackbird bird management tool (Chapter 10, this volume). The U.S. Department of Agriculture offered candidate sunflower producers US$375.00/ha to plant WCSP with histories of elevated blackbird damage. The WCFP produced an average of 1,290 kg/ha and birds removed 435 kg/ha, valued at US$160.95/ha. Hagy et al. (2008) concluded that the cost–benefit ratio was 3.4:1, indicating economic inefficiency.

We conclude that over the last four decades scientists have been cognizant of the need to include costs and benefits of bird damage management techniques. For a technique to be cost-effective, the cost of the control measure must be less than the anticipated monetary loss (Dolbeer 1981). Obviously, a producer growing a valuable crop can afford to expend more resources to protect the crop. For example, confectionery varieties of sunflower are more valuable than oilseed varieties and thus might warrant extra protection from depredating blackbirds.

These analyses are important for individual growers. However, economists are often asked what wildlife damage means to the economic health of the entire industry. Obviously, if bird damage to a particular crop is widespread and severe, the entire industry and associated macroeconomy could be impacted. In the following section, we provide background and a case history on how loss of production due to bird damage can damage an economy.

12.3 ESTIMATES OF INDIRECT OR SECONDARY DAMAGE

Direct (i.e., primary) damages can generate indirect (i.e., secondary) impacts due to economic factors that create linkages to established economic sectors (Figure 12.2). Regional economic models (also known as *IO models*) attempt to quantify the impacts on output as a result of input changes in a regional economy (Richardson 1972; Treyz et al. 1991). The model then uses existing estimates of direct impacts as inputs into the regional economic model to quantify the resulting indirect impacts, thereby calculating the total effect on macroeconomic indicators like employment and gross domestic product (GDP) in a specified regional economy. A dynamic regional economic model has been developed to generate annual forecasts and simulate behavioral responses to compensation, price, and other economic factors (REMI: Model Documentation—Version 9.5; Treyz et al. 1991). For example, when birds consume sunflower, fewer sunflower enter the supply chain and as a result sunflower is not processed into products (Elser et al. 2016).

To capture this in a regional economic model, two forecasts are created (Figure 12.2). The first forecast is the control or baseline forecast in which no bird damage to sunflower has occurred and the model projects economic conditions within a region on the basis of trends in historical data. The second forecast is the alternate forecast in which bird damage to sunflower has reduced the amount of sunflower into the supply chain and the model must account for changes in variables such as industry-specific income, value added, and employment. The model then compares the two forecasts to determine the overall impact to the regional economy.

12.4 CASE STUDY—SUNFLOWER DAMAGE

12.4.1 Introduction

From 2001 to 2013 (except 2004), the National Sunflower Association (Mandan, ND) sponsored a comprehensive production survey of physiologically mature sunflower (*Helianthus annuus*) fields in the Canadian province of Manitoba and eight states in the United States (Kandel and Linz 2015). Trained teams of surveyors randomly stopped at one sunflower field for every 4,047–6,070 ha. Yield was based on plant stand, head size, seed size, percent filled seeds, center seed-set, and percent loss due to bird feeding at two random locations within the field. Loss due to bird damage was estimated based on sample charts with examples of various levels of bird damage.

12.4.2 Methods

The data were pooled during the most recent 5 years (2009 to 2013) in the United States for statistical analysis (Kandel and Linz 2015). The overall percentage of sunflower damaged and percentage of oilseed and confectionery sunflower hybrids damaged did not differ across the five study years, hence study years were combined for further analyses. Confectionery and oilseed hybrids, however, produce achenes that are fundamentally different in oil content, size, and hull thickness, so the damage data were presented for both variety types. Additionally, confectionery achenes are also sold at a premium over oilseeds (U.S. Department of Agriculture 2016). Total economic impact was calculated using a model of the regional economy (i.e., IO model) that predicts how a change in one industry can affect revenue and employment throughout the economy.

12.4.3 Results

Across all eight states, mean oilseed yield was 1,456 kg/ha and annual blackbird damage was 2.59%, whereas confectionery fields yielded an average of 1,420 kg/ha and blackbirds damaged

THE ECONOMIC IMPACT OF BLACKBIRD DAMAGE TO CROPS

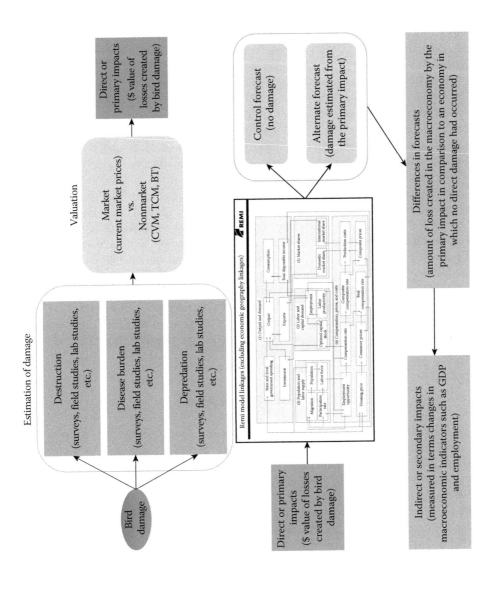

Figure 12.2 Determination of direct (primary) and indirect (secondary) economic impacts.

1.66% of the crop. Overall, blackbird damage to oilseed and confectionery varieties was valued at US$13.26 million and US$4.3 million annually, respectively. The average annual (direct + indirect) economic impact for bird damage to sunflower production in the eight study states was US$29.5 million and reduced employment by 14 jobs (U.S. Department of Agriculture 2015).

12.4.4 Summary

From our review of direct-impact studies, we conclude that blackbird prebreeding populations can sometimes cause significant damage to spring-seeded crops, especially rice (Meanley 1971; Wilson et al. 1989). In fact, severe blackbird damage to newly planted rice can result in a total loss and require that the crop be replanted (Wilson et al. 1989).

The postbreeding blackbird population, which increases about 45% after nesting, likely exceeds 350 million individuals (Peer et al. 2003; Rosenberg et al. 2016). These birds typically damage 1%–2% of ripening grain crops, but most of that damage is within 8 km of a roost where a few farmers suffer most of the damage (Dolbeer 1981; Otis and Kilburn 1988; Wywialowski 1996). These losses would seem inconsequential if damage was distributed evenly; however, bird damage becomes economically significant if individual producers lose >5% of their crop (Dolbeer 1980; Linz and Homan 2011).

Historically, researchers obtained the market value of the affected crop from government publications and subtracted the value of the damage to determine the estimated direct impact to growers (e.g., Dolbeer 1981; Cumming et al. 2005; Hagy et al. 2008; Linz and Homan 2011; Anderson et al. 2013). Direct estimates of bird damage provide valuable information on the impact of damage to individual growers and associated industries. However, additional information on the impacts of reduced production rippling through the economy can be obtained by using regional economic models. Modelling impacts in this way can translate the primary impacts of birds into regional impacts on revenue and jobs, expanding the general public's perception of the potential benefits of preventing or combatting bird damage. These indirect impacts not only help estimate the total impact of bird damage but also help engage a broader audience by highlighting the implications of bird damage for local communities and economies.

12.5 RESEARCH NEEDS

Our review of the literature has revealed an incomplete understanding of the economic damages and control costs arising from blackbirds. Further, recent objective surveys assessing bird damage to corn and rice are not available. Given that these surveys are costly, bioenergetics models and population estimates should be updated on a periodic basis to provide valuable information (see Weatherhead et al. 1982; Peer et al. 2003). These improved data and associated research insights need to be integrated into future management decisions to identify economically efficient (or at least the most cost-effective) management strategies for birds. Lastly, regional economic models should be used to rigorously link primary damage impacts to the appropriate economic sector in order to estimate secondary impacts.

To mitigate damage caused by blackbirds, substantial resources have been committed to management and control efforts. These efforts impose both direct costs (in terms of outlays of actual dollars on lethal and nonlethal control efforts) as well as indirect costs (in terms of lost time and resources devoted to controlling birds). Management and control costs, however, are categorically different than damages inflicted by birds. These two forms of expense should therefore be recorded separately as they are accounted for differently in regional economic models.

REFERENCES

American Birding Association. 2016. The ABS Checklist; Version 7.9.0. http://listing.aba.org/checklist/abachecklist_v7.9.0.pdf (accessed March 12, 2017).

Anderson, A., C. A. Lindell, K. M. Moxcey, W. F. Siemer, G. M. Linz, P. D. Curtis, J. E. Carroll, et al. 2013. Bird damage to select fruit crops: The cost of damage and benefits of control in five states. *Crop Protection* 52:103–109.

Besser, J. F. 1985. *A grower's guide to reducing bird damage to US agriculture crops*. Bird Damage Research Report 340. U.S. Fish and Wildlife Service Denver Wildlife Research Center, Denver, CO.

Besser, J. F., and D. J. Brady. 1986. *Bird damage to ripening field corn increases in the United States from 1971 to 1981*. Fish and Wildlife Leaflet 7, U.S. Fish and Wildlife Service, Washington, DC.

Blackwell, B. F., E. Huszar, G. M. Linz, and R. A. Dolbeer. 2003. Lethal control of red-winged blackbirds to manage damage to sunflower: An economic evaluation. *Journal of Wildlife Management* 67:818–828.

Cummings, J. L., J. L. Guarino, and C. E. Knittle. 1989. Chronology of blackbird damage to sunflowers. *Wildlife Society Bulletin* 17:50–52.

Cummings J. L., C. E. Knittle, and J. L. Guarino. 1986. Evaluating a pop-up scarecrow coupled with propane exploder for reducing blackbird damage to ripening sunflower. *Vertebrate Pest Conference* 12:286–291.

Cummings, J. S. Shwiff, and S. Tupper. 2005. Economic impacts of blackbird damage to the rice industry. *Wildlife Damage Management Conference* 11:317–322.

Decker, D. G., M. L. Avery, and M. O. Way. 1990. Reducing blackbird damage to newly planted rice with a nontoxic clay-based seed coating. *Vertebrate Pest Conference* 14:327–331.

DeGrazio, J. W., J. F. Besser, T. J. DeCino, J. L. Guarino, and E. W. Schafer, Jr. 1972. Protecting ripening corn from blackbirds by broadcasting 4-aminopyridine baits. *Journal of Wildlife Management* 36:1316–1320.

DeGrazio, J. W., T. J. DeCino, J. L. Guarino, and R.I. Starr. 1971. Use of 4-aminopyridine to protect ripening corn from blackbirds. *Journal of Wildlife Management* 35:565–569.

Dolbeer, R. A. 1981. Cost-benefit determination of blackbird damage control for cornfields. *Wildlife Society Bulletin* 9:44–51.

Dolbeer, R. A. 1990. Ornithology and integrated pest management: Red-winged blackbirds *Agelaius phoeniceus* and corn. *Ibis* 132:309–322.

Dolbeer, R. A., N. R. Holler, and D. W. Hawthorne. 1994. Identification and assessment of wildlife damage: An overview. In *The handbook: Prevention and control of wildlife damage*. eds. S. E. Hygnstrom, R. M. Timm, G. E. Larson, A1–A18. University of Nebraska, Lincoln, NE.

Dolbeer, R. A., P. P. Woronecki, A. R. Stickley, Jr., and S. B. White. 1978. Agricultural impact of a winter population of blackbirds and starlings. *Wilson Journal of Ornithology* 90:31–44.

Elser, J. L., A. Anderson, C. A. Lindell, N. Dalsted, A. Bernasek, and S. A. Shwiff. 2016. Economic impacts of bird damage and management in U.S. sweet cherry production. *Crop Protection* 83:9–14.

Guarino, J. L. 1984. Current status of research on the blackbird-sunflower problem in North Dakota. *Vertebrate Pest Conference* 11:211–216.

Hagy, H. M., G. M. Linz, and W. J. Bleier. 2008. Optimizing the use of decoy plots for blackbird control in commercial sunflower. *Crop Protection* 27:1442–1447.

Hothem, R. L., R. W. DeHaven, and S. D. Fairaizl. 1988. *Bird damage to sunflower in North Dakota, South Dakota, and Minnesota, 1979–1981*. Fish and Wildlife Technical Report 15. U.S. Fish and Wildlife Service, Washington, DC.

Hulke, B. S., and L. W. Kleingartner. 2014. Sunflower. In: *Yield Gains in Major U.S. Field Crops*, eds. S. Smith, B. Diers, J. Specht, and B. Carver, pp. 433–457. Soil Science Society of America Special Publication 33, American Society of Agronomy, Crop Science Society of America, and Soil Science Society of America. Madison, WI.

Kandel, H., and G. M. Linz. 2016. Bird damage is an important economic agronomic factor influencing sunflower production. *Wildlife Damage Management Conference* 16:75–82.

Klosterman, M. E., G. M. Linz, A. A. Slowik, and H. J. Homan. 2013. Comparisons between blackbird damage to corn and sunflower in North Dakota. *Crop Protection* 53:1–5.

Linz, G. M., E. H. Bucher, S. B. Canavelli, E. Rodriguez, and M. L. Avery. 2015. Limitations of population suppression for protecting crops from bird depredation: A review. *Crop Protection* 76:46–52.

Linz, G. M., and H. J. Homan. 2011. Use of glyphosate for managing invasive cattail (*Typha* spp.) to protect crops near blackbird (Icteridae) roosts. *Crop Protection* 30:98–104.

Linz, G. M., H. J. Homan, S. W. Werner, H. M. Hagy, and W. J. Bleier. 2011. Assessment of bird management strategies to protect sunflower. *BioScience* 61:960–970.

Lowther, P. E. 1993. Brown-headed cowbird (*Molothrus ater*). No. 47. *The birds of North America*, ed. P. G. Rodewald. Cornell Laboratory of Ornithology, Ithaca, NY. https://birdsna.org/Species-Account/bna/species/bnhcow (accessed November 12, 2016).

Meanley, B. 1971. *Blackbirds and the southern rice crop*. Resource Publication *100*. U.S. Department of Interior, U.S. Fish and Wildlife Service, Washington, DC. http://pubs.usgs.gov/rp/100/report.pdf (accessed June 28, 2016).

Otis, D. L., and C. M. Kilburn. 1988. Influence of environmental factors on blackbird damage to sunflower. Technical Report No. 16, U.S. Fish and Wildlife Service, Department of the Interior, Washington, DC.

Peer, B. D., and E. K. Bollinger. 1997. Common grackle (*Quiscalus quiscula*). No. 197. *The birds of North America*, ed. P. G. Rodewald. Cornell Laboratory of Ornithology, Ithaca, NY. https://birdsna.org/Species-Account/bna/species/comgra (accessed November 12, 2016).

Peer, B. D., H. J. Homan, G. M. Linz, and W. J. Bleier. 2003. Impact of blackbird damage to sunflower: Bioenergetics and economic models. *Ecological Applications* 13:248–256.

Pimentel, D., R. Zuniga, and D. Morrison. 2005. Update on the environmental and economic costs associated with alien-invasive species in the United States. *Ecological Economics* 52:273–288.

Ralston, S. T., G. M. Linz, W. J. Bleier, and H. J. Homan. 2007. Cattail distribution and abundance in North Dakota. *Journal of Aquatic Plant Management* 45:21–24.

Richardson, H. W. 1972. *Input–output and regional economics*. Wiley, New York, NY.

Rosenberg, K. V., J. A. Kennedy, R. Dettmers, R. P. Ford, D. Reynolds, C. J. Beardmore, P. J. Blancher, et al. 2016. *Partners in Flight Landbird Conservation Plan: 2016 Revision for Canada and Continental United States*. Partners in Flight Science Committee. http://www.partnersinflight.org/ (accessed September 25, 2016).

Stickley, A. R., Jr., R. T. Mitchell, J. L. Seubert, C. R. Ingram, and M. L. Dyer. 1976. Large-scale evaluation of blackbird frightening agent 4-aminopyridine in corn. *Journal of Wildlife Management* 40:123–131.

Stone, C. P., D. F. Mott, J. F. Besser, and J. W. DeGrazio. 1972. Bird damage to corn in the United States in 1970. *Wilson Bulletin* 84:101–105.

Treyz, G. I., D.S. Rickman, and G. Shao. 1991. The REMI economic-demographic forecasting and simulation model. *International Regional Science Review* 14:221–253.

Twedt, D. J., and R. D. Crawford. 1995. Yellow-headed blackbird (*Xanthocephalus xanthocephalus*). No. 192. *The birds of North America*, ed. P.G. Rodewald. Cornell Laboratory of Ornithology, Ithaca, NY. http://bna.birds.cornell.edu/bna/species/1921995 (accessed June 23, 2016).

U.S. Department of Agriculture. 2015. *Blackbird damage to the sunflower crop in ND, SD, KS, CO, MN, TX, NE, and VT*. U.S. Department of Agriculture, Animal and Plant Health Inspection Service, Wildlife Services, National Wildlife Research Center, Fort Collins, CO.

U.S. Department of Agriculture. 2016. *Quick Stats 2.0*. U.S. Department of Agriculture, National Agricultural Statistics Service, Washington, DC. https://quickstats.nass.usda.gov/ (accessed September 25, 2016).

Weatherhead, P. J., H. Greenwood, S. H. Tinker, and J. R. Bider. 1980. Decoy traps and the control of blackbird populations. *Phytoprotection* 61:65–71.

Weatherhead, P. J., S. Tinker, and H. Greenwood. 1982. Indirect assessment of avian damage to agriculture. *Journal of Applied Ecology* 19:773–782.

Wilson, E. A., E. A. LeBoeuf, K. M. Weaver, and D. J. LeBlanc. 1989. Delayed seeding for reducing blackbird damage to sprouting rice in southwestern Louisiana. *Wildlife Society Bulletin* 17:165–171.

Wywialowski, A. P. 1996. Wildlife damage to field corn in 1993. *Wildlife Society Bulletin* 24:264–271.

Yasukawa, K., and W. A. Searcy. 1995. Red-winged blackbird (*Agelaius phoeniceus*). No. 184. *The birds of North America*, ed. P. G. Rodewald. Cornell Lab of Ornithology, Ithaca, NY. https://birdsna.org/Species-Account/bna/species/rewbla (accessed September 25, 2016).

CHAPTER 13

The Future of Blackbird Management Research

Page E. Klug
National Wildlife Research Center
Bismarck, North Dakota

CONTENTS

13.1	Blackbird Ecology	219
13.2	Management Tools	221
	13.2.1 Lethal Control	221
	13.2.2 Chemical Repellents	222
	13.2.3 Frightening Devices	223
	13.2.4 Evading Strategies	225
13.3	Economics and Human Dimensions	227
13.4	Conclusions	228
References		229

Human society values birds for their intrinsic and aesthetic value as well as the ecosystem services they provide as pollinators, consumers of pests, and distributors of nutrients and seeds (Wenny et al. 2011). At the same time, conflict between birds and humans is an age-old phenomenon that has persisted as society has transformed and the scale of agriculture has expanded (Conover 2002). Managing conflict between birds and agriculture is challenging for many reasons. Foremost, the need to consider both human welfare and conservation of protected bird species is paramount, with nonlethal management methods preferred to lethal measures from societal, economical, and ecological standpoints (Miller 2007; Linz et al. 2015). Second, methods must be effective, practical, and economical for agricultural implementation. Finally, management methods must overcome characteristics that make birds difficult to manage including uncertainty in population estimates, fecundity, mobility, and adaptive behaviors. All challenges are compounded when attempting to establish management methods that fit within modern agricultural practices, while simultaneously supporting conservation efforts to protect wildlife.

Labor-saving devices and methodologies resulting from agricultural advances in mechanical, chemical, genetic, and information technologies have facilitated a shift to larger crop fields, a broader range of suitable habitat for a variety of crops, and consolidated farms in North America (MacDonald et al. 2013). This shift to large, less labor-intensive farms has supported the ability to feed an ever-increasing human population but has complicated the relationship between humans and wildlife. Modern agriculture directly impacts wildlife by altering natural habitat, resulting in the increase of species able to thrive in agricultural landscapes and the decline of species unable to adapt. Thus, agriculture often provides increased carrying capacity for species responsible for agricultural damage (Van Vuren and Smallwood 1996). However, changes in harvest efficiency have resulted in less crop waste and reduced availability of high-energy foods available to birds postharvest, potentially placing common farmland birds at risk of decline (Krapu et al. 2004; Galle et al. 2009). Nevertheless, vertebrate species able to adapt to the agricultural landscape often reach pest levels, resulting in producers seeking tools to reduce damage, tools that have not necessarily advanced in concert with modern agriculture.

Red-winged blackbirds (150 million; *Agelaius phoeniceus*), brown-headed cowbirds (120 million; *Molothrus ater*), common grackles (69 million; *Quiscalus quiscula*), and yellow-headed blackbirds (15 million; *Xanthocephalus xanthocephalus*) are among the most numerous birds in North America (Rosenberg et al. 2016). This book has identified conflicts between blackbirds and agricultural commodity groups including livestock, rice, corn, sunflower, and numerous specialty crops (Dolbeer 1990; Cummings et al. 2005; Anderson et al. 2013; Klosterman et al. 2013; Figure 13.1). Continued progress in development of blackbird management methods and acquisition of baseline knowledge as to its impacts on blackbird populations are needed at local, regional, and national scales.

In this chapter, I evaluate gaps in knowledge and potential research directions. I address the following topics: (1) blackbird biology at the species, population, and community levels; (2) the influence of changing landscapes on blackbirds and agricultural damage in terms of agricultural practices, habitat, and climate change; (3) the limitations of lethal and nonlethal management tools (i.e., repellents, frightening devices, and evading strategies) and how research can optimize techniques or facilitate new tool discovery; and (4) economic evaluation of management and human dimensions.

Figure 13.1 Evidence of blackbird damage to sunflower in which expelled shells are left on the back of the downward-facing sunflower head. (Courtesy of Conor Egan/USDA Wildlife Services.)

13.1 BLACKBIRD ECOLOGY

Although the red-winged blackbird is one of the most studied wildlife species, much is left to understand about its biology and the biology of other blackbird species. The majority of blackbird literature focuses on mating systems, sexual selection, and breeding behavior (Searcy and Yasukawa 1995; Beletsky 1996; Beletsky and Orians 1996), with additional focus on avian communication and social bonds of species with both territorial and colonial behaviors (Beletsky 1996). Beyond the breeding season, most research has been conducted in the context of blackbirds as pests when large roosts or flocks come into conflict with human society (Conover 2002). Searcy and Yasukawa (1995) listed gaps in our knowledge of red-winged blackbirds, including several that influence management in relation to agriculture. I concur that little is known about blackbird physiology in relation to migration, behavior of independent young birds, and overall effect of species and subsets of populations on agriculture and human health and safety. Brown-headed cowbirds have been the focus of much research due to their unique nest parasitism behavior, potential influence on birds of conservation concern, and agricultural crop damage, but many data gaps exist for cowbirds as well as other less studied blackbirds.

The impact of yellow-headed blackbirds, common grackles, and brown-headed cowbirds on agriculture are thought to be substantially less than red-winged blackbirds due to factors such as smaller population sizes, habitat use, feeding habits, or earlier molt and migration (Besser 1985; Twedt et al. 1991; Homan et al. 1994; Peer et al. 2003; Twedt and Linz 2015). Research has mainly focused on management tools to address damage from red-winged blackbirds (e.g., Dolbeer 1990; Linz et al. 2011; U.S. Department of Agriculture 2015), but the impact of other species holds potential to change as avian populations respond to habitat and climate change (Homan et al. 1994). Additionally, tools aimed at red-winged blackbirds may negatively impact species with small or declining populations (e.g., Brewer's blackbirds [*Euphagus cyanocephalus*] and rusty blackbirds [*Euphagus carolinus*]) or may impact the continental population of red-winged blackbirds (Greenberg et al. 2011; Sauer et al. 2014). Understanding the importance of the southern United States as overwintering habitat and the Prairie Pothole Region (Bird Conservation Region [BCR] 11) as a stronghold for breeding blackbirds experiencing continental declines is necessary to assure protection of a native species and to maintain a balance between human and wildlife well-being (Weatherhead 2005; Strassburg et al. 2015; Chapter 7, this volume). Monitoring changes in both winter roost and breeding numbers is essential, and evaluating possible factors influencing abundance and distribution should be a research focus. Updated take models using accurate demographic information are necessary, given declining blackbird populations, specifically brown-headed cowbirds, where aggressive population reduction may not be warranted (Peer et al. 2003; Chapter 5, this volume). Thus, the status of blackbird populations must be addressed at multiple scales and the influence of management on demography explored throughout their annual cycle, especially considering the impact of habitat and climate change (e.g., Blackwell and Dolbeer 2001).

Agricultural stakeholders have voiced concerns about limitations for effective bird damage management and identified three critical research needs: (1) development of national management plans for each blackbird species; (2) development of management tools, including species-specific lethal methods and chemical, auditory, and visual repellents; and (3) research on blackbird biology in relation to damage and avian-borne diseases (U.S. Department of Agriculture 2008). Any program to manage wildlife must be in compliance with the National Environmental Protection Act and the Endangered Species Act, which require research-based information on ecosystem impacts. Thus, to justify management actions, baseline biological information is needed for all blackbirds, with attention also given to nontarget animals.

Research is needed to understand blackbird population dynamics, optimal deployment of management tools, and relationship to crop damage. Studies evaluating blackbird response to climate, habitat, and management at finer scales than publicly available data (e.g., North American Breeding

Bird Survey) would give a better understanding of population trends and impact of management within regions of concern (e.g., overwintering, migration, and breeding grounds; Chapter 6, this volume). For instance, birds may alter migration timing or location of overwintering sites with a warming climate (Van Buskirk et al. 2009). At the same time the proliferation of concentrated animal feedlots (i.e., concentrated, high-energy food) and changes in crop varieties (i.e., genetically modified crops that reduce waste and weed seeds) have altered food distribution and availability, creating complex situations with unknown impact on bird populations and behavior (Gibbons et al. 2006). Regional monitoring programs could elucidate how blackbird populations are changing in concert with land cover or how climate change may be impacting migration timing and onset of breeding and the ultimate impact on crop damage (Nelms et al. 1994).

Information about how molt patterns influence the timing of migration and how that may be affected by climate change is important, especially for yellow-headed blackbirds, where cold sensitivity is a factor in early emigration (Chapter 3, this volume). Additionally, the molt pattern of common grackles has yet to be described in relation to impact on agriculture (Chapter 4, this volume). Although diet and molt pattern in relation to agricultural damage have been evaluated for red-winged and yellow-headed blackbirds, updated data would elucidate changes occurring with changing habitat, climate, and agricultural practices (Linz et al. 1983; Twedt et al. 1991; Twedt and Linz 2015; Chapter 2, this volume). Further investigations into migration, molt patterns, and food habits can be evaluated using stable isotope markers to understand the full annual cycle of blackbirds at a continental scale (Werner et al. 2016). An understanding of species' biology, such as molt, migration, habitat use, diet, dispersal, survival, and reproductive success, could link different periods in the annual cycle and lead to new approaches for managing conflict with blackbirds.

A changing climate will impact not only the phenology of avian populations and natural habitat but also crop phenology and crop variety. Thus, the synergy among climate, land use, and avian populations should be explored (Forcey et al. 2015; Chapter 6, this volume). Changes in the type, amount, and distribution of woody vegetation could impact blackbird populations, especially grackles and brown-headed cowbirds (Rothstein 1994; Peer and Bollinger 1997; Wehtje 2003). Increased abundance of grackles in North Dakota has been linked to warmer temperatures (Forcey et al. 2015), but the reasons behind their range expansion in the West deserves further attention (Marzluff et al. 1994). While blackbirds may respond to loss of forested habitat at their overwintering sites in the southeastern United States, red-winged blackbirds and yellow-headed blackbirds may respond more to oscillations between wet and dry years at their breeding sites (BCR 11) due to their dependence on wetlands. Thus, regional climate projection models in conjunction with land-use data could forecast the impact of climate on blackbirds and help assess future needs and allocation of management (Forcey et al. 2015).

Although the relationship of blackbirds to local and regional habitat is imperative, response to habitat along continental migration pathways should also be emphasized. As technology for tracking individual birds becomes more sophisticated (Bridge et al. 2011), dispersal and migration patterns for each blackbird species and subsets of their populations (i.e., age class and sex) can be evaluated to complement previous estimates (Dolbeer 1978, 1982; Moore and Dolbeer 1989; Homan et al. 2004). With the exception of brown-headed cowbirds (Dufty 1982; Rothstein et al. 1984; Goguen and Mathews 2001), few tracking studies have been conducted to evaluate sociality, habitat use, survival, and migration in blackbirds (Homan et al. 2004). Foremost, the importance of various habitats used during the annual cycle and its impact on physical condition, migration timing, and reproductive success has not been addressed but could elucidate how management during winter (i.e., rice), migration (i.e., concentrated animal feedlots), and postbreeding (i.e., corn, rice, sunflower) seasons are interconnected (e.g., Marra et al. 2015). Movement ecology throughout the annual cycle is also fundamental to understanding population status and the impact of management targeting a specific region, species, sex, or age class.

THE FUTURE OF BLACKBIRD MANAGEMENT RESEARCH

Understanding the survival of blackbirds by species, age class, and sex is crucial to determining impacts of management in relation to other sources of mortality and natural regulation of populations (Fankhauser 1971; Bray et al. 1979; Stehn 1989). Hatch-year blackbirds hold potential to inflict damage to crops, given that fledglings are the driver behind the annual population numbers of red-winged blackbirds increasing from an estimated 170 million at the start of nesting to 328 million postbreeding (Chapter 8, this volume). As chemosterilant technologies advance, the feasibility of species-specific reproductive inhibition techniques for regionally managing blackbird populations should be explored under the limits of biological and economic feasibility as well as environmental regulations (Fagerstone et al. 2010). Assessing postfledging ecology would also improve management tool distribution, management tool effectiveness, and demographic models for this age class (Chapter 11, this volume). Research projects focused on migration and dispersal of population subsets could improve bioenergetic and economic models for estimating species-specific, region-wide crop damage and impact of management (Peer et al. 2003).

13.2 MANAGEMENT TOOLS

Many management tools, in some form, have been in existence for millennia (Benson 1937; Warnes 2016). Traps, poisons, and scarecrows have been used since prehistoric times, continue to be used today, and hold potential for the future (Conover 2002). Historically, farmers were able to protect resources within a given distance of their domicile and could dedicate significant time to the task. Today, the limited range of most tools is dwarfed by the size of the field to be protected, thus reducing their efficacy. Regardless, agricultural producers still use various techniques to disperse blackbirds, including repellents, decoy crops, firearms, propane cannons, pyrotechnics, and habitat management (Linz et al. 2011). In addition to inconsistent results, methods are often labor intensive and cost prohibitive, especially at the broad scales seen in current agriculture. Integrated pest management is often touted to optimize management, but few studies evaluate the combined effectiveness of methods (Avery 2002).

13.2.1 Lethal Control

Major challenges exist in attempts to benefit agriculture by lethal control of blackbirds, including large continental population sizes, magnitude of natural annual turnover, compensatory factors of increased survival and reproductive success, and migration dynamics (Chapter 7, this volume). Numerous programs have been implemented to reduce blackbird numbers, with limited reduction in crop damage (Linz 2013). First, blackbirds inflicting or about to inflict crop damage may be taken legally in the United States without a permit under an existing depredation order for blackbirds, cowbirds, grackles, crows, and magpies (50 CFR 21.43), but this small-scale control only functions to temporarily scare birds from a localized area. On a broader scale, Blackwell et al. (2003) showed that the cost of annually removing up to 2 million red-winged blackbirds during spring migration would not result in substantial damage reduction during the late-summer sunflower maturity. Additionally, the estimated sustainable allowed take of female red-winged blackbirds should range between 392,000 and 783,000 for BCR 11 (Chapter 11, this volume). Given evidence that the number of birds allowed for a sustainable take is considerably less than the estimated number needed to reduce crop damage, the U.S. Fish and Wildlife Service and public sentiment will likely not support broad-scale lethal control. The need to develop methods for culling large numbers of blackbirds is limited in both feasibility and cost-effectiveness; therefore nonlethal methods should be emphasized (Linz et al. 2015; Chapter 7, this volume).

13.2.2 Chemical Repellents

Chemical repellents have the potential to be a cost-effective method to protect large, commercial fields if used in conjunction with other tools to disperse birds, such as frightening devices, evading strategies, and habitat management (Avery 2002; Hagy et al. 2008; Linz and Homan 2011). Although a variety of chemicals have been tested for repellency (Chapter 8, this volume), registered repellents are restricted to nonlethal formulations shown to be safe for the environment and food consumption. Thus, one avenue of research is the continued evaluation of naturally occurring compounds and formulations, including mixtures of repellents and visual deterrents (Avery 2002). For instance, Werner et al. (2014a) found that the addition of nontoxic visual cues added to anthraquinone (AQ) formulations may enhance avian repellency at lower repellent concentrations. Although this is promising in that EPA registrations of repellents are more likely at lower chemical concentrations, execution of this approach in the field needs to be explored for each crop to maximize efficacy and minimize cost.

AQ-based repellents have shown >80% repellency in the lab (Avery et al. 1997; Werner et al. 2009), but translating efficacy from the lab to field is a challenge at the scale of commercial agriculture (Dolbeer et al. 1998; Kandel et al. 2009; Werner et al. 2011, 2014b; Niner et al. 2015). Issues arise when applying any repellent to all major food crops impacted by blackbirds including rice, corn, and sunflower (Werner et al. 2005; Carlson et al. 2013; Werner et al. 2014b). For example, one obstacle to using AQ in ripening sunflower is applying sufficient repellent directly on the face of the sunflower to repel birds while simultaneously minimizing AQ residues on harvested seed. As sunflower matures, the head faces down, making the preferred aerial application problematic given that blackbirds must ingest the repellent to be effective (Avery et al. 1997). Therefore, research should focus on developing application strategies such as ground rigs equipped with drop nozzles to apply chemicals directly to the sunflower face (Mullally 2010; Wunsch et al. 2016; Figure 13.2). Even with effective application technology, achenes will only be partially treated because most of each achene is concealed within the sunflower head or protected by disk flowers (Figure 13.3). However, reduced achene coverage may be sufficient given that birds must remove the adulterated disk flowers or manipulate exposed seed during consumption. Corn and rice have similar application issues, with the target seed being protected by vegetative components of the plant. Understanding avian feeding behavior on crops with varying repellent coverages will provide application details for improved effectiveness, given that repellent coverage is variable and often <100% at the plant scale (Avery 1985).

Researchers should explore crop-specific feeding behavior of blackbirds at various scales ranging from the individual plant and field to the diverse agricultural landscape. Identifying the behavior of each blackbird species and population subsets responsible for damage will inform repellent application and increase cost-effectiveness through precision agriculture. For example, research on blackbird

Figure 13.2 Small-plot ground rig equipped with 360 Undercover® drop nozzles (360 Yield Center, LLC; Morton, IL) to apply avian repellent under the crop canopy and increase application to targeted area (e.g., sunflower face or corn husk). Use of trade names does not imply endorsement by the U.S. government. (Courtesy of Page Klug/USDA Wildlife Services; and 360 Yield Center, https://360yieldcenter.com/products/360-undercover/.)

THE FUTURE OF BLACKBIRD MANAGEMENT RESEARCH

Figure 13.3 Agricultural crops are often difficult to protect with avian repellents due to the growth form of the plant acting to decrease the amount of repellent on the ingested seed. For example, corn is protected by a husk, rice is protected by awls, and, as pictured here, sunflower is protected by disk flowers and seed husks. (Courtesy of Page Klug/USDA Wildlife Services.)

foraging, habitat use, and flocking behavior could inform the temporal and spatial distribution of repellent at the field scale (Avery 1989). A repellent with a visual cue could be applied with a drop-nozzle–equipped ground rig in areas where birds are likely to learn the negative effects of the repellent-treated crop, and the remainder of the field could be treated aerially, reducing cost. It is important to understand how repellent should be distributed on the landscape as a function of realized damage and the level of partial repellent treatment needed to maintain repellent cost-effectiveness. The use of chemical repellents involves considerable expense in production and application; thus cost–benefit studies must be done to ensure application only in favorable situations (Dolbeer 1981).

In addition to evaluating spatial distribution of damage and repellent application, timing during the growing season must be considered (Bridgeland 1979). The functional cue to which blackbirds respond for onset of damage and food selection in varying crops (i.e., rice, corn, and sunflower) needs to be further addressed. The presence of insects and weeds in fields has been thought to influence the establishment of feeding areas, and this has not been evaluated in relation to cues derived from the crops themselves (Linz et al. 2011). Although vision is a large part of how birds sense their environment, the role of gustatory, olfactory, and chemesthetic senses must also be addressed and may be differentially important or work in concert at varying scales from selection of roosts and crop fields to selection of seeds (Mah and Nuechterlein 1991; Mason et al. 1991; Avery and Mason 1997).

13.2.3 Frightening Devices

Frightening devices have a long history in the management of human–wildlife conflict and hold the possibility for effective hazing of blackbirds in agricultural fields (Bomford and O'Brien 1990; Gilsdorf et al. 2002; Chapter 9, this volume). Factors limiting the success of frightening devices include bird behaviors such as limited mobility during feather molt, strong fidelity to established feeding areas, and habituation to nonrandom noise (Washburn et al. 2006). The limitations of the devices themselves include extent of effectiveness in space and time, immobility, and labor intensity (Linz and Hanzel 2015). Research is needed to develop frightening devices that can respond to the needs of broad-scale agriculture.

Well-designed studies focused on blackbirds are needed; there are few published reports of frightening devices that include testing against blackbirds under field conditions. Modifications to current frightening devices such as propane cannons and pyrotechnics are necessary to increase efficacy and include variation in directionality and timing. Lethal reinforcement is often referenced to limit habituation; however, limited scientific evidence is available to support this contention and differences may exist depending on species (Washburn et al. 2006; Baxter and Allan 2008; Seamans et al. 2013; Chapter 9, this volume). Evaluation of cost-effectiveness is scant in relation to the sheer number of frightening devices on the market, and resources for objective testing of products are limited. Therefore, a strong understanding of the biology of the animal and environmental conditions in which the frightening device would be deployed are necessary for thoughtful selection of devices to be evaluated.

Species-specific frightening devices may be beneficial, especially for the few species that cause the majority of damage (Swaddle et al. 2016). Introduced noise at frequencies interrupting avian communication holds the potential to deter birds from areas of concern. The technology has been shown to be successful in reducing feeding rate in captivity and in reducing bird activity in airfields (Mahjoub et al. 2015; Swaddle et al. 2016) but has yet to be evaluated in agricultural settings. Swaddle et al. (2016) suggested that if birds are not displaced from agricultural areas, the "sonic net" may influence antipredator behavior by masking alarm and predator calls, causing increased vigilance and decreased feeding (Lima and Bednekoff 1999). These sonic nets are appealing in that habituation is decreased, but limitations in spatial extent are evident along with power source restrictions. The effectiveness of disruptive sound for deterring birds is species-specific and may vary with environment but is worth pursuing.

Another promising technology in wildlife damage management is unmanned aircraft systems (UAS), which have already been deployed by producers to protect agricultural fields (BBC 2014; Kerzman 2015) and are being evaluated for use in wildlife and agricultural monitoring (Christie et al. 2016; Figure 13.4). A main benefit to UAS is the ability to overcome mobility limitations of stationary devices and to create a dynamic object. Research is needed to evaluate the feasibility of UAS to mitigate bird damage by evaluating avian physiological and behavioral responses and potential habituation or tolerance (e.g., Ditmer et al. 2015). Researchers also need to establish best practices (i.e., color, size, shape, approach, altitude, and speed) for entities looking to buy and incorporate UAS in blackbird hazing. The potential efficacy of UAS as hazing tools will depend on bird

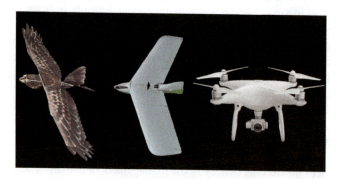

Figure 13.4 Unmanned aircraft systems (UAS) hold potential for use in wildlife and agricultural monitoring as well as frightening devices to reduce the impact of pest species. The potential efficacy of UAS as hazing tools will likely depend on bird detection and response to the flight dynamics. Research is needed to understand avian response to UAS platforms, such as multirotor quadcopters, traditional fixed-wing models or fixed-wing models shaped like a predator. Use of trade names does not imply endorsement by the U.S. government. (Courtesy of Page Klug/USDA Wildlife Services, HobbyKing.com®, https://hobbyking.com/en_us/eagle-epp-slow-flyer-1430mm-w-motor-kit.html; and DJI Technology Co., Ltd.®, http://www.dji.com/products/drones#consumer-nav.)

detection and response to UAS design and flight dynamics. Avoidance responses might be enhanced by designing vehicles based on a perceptual model of red-winged blackbird visual capabilities, so as to enhance detection under varying ambient conditions and responses to UAS during hazing (Blackwell et al. 2012). As technology continues to advance, UAS is a rich area for research with the potential for completely autonomous flight, which would act to substantially decrease labor by removing the need for a human operator and allow the aircraft to deploy when necessary in time and space (Grimm et al. 2012).

Current limitations of UAS as hazing devices include FAA regulations as well as a lack of onboard bird detection systems (Ampatzidis et al. 2015). Thus, signal processing research is needed to improve technology for identifying animal presence or abundance through real-time audio or visual monitoring (Pijanowski et al. 2011; Pérez-García 2012). Labor-saving approaches in wildlife monitoring would allow for measures of blackbird activity and, along with the distribution of crop damage, would allow a better understanding of factors that influence regional dynamics and rigorous testing of methods at the landscape scale. Another benefit of identifying birds in real time would be the ability to develop a detector for initiating scare devices or deploying an autonomous UAS when a nuisance species enters a protected area (Gilsdorf et al. 2002; Ampatzidis et al. 2015). Combining UAS technology with a primary repellent (e.g., methyl anthranilate) may also function to reduce habituation and increase negative connotation with the UAS, if the system released a primary repellent only when a pest scenario arose (Ampatzidis et al. 2015). Difficulty arises in deploying networks that can identify the presence of pest animals at broad landscape scales and is further complicated by topographically complex landscapes and fast-moving, small-bodied organisms. Until research in signal processing advances, use of automated UAS would include predetermined paths to patrol areas harboring the majority of damage (Grimm et al. 2012). Predetermined paths run the risk of habitation, but paths could be designed to vary in space and time and focus on areas of high risk.

13.2.4 Evading Strategies

Habitat management plays a fundamental role in reducing carrying capacity of blackbirds (Linz and Homan 2011; Chapter 10, this volume). The availability of nesting or roosting habitat as a function of water availability (hence cattail stands [*Typha* spp.]) is a likely factor limiting blackbird populations, given that seed-based food is abundant preharvest on agricultural landscapes such as in the Prairie Pothole Region. Management strategies for reducing damage should consider weather effects in addition to broad-scale landscape, given that such factors have been shown to contribute to blackbird relative abundance by impacting wetland habitat (Forcey et al. 2015). Cost-effective and environmentally safe methods to restore wetlands and reduce the dominance of invasive cattails and its impact on avian abundance need to be explored further and include traditional management such as burning, grazing, disking, and herbicides as well as studies exploring the utility of biological control (Linz et al. 2003; Kostecke et al. 2004). Distributing birds across the landscape by managing cattail stands has been shown to be a valuable approach to reducing damage experienced by producers while conserving valued wildlife and thus should be promoted (Linz and Homan 2011).

The use of crop varieties resistant to damage by blackbirds has also shown promise and is worthy of future development, especially in the era of genetic engineering. For example, Dolbeer et al. (1986, 1995) showed for both sweet and field corn that varieties with thicker, longer husks that extend beyond the ear tip have less damage than ears with lesser husks. Research in rice has also shown that modifications to plant morphology (e.g., awns and long, erect flag leaves) could increase resistance to bird depredation (Avery 1979; Abifarin 1984; Bullard 1988). Classical sunflower breeding techniques have been used to develop bird-resistant hybrids with limited utility, given that traits thought to be resistant to birds such as thick, white, fibrous hulls and increased chlorogenic acid and anthocyanin in the hull are related to unacceptable oil content and agronomic yield (Dolbeer et al. 1986; Parfitt and Fox 1986;

Mah et al. 1990; Mason et al. 1991). Although genetic engineering holds potential for corn and rice, regulations for genetically modified sunflower seed are strict due to potential for gene flow between cultivated and wild sunflower (*Helianthus annuus*) in North America (Burke et al. 2002; Cantamutto and Poverene 2007). Thus, sunflower breeders interested in developing bird-resistant hybrids may instead focus on double-haploid technology in which desired cultivars can be developed much faster compared to conventional breeding methods (Jan et al. 2011; Linz et al. 2011). In addition to this, a new frontier in genome engineering with CRISPR-Cas9 technology provides opportunities for incorporating bird-resistance into various crops without the presence of foreign DNA (Doudna and Charpentier 2014).

When implementing management tools to disperse or discourage blackbirds from feeding on a crop, alternative sources of foods are necessary to improve efficacy (Avery 2002). Wildlife conservation food plots (WCFP; also known as *diversionary feeding*, *decoy plots*, and *supplemental*, *lure*, or *trap crops*) are used to entice animals away from situations in which they are viewed as pests and have the potential to be a socially acceptable conservation action to avoid pest scenarios while providing wildlife habitat (Kubasiewicz et al. 2016; Chapter 10, this volume). The few studies that have assessed efficacy of WCFP for blackbirds indicate juxtaposition of WCFP and other less valuable crops is an important factor (Hagy et al. 2008; Linz et al. 2011; Klosterman et al. 2013). Limitations to implementing WCFP include finding an alternative food that blackbirds would prefer over an abundant and calorically dense agricultural crop, siting of WCFP, and cost-effectiveness for producers. A perennial sunflower variety may be developed that could be used as an alternative food source for birds and reduce the cost of WCFP (Kantar et al. 2014; Linz et al. 2014). Planting diversity, crop varieties, plant spacing, planting times, field size, and plot locations are research avenues that can be explored to increase the cost-effectiveness of WCFP (Cummings et al. 1987; Hagy et al. 2008).

Risk factors at the landscape and farm scale need to be evaluated with the potential of habitat manipulation to minimize risk or to identify where not to grow a susceptible crop (Lindell et al. 2016). Bird damage to agricultural crops has been shown to be greater on the edge (Fleming et al. 2002), near tall trees on an otherwise open habitat (Schäckermann et al. 2014), and near cattail marshes (Dolbeer 1980; Otis and Kilburn 1988; Figure 13.5). Additionally, research on how to use a less valuable crop (e.g., corn) as an alternate food source to protect a more valuable crop (e.g., sunflower) might be useful in some situations. To effectively manage bird damage, information is needed as to the influence of habitat composition and cover (e.g., target crop, alternate crops, wetlands, grassland,

Figure 13.5 In the Prairie Pothole Region of North America, blackbirds roost in cattail marshes with flight lines emanating from roosting to feeding areas. Bird damage to agricultural crops has been shown to be greater on habitat edges, near tall trees on an otherwise open habitat, and near roosting habitat such as cattail marshes, all of which are evident in this picture. (Courtesy of USDA Wildlife Services.)

and woodlots; Hagy et al. 2008; Linz et al. 2011; Forcey et al. 2015), timing and synchronization of planting and harvest (Wilson et al. 1989; Samanci 1995; Killi et al. 2004; Alizadeh 2009), and within-field characteristics such as weed and insect abundance, field size and shape, crop density, and short-stature sunflower (Otis and Kilburn 1988; Linz et al. 2011; Trostle et al. 2013). Studies evaluating bird abundance and distribution of crop damage as a function of landscape can inform cropping strategies, location of WCFP, and habitat management implementation (Cummings et al. 1987; Hagy et al. 2008).

In addition to understanding the spatial distribution of damage across the landscape, we must also consider the timing of management tool deployment. Understanding the growth stage at which visual cues of sunflower, rice, and corn indicate palatability to a blackbird would inform the growth stage to apply a tool (Cummings et al. 1989; Wilson et al. 1989; Dolbeer 1990). Understanding how blackbirds perceive their environment and select habitat is vital to being able to influence birds to avoid valued agricultural crops and instead use alternative forage (e.g., Hagy et al. 2008). Future research aimed at understanding the characteristics of a plant or field that make it susceptible to damage will help direct the spatial distribution of management tools, identify high risk areas, and help develop or optimize management tools (Cummings et al. 1989; Dolbeer 1990; Okurut-Akol et al. 1990; Somers and Morris 2002).

13.3 ECONOMICS AND HUMAN DIMENSIONS

A better understanding of economic damage from each blackbird species and the cost of control are needed in all impacted commodities (i.e., livestock, rice, corn, sunflower; Chapter 12, this volume). Estimates of crop damage are the baseline value upon which the cost-effectiveness of a management program can be evaluated (e.g., Dolbeer 1981); therefore, accurate estimates of damage are necessary for making sound decisions on management strategies. Damage estimates at regional scales could be enhanced by using remotely sensed data or using UAS to monitor crop damage (Anderson and Gaston 2013). For example, a normalized difference water index may be able to signal areas with high bird damage in sunflower. Near sunflower harvest, the vegetative parts of the plants are desiccated but the sunflower seeds still contain water. Consequently, heads with reduced seeds would have lower water content, thus signaling damage (Figure 13.6). Alternatively, bioenergetics and economic models along with population estimates of blackbird species are a labor-saving method to estimate damage and should be routinely updated and integrated into management strategies for impacted commodities such as rice, which has not yet been evaluated using this tool (Weatherhead et al. 1982; Peer et al. 2003).

Research is also needed to survey producers about blackbird abundance, crop damage, management tools, and socioeconomic standing to provide a better understanding of varying attitudes and factors influencing producer tolerance and response to damage (Conover 1998; Jacobson et al. 2003). Small-scale farmers or those attempting to initiate a new crop may be hit the hardest economically and thus may be more ardent about finding solutions to reduce bird damage. Conversely, a percentage of producers see no need to control birds or use management tools, and understanding the characteristics of individuals with this viewpoint would inform how to best reach out to concerned producers (Conover 2002). Likewise, producers implementing organic methods are increasing (U.S. Department of Agriculture 2016) and require a different suite of bird management tools than traditional farmers, which may provide opportunity for developing nontraditional approaches to human–wildlife conflict.

Multidisciplinary approaches to understanding conflict between blackbirds and agricultural producers could be developed by combining ecological, socioeconomic, and consumer marketing approaches. For example, consumer interest in food production practices such as eco-labels has shown to increase the market value of fruit crops (Oh et al. 2015). Although connections between producer and consumer are less direct in commodities such as rice, corn, and sunflower compared to fruits and vegetables, small-scale or organic producers may find marketing "bird-friendly" practices beneficial

Figure 13.6 Severely damaged sunflower head close to harvest. Vegetative parts of the sunflower plant are dry near harvest, but the sunflower seeds still contain water. Thus, a normalized difference water index collected through remotely sensed imagery may be able to signal areas of high sunflower damage. (Courtesy of Conor Egan/USDA Wildlife Services.)

(Jacobson et al. 2003). Discovery and testing of nonlethal management tools (e.g., WCFP) could first be tested on small-scale production areas such as organic farms and scaled up to traditional broad-scale agriculture. For example, marketing of bird-friendly products by commercial birdseed companies and avian conservation groups could subsidize producers participating in a WCFP program. Diverse WCFP in terms of crop and hybrid variety (e.g., sunflower, millet, and safflower) could be planted and harvested for sale as bird-friendly birdseed mixes or kept as overwintering habitat for nontarget animals. Such nontraditional approaches could stimulate discussion among producers, government agencies, and conservationists to develop positive attitudes and mechanisms for coexistence (Conover 2002).

13.4 CONCLUSIONS

Strategies to allow humans and wildlife to coexist will remain vital as habitat loss and fragmentation increase in concert with challenges from climate change and human population growth. As human society and the culture of agriculture evolve, so too will approaches to managing conflict between humans and wildlife. Today, local problems are shaped by global phenomena and potential solutions to local problems have far-reaching implications. Thus, optimizing current tools and developing new methods are necessary for effectively managing conflicts between blackbirds and agricultural producers.

REFERENCES

Abifarin, A. O. 1984. The importance of rice awns in the reduction of bird damage. *West African Rice Development Association Technical Newsletter* 5:27–28.

Alizadeh, E. 2009. Effects of sunflower cultivars and different sowing dates on the damage rate caused by birds, in particular house sparrow *Passer domesticus*. *Podoces* 4:108–114.

Ampatzidis, Y., J. Ward, and O. Samara. 2015. *Autonomous system for pest bird control in specialty crops using unmanned aerial vehicles*. ASABE Annual International Meeting. ASABE, New Orleans, LA.

Anderson, A., C. A. Lindell, K. M. Moxcey, W. F. Siemer, G. M. Linz, P. D. Curtis, J. E. Carroll, et al. 2013. Bird damage to select fruit crops: The cost of damage and benefits of control in five states. *Crop Protection* 52:103–109.

Anderson, K., and K. J. Gaston. 2013. Lightweight unmanned aerial vehicles will revolutionize spatial ecology. *Frontiers in Ecology and the Environment* 11:138–146.

Avery, M. L. 1979. Food preferences and damage levels of some avian rice field pests in Malaysia. *Bird Control Seminar* 8:161–166.

Avery, M. L. 1985. Application of mimicry theory to bird damage control. *Journal of Wildlife Management* 49:1116–1121.

Avery, M. L. 1989. Experimental evaluation of partial repellent treatment for reducing bird damage to crops. *Journal of Applied Ecology* 26:433–439.

Avery, M. L. 2002. Avian repellents. *Encyclopedia of Agrochemicals*, eds. J. R. Plimmer, D. W. Gammon, and N. N. Ragsdale, 1–8. Wiley, New York, NY.

Avery, M. L., J. S. Humphrey, and D. G. Decker. 1997. Feeding deterrence of anthraquinone, anthracene, and anthrone to rice-eating birds. *Journal of Wildlife Management* 61:1359–1365.

Avery, M. L., and J. R. Mason. 1997. Feeding responses of red-winged blackbirds to multisensory repellents. *Crop Protection* 16:159–164.

Baxter, A. T., and J. R. Allan. 2008. Use of lethal control to reduce habituation to blank rounds by scavenging birds. *Journal of Wildlife Management* 72:1653–1657.

BBC. 2014. *Suffolk farm uses drone to scare pigeons off rape crop.* http://www.bbc.com/news/uk-england-suffolk-27069825 (accessed September 1, 2016).

Beletsky, L. D. 1996. *The red-winged blackbird: The biology of a strongly polygynous songbird.* Academic Press, San Diego, CA.

Beletsky, L. D., and G. H. Orians. 1996. *Red-winged blackbirds: Decision making and reproductive success.* University of Chicago Press, Chicago, IL.

Benson, A. B. (ed.) 1937. *Peter Kalm's travels in North America.* Dover Publications Inc., New York, NY.

Besser, J. F. 1985. *A grower's guide to reducing bird damage to US agriculture crops.* Bird Damage Research Report 340. U.S. Fish and Wildlife Service Denver Wildlife Research Center, Denver, CO.

Blackwell, B. F., T. L. DeVault, T. W. Seamans, S. L. Lima, P. Baumhardt, and E. Fernández-Juricic. 2012. Exploiting avian vision with aircraft lighting to reduce bird strikes. *Journal of Applied Ecology* 49:758–766.

Blackwell, B. F., and R. A. Dolbeer. 2001. Decline of the red-winged blackbird population in Ohio correlated to changes in agriculture (1965–1996). *Journal of Wildlife Management* 65:661–667.

Blackwell, B. F., E. Huszar, G. M. Linz, and R. A. Dolbeer. 2003. Lethal control of red-winged blackbirds to manage damage to sunflower: An economic evaluation. *Journal of Wildlife Management* 67:818–828.

Bomford, M., and P. H. O'Brien. 1990. Sonic deterrents in animal damage control: A review of device tests and effectiveness. *Wildlife Society Bulletin* 18:411–422.

Bray, O. E., A. M. Gammell, and D. R. Anderson. 1979. Survival of yellow-headed blackbirds banded in North Dakota. *Bird-Banding* 50:252–255.

Bridge, E. S., K. Thorup, M. S. Bowlin, P. B. Chilson, R. H. Diehl, R. W. Fleron, P. Hartl, et al. 2011. Technology on the move: Recent and forthcoming innovations for tracking migratory birds. *BioScience* 61:689–698.

Bridgeland, W. 1979. Timing bird control applications in ripening corn. *Bird Control Seminar* 8:222–228.

Bullard, R. W. 1988. Characteristics of bird-resistance in agricultural crops. *Vertebrate Pest Conference* 13:305–309.

Burke, J. M., K. A. Gardner, and L. H. Rieseberg. 2002. The potential for gene flow between cultivated and wild sunflower (*Helianthus annuus*) in the United States. *American Journal of Botany* 89:1550–1552.

Cantamutto, M., and M. Poverene. 2007. Genetically modified sunflower release: Opportunities and risks. *Field Crops Research* 101:133–144.

Carlson, J. C., S. K. Tupper, S. J. Werner, S. E. Pettit, M. M. Santer, and G. M. Linz. 2013. Laboratory efficacy of an anthraquinone-based repellent for reducing bird damage to ripening corn. *Applied Animal Behaviour Science* 145:26–31.

Christie, K. S., S. L. Gilbert, C. L. Brown, M. Hatfield, and L. Hanson. 2016. Unmanned aircraft systems in wildlife research: Current and future applications of a transformative technology. *Frontiers in Ecology and the Environment* 14:241–251.

Conover, M. R. 1998. Perceptions of American agricultural producers about wildlife on their farms and ranches. *Wildlife Society Bulletin* 26:597–604.

Conover, M. R. 2002. *Resolving human-wildlife conflicts: The science of wildlife damage management.* CRC Press/Taylor & Francis, Boca Raton, FL.

Cummings, J. L., J. L. Guarino, C. E. Knittle, and W. C. Royall. 1987. Decoy plantings for reducing blackbird damage to nearby commercial sunflower fields. *Crop Protection* 6:56–60.

Cummings, J. L., J. L. Guarino, C. E. Knittle, and W. C. Royall. 1989. Chronology of blackbird damage to sunflowers. *Wildlife Society Bulletin* 17:50–52.

Cummings, J. L., Shwiff, S., and S. Tupper. 2005. Economic impacts of blackbird damage to the rice industry. *Wildlife Damage Management Conference* 11:317–322.

Ditmer, M. A., J. B. Vincent, L. K. Werden, J. C. Tanner, T. G. Laske, P. A. Iaizzo, D. L. Garshelis, et al. 2015. Bears show a physiological but limited behavioral response to unmanned aerial vehicles. *Current Biology* 25:2278–2283.

Dolbeer, R. A. 1978. Movement and migration patterns of red-winged blackbirds: A continental overview. *Bird-Banding* 49:17–34.

Dolbeer, R. A. 1980. *Blackbirds and corn in Ohio.* Resource Publication 136. U.S. Department of the Interior Fish and Wildlife Service, Washington, DC.

Dolbeer, R. A. 1981. Cost-benefit determination of blackbird damage control for cornfields. *Wildlife Society Bulletin* 9:44–51.

Dolbeer, R. A. 1982. Migration patterns for sex and age classes of blackbirds and starlings. *Journal of Field Ornithology* 53:28–46.

Dolbeer, R. A. 1990. Ornithology and integrated pest management: Red-winged blackbirds *Agelaius phoeniceus* and corn. *Ibis* 132:309–322.

Dolbeer, R. A., T. W. Seamans, B. F. Blackwell, and J. L. Belant. 1998. Anthraquinone formulation (Flight Control) shows promise as avian feeding repellent. *Journal of Wildlife Management* 62:1558–1564.

Dolbeer, R. A., P. P. Woronecki, and T. W. Seamans. 1995. Ranking and evaluation of field corn hybrids for resistance to blackbird damage. *Crop Protection* 14:399–403.

Dolbeer, R. A., P. P. Woronecki, and R. A. Stehn. 1986. Resistance of sweet corn to damage by blackbirds and starlings. *Journal of American Society of Horticultural Science* 111:306–311.

Doudna, J. A, and E. Charpentier. 2014. The new frontier of genome engineering with CRISPR-Cas9. *Science* 346:1258096.

Dufty, A. M. 1982. Movements and activities of radio-tracked brown-headed cowbirds. *Auk* 99:316–327.

Fagerstone, K. A., L. A. Miller, G. Killian, and C. A. Yoder. 2010. Review of issues concerning the use of reproductive inhibitors, with particular emphasis on resolving human-wildlife conflicts in North America. *Integrative Zoology* 5:15–30.

Fankhauser, D. P. 1971. Annual adult survival rates of blackbirds and starlings. *Bird-Banding* 42:36–42.

Fleming, P. J. S., A. Gilmour, and J. A. Thompson. 2002. Chronology and spatial distribution of cockatoo damage to two sunflower hybrids in south-eastern Australia, and the influence of plant morphology on damage. *Agriculture, Ecosystems and Environment* 91:127–137.

Forcey, G. M., W. E. Thogmartin, G. M. Linz, P. C. McKann, and S. C. Crimmins. 2015. Spatially explicit modeling of blackbird abundance in the Prairie Pothole Region. *Journal of Wildlife Management* 79:1022–1033.

Galle, A. M., G. M. Linz, H. J. Homan, and W. J. Bleier. 2009. Avian use of harvested crop fields in North Dakota during spring migration. *Western North American Naturalist* 69:491–500.

Gibbons, D. W., D. A. Bohan, P. Rothery, R. C. Stuart, A. J. Haughton, R. J. Scott, J. D. Wilson, et al. 2006. Weed seed resources for birds in fields with contrasting conventional and genetically modified herbicide-tolerant crops. *Proceedings of the Royal Society of London B: Biological Sciences* 273:1921–1928.

Gilsdorf, J. M., S. E. Hygnstrom, and K. C. VerCauteren. 2002. Use of frightening devices in wildlife damage management. *Integrated Pest Management Reviews* 7:29–45.

Goguen, C. B., and N. E. Mathews. 2001. Brown-headed cowbird behavior and movements in relation to livestock grazing. *Ecological Applications* 11:1533–1544.

Greenberg, R., D. W. Demarest, S. M. Matsuoka, C. Mettke-Hofmann, D. Evers, P. B. Hamel, and K. A. Hobson. 2011. Understanding declines in rusty blackbirds. *Studies in Avian Biology* 41:107–125.

Grimm, B. A., B. A. Lahneman, P. B. Cathcart, R. C. Elgin, G. L. Meshnik, and J. P. Parmigiani. 2012. Autonomous unmanned aerial vehicle system for controlling pest bird population in vineyards. *ASME 2012 International Mechanical Engineering Congress and Exposition* 4:499–505.

Hagy, H. M., G. M. Linz, and W. J. Bleier. 2008. Optimizing the use of decoy plots for blackbird control in commercial sunflower. *Crop Protection* 27:1442–1447.

Homan, H. J., G. M. Linz, and W. J. Bleier. 1994. Effect of crop phenology and habitat on the diet of common grackles (*Quiscalus quiscula*). *American Midland Naturalist* 131:381–385.

Homan, H. J., G. M. Linz, R. M. Engeman, and L. B. Penry. 2004. Spring dispersal patterns of red-winged blackbirds, *Agelaius phoeniceus*, staging in eastern South Dakota. *Canadian Field-Naturalist* 118:201–209.

Jacobson, S. K., K. E. Sieving, G. A. Jones, and A. Van Doorn. 2003. Assessment of farmer attitudes and behavioral intentions toward bird conservation on organic and conventional Florida farms. *Conservation Biology* 17:595–606.

Jan, C. C., L. Qi, B. Hulke, X. Fu. 2011. *Present and future plans of the sunflower "Doubled Haploid" project.* National Sunflower Association Research Forum, Fargo, ND. http://www.sunflowernsa.com/research/searchable-database-of-forum-papers/ (accessed September 1, 2016).

Kandel, H., B. Johnson, C. Deplazes, G. M. Linz, and M. M. Santer. 2009. *Sunflower treated with Avipel (Anthraquinone) bird repellent.* National Sunflower Association Research Forum, Fargo, ND. http://www.sunflowernsa.com/research/searchable-database-of-forum-papers/ (accessed September 1, 2016).

Kantar, M. B., K. Betts, J.-M. S. Michno, J. J. Luby, P. L. Morrell, B. S. Hulke, R. M. Stupar, et al. 2014. Evaluating an interspecific *Helianthus annuus* × *Helianthus tuberosus* population for use in a perennial sunflower breeding program. *Field Crops Research* 155:254–264.

Kerzman, J. 2015. Drones & sunflower. *The Sunflower Magazine* 41(5):20–22.

Killi, F., B. Kilic, and K Goner. 2004. The effect of different planting dates on the extent of bird damage in confection and oilseed sunflowers. *Journal of Agronomy* 3:36–39.

Klosterman, M. E., G. M. Linz, A. A. Slowik, and H. J. Homan. 2013. Comparisons between blackbird damage to corn and sunflower in North Dakota. *Crop Protection* 53:1–5.

Kostecke, R. M., L. M. Smith, and H. M. Hands. 2004. Vegetation response to cattail management at Cheyenne Bottoms, Kansas. *Journal of Aquatic Plant Management* 42:39–45.

Krapu, G. L., D. A. Brandt, and R. R. Cox Jr. 2004. Less waste corn, more land in soybeans, and the switch to genetically modified crops: Trends with important implications for wildlife management. *Wildlife Society Bulletin* 32:127–136.

Kubasiewicz, L. M., N. Bunnefeld, A. I. T. Tulloch, C. P. Quine, and K. J. Park. 2016. Diversionary feeding: An effective management strategy for conservation conflict? *Biodiversity and Conservation* 25:1–22.

Lima, S. L., and P. A. Bednekoff. 1999. Back to the basics of antipredatory vigilance: Can nonvigilant animals detect attack? *Animal Behaviour* 58:537–543.

Lindell, C. A., K. M. M. Steensma, P. D. Curtis, J. R. Boulanger, J. E. Carroll, C. Burrows, D. P. Lusch, et al. 2016. Proportions of bird damage in tree fruits are higher in low-fruit-abundance contexts. *Crop Protection* 90:40–48.

Linz, G. M. 2013. *Blackbird population management to protect sunflower: A history.* National Sunflower Association Research Forum, Fargo, ND. http://www.sunflowernsa.com/research/searchable-database-of-forum-papers/ (accessed September 1, 2016).

Linz, G. M., S. B. Bolin, and F. J. Cassel. 1983. Postnuptial and postjuvenal molts of red-winged blackbirds in Cass County, North Dakota. *Auk* 100:206–209.

Linz, G. M., E. H. Bucher, S. B. Canavelli, E. Rodriguez, and M. L. Avery. 2015. Limitations of population suppression for protecting crops from bird depredation: A review. *Crop Protection* 76:46–52.

Linz, G. M., and J. J. Hanzel. 2015. Sunflower and bird pests. In *Sunflower: Chemistry, production, processing, and utilization*, eds. E. M. Force, N. T. Dunford, and J. J. Salas, 175–186. AOCS Press, Urbana, IL.

Linz, G. M., and H. J. Homan. 2011. Use of glyphosate for managing invasive cattail (*Typha* spp.) to protect crops near blackbird (Icteridae) roosts. *Crop Protection* 30:98–104.

Linz, G. M., H. J. Homan, L. B. Penry, and P. Mastrangelo. 2003. Reducing blackbird-human conflicts in agriculture and feedlots: New methods for an integrated management approach. In *Management of North American blackbirds: Special symposium of the wildlife society ninth annual conference*, ed. G. M. Linz, 21–24. National Wildlife Research Center, Fort Collins, CO.

Linz, G. M., H. J. Homan, S. J. Werner, H. M. Hagy, and W. J. Bleier. 2011. Assessment of blackbird management strategies to protect sunflower. *BioScience* 61:960–970.

Linz, G. M., B. S. Hulke, M. B. Kantar, H. J. Homan, R. M. Stupar, and D. L. Wyse. 2014. Potential use of perennial sunflower to reduce blackbird damage to sunflower. *Vertebrate Pest Conference* 26:356–359.

MacDonald, J. M., P. Korb, and R. A. Hoppe. 2013. *Farm size and the organization of U.S. crop farming, ERR-152*. U.S. Department of Agriculture, Economic Research Service, Washington, DC. http://www.ers.usda.gov/publications/pub-details/?pubid=45110 (accessed November 17, 2016).

Mah, J., G. M. Linz, and J. J. Hanzel. 1990. Relative effectiveness of individual sunflower traits for reducing red-winged blackbird depredation. *Crop Protection* 9:359–362.

Mah, J., and G. L. Nuechterlein. 1991. Feeding behavior of red-winged blackbirds on bird-resistant sunflowers. *Wildlife Society Bulletin* 19:39–46.

Mahjoub, G., M. K. Hinders, and J. P. Swaddle. 2015. Using a "sonic net" to deter pest bird species: Excluding European starlings from food sources by disrupting their acoustic communication. *Wildlife Society Bulletin* 39:326–333.

Marra, P. P., E. B. Cohen, S. R. Loss, J. E. Rutter, and C. M. Tonra. 2015. A call for full annual cycle research in animal ecology. *Biology Letters* 11:20150552.

Marzluff, J. M., R. B. Boone, and G. W. Cox. 1994. Historical changes in populations and perceptions of native pest bird species in the west. *Studies in Avian Biology* 15:202–220.

Mason, J. R., G. Nuechterlein, G. Linz, R. A. Dolbeer, and D. L. Otis. 1991. Oil concentration differences among sunflower achenes and feeding preferences of red-winged blackbirds. *Crop Protection* 10:299–304.

Miller, J. E. 2007. Evolution of the field of wildlife damage management in the United States and future challenges. *Human-Wildlife Conflict* 1:13–20.

Moore, W. S., and R. A. Dolbeer. 1989. The use of banding recovery data to estimate dispersal rates and gene flow in avian species: Case studies in the red-winged blackbird and common grackle. *Condor* 91:242–253.

Mullally, S. 2010. 'High Boys' in sunflower: Another look. *The Sunflower Magazine* 36(5):24–25.

Nelms, C. O., W. J. Bleier, D. L. Otis, and G. M. Linz. 1994. Population estimates of breeding blackbirds in North Dakota, 1967, 1981–1982 and 1990. *American Midland Naturalist* 132:256–263.

Niner, M. D., G. M. Linz, and M. E. Clark. 2015. Evaluation of 9,10 anthraquinone application to pre-seed set sunflowers for repelling blackbirds. *Human-Wildlife Interactions* 9:4–13.

Oh, C., Z. Herrnstadt, and P. H. Howard. 2015. Consumer willingness to pay for bird management practices in fruit crops. *Agroecology and Sustainable Food Systems* 39:782–797.

Okurut-Akol, F. H., R. A. Dolbeer, and P. P. Woronecki. 1990. Red-winged blackbird and starling feeding responses on corn earworm-infested corn. *Vertebrate Pest Conference* 14:296–301.

Otis, D. L., and C. M. Kilburn. 1988. *Influence of environmental factors on blackbird damage to sunflower*. Technical Report No. 16. U.S. Fish and Wildlife Service, Washington, DC.

Parfitt, D. E., and G. J. Fox. 1986. Genetic sources of resistance to blackbird predation in sunflower. *Canadian Journal of Plant Science* 66:19–23.

Peer, B. D., and E. K. Bollinger. 1997. Common grackle (*Quiscalus quiscula*). No. 271. *The birds of North America*, ed. P. G. Rodewald. Cornell Lab of Ornithology, Ithaca, NY. https://birdsna.org/Species-Account/bna/species/rewbla (accessed September 25, 2016).

Peer, B. D., H. J. Homan, G. M. Linz, and W. J. Bleier. 2003. Impact of blackbird damage to sunflower: Bioenergetic and economic models. *Ecological Applications* 13:248–256.

Pérez-García, J. M. 2012. The use of digital photography in censuses of large concentrations of passerines: The case of a winter starling roost-site. *Revista Catalana d'Ornitologia* 28:28–33.

Pijanowski, B. C., L. J. Villanueva-Rivera, S. L. Dumyahn, A. Farina, B. L. Krause, B. M. Napoletano, S. H. Gage, et al. 2011. Soundscape ecology: The science of sound in the landscape. *BioScience* 61:203–216.

Rosenberg, K. V., J. A. Kennedy, R. Dettmers, R. P. Ford, D. Reynolds, C. J. Beardmore, P. J. Blancher, et al. 2016. *Partners in Flight Landbird Conservation Plan: 2016 Revision for Canada and Continental United States.* Partners in Flight Science Committee. http://www.partnersinflight.org/ (accessed September 25, 2016).

Rothstein, S. I. 1994. The cowbird's invasion of the far west: History, causes and consequences experienced by host species. *Studies in Avian Biology* 15:301–315.

Rothstein, S. I., J. Verner, and E. Steven. 1984. Radio-tracking confirms a unique diurnal pattern of spatial occurrence in the parasitic brown-headed cowbird. *Ecology* 65:77–88.

Samanci, B. 1995. The effect of different planting dates on the extent of bird damage in sunflower. *Turkish Journal of Agriculture and Forestry* 19:207–211.

Sauer, J. R., J. E. Hines, J. E. Fallon, K. L. Pardieck, D. J. Ziolkowski, Jr., and W. A. Link. 2014. *The North American Breeding Bird Survey, Results and Analysis 1966–2013. Version 01.30.2015.* U.S. Geological Survey, Patuxent Wildlife Research Center, Laurel, MD.

Schäckermann, J., N. Weiss, H. von Wehrden, and A. M. Klein. 2014. High trees increase sunflower seed predation by birds in an agricultural landscape of Israel. *Frontiers in Ecology and Evolution* 2:1–35.

Seamans, T. W., B. F. Blackwell, and T. L. DeVault. 2013. Brown-headed cowbird (*Molothrus ater*) response to pyrotechnics and lethal removal in a controlled setting. *Wildlife Damage Management Conference* 15:56–62.

Searcy, W. A., and K. Yasukawa. 1995. *Polygyny and sexual selection in red-winged blackbirds.* Princeton University Press, Princeton, NJ.

Somers, C. M., and R. D. Morris. 2002. Birds and wine grapes: Foraging activity causes small-scale damage patterns in single vineyards. *Journal of Applied Ecology* 39:511–523.

Stehn, R. A. 1989. *Adult survival rate of red-winged blackbirds.* Bird Damage Research Report 371. U.S. Fish and Wildlife Service, Denver Wildlife Research Center, Denver, CO.

Strassburg, M., S. M. Crimmins, G. M. Linz, P. C. McKann, W. E. Thogmartin. 2015. Winter habitat associations of blackbirds and starlings wintering in the south-central United States. *Human-Wildlife Interactions* 9:171–179.

Swaddle, J., D. Moseley, M. Hinders, and E. P. Smith. 2016. A sonic net excludes birds from an airfield: Implications for reducing bird strike and crop losses. *Ecological Applications* 26:339–345.

Trostle, C., D. Pietsch, A. Schlegel, and P. Evans. 2013. *Height, yield, and oil content of short-stature sunflower (Helianthus annuus) vs. conventional height sunflower in the Southern Great High Plains.* National Sunflower Association Research Forum, Fargo, ND. https://www.sunflowernsa.com/Research/Research-Forum-PowerPoint-Presentations-Since-2008/2013/ (accessed September 1, 2016).

Twedt, D. J., W. J. Bleier, and G. M. Linz. 1991. Geographic and temporal variation in the diet of yellow-headed blackbirds. *Condor* 93:975–986.

Twedt, D. J., and G. M. Linz. 2015. Flight feather molt in yellow-headed blackbirds (*Xanthocephalus xanthocephalus*) in North Dakota. *Wilson Journal of Ornithology* 127:622–629.

U.S. Department of Agriculture. 2008. *Recommendations of conferees for management of blackbird, starling, and corvid impacts on municipalities, airports, crops and livestock.* Management of Blackbird, Starling, and Corvid Conference, Nashville, Tennessee, USA. Animal and Plant Health Inspection Service, Wildlife Services, Washington, DC. https://www.aphis.usda.gov/wildlife_damage/nwrc/publications/Linz_Management%20of%20blackbird%20starling%20and%20corvid%20conference.pdf (accessed November 28, 2016).

U.S. Department of Agriculture. 2015. *Managing blackbird damage to sprouting rice in southwestern Louisiana. Environmental Assessment.* Animal and Plant Health Inspection Service, Wildlife Services, Washington, DC.

U.S. Department of Agriculture. 2016. *Organic market overview.* Economic Research Service, Washington DC. http://www.ers.usda.gov/topics/natural-resources-environment/organic-agriculture/organic-market-overview/ (accessed November 15, 2016).

Van Buskirk, J., R. S. Mulvihill, and R. C. Leberman. 2009. Variable shifts in spring and autumn migration phenology in North American songbirds associated with climate change. *Global Change Biology* 15:760–771.

Van Vuren, D., and K. S. Smallwood. 1996. Ecological management of vertebrate pests in agricultural systems. *Biological Agriculture and Horticulture* 12:39–62.

Warnes, K. 2016. Scarecrows historically speaking. http://historybecauseitshere.weebly.com/scarecrows-historically-speaking.html (accessed September 1, 2016).

Washburn, B. E., R. B. Chipman, and L. C. Francoeur. 2006. Evaluation of bird response to propane exploders in an airport environment. *Vertebrate Pest Conference* 22:212–215.

Weatherhead, P. J. 2005. Long-term decline in a red-winged blackbird population: Ecological causes and sexual selection consequences. *Proceedings of the Royal Society of London B: Biological Sciences* 272:2313–2317.

Weatherhead, P. J., S. Tinker, and H. Greenwood. 1982. Indirect assessment of avian damage to agriculture. *Journal of Applied Ecology* 19:773–782.

Wehtje, W. 2003. The range expansion of the great-tailed grackle (*Quiscalus mexicanus* Gmelin) in North America since 1880. *Journal of Biogeography* 30:1593–1607.

Wenny, D. G., T. L. DeVault, M. D. Johnson, D. Kelly, C. H. Sekercioglu, D. F. Tomback, and C. J. Whelan. 2011. The need to quantify ecosystem services provided by birds. *Auk* 128:1–14.

Werner, S. J., J. C. Carlson, S. T. Tupper, M. S. Santer, and G. M. Linz. 2009. Threshold concentrations of an anthraquinone-based repellent for Canada geese, red-winged blackbirds, and ring-necked pheasants. *Applied Animal Behaviour Science* 121:190–196.

Werner, S. J., S. T. DeLiberto, S. E. Pettit, and A. M. Mangan. 2014a. Synergistic effect of an ultraviolet feeding cue for an avian repellent and protection of agricultural crops. *Applied Animal Behaviour Science* 159:107–113.

Werner, S. J., K. A. Hobson, S. L. Van Wilgenberg, and J. W. Fischer. 2016. Multi-Isotopic (δ^2H, δ^{13}C, δ^{15}N) tracing of molt origin for red-winged blackbirds associated with agro-ecosystems. *PLoS One* 11:e0165996.

Werner, S. J., H. J. Homan, M. L. Avery, G. M. Linz, E. A. Tillman, A. A. Slowik, R. J. Byrd, et al. 2005. Evaluation of Bird Shield™ as a blackbird repellent in ripening rice and sunflower fields. *Wildlife Society Bulletin* 33:251–257.

Werner, S. J., G. M. Linz, J. C. Carlson, S. E. Pettit, S. T. Tupper, and M. M. Santer. 2011. Anthraquinone-based bird repellent for sunflower crops. *Applied Animal Behaviour Science* 129:162–169.

Werner, S. J., S. K. Tupper, S. E. Pettit, J. W. Ellis, J. C. Carlson, D. A. Goldade, N. M. Hofmann, et al. 2014b. Application strategies for an anthraquinone-based repellent to protect oilseed sunflower crops from pest blackbirds. *Crop Protection* 59:63–70.

Wilson, E. A., E. A. LeBoeuf, K. M. Weaver, and D. J. LeBlanc. 1989. Delayed seeding for reducing blackbird damage to sprouting rice in southwestern Louisiana. *Wildlife Society Bulletin* 17:165–171.

Wunsch, M. J., M. Schaefer, B. Kraft, J. Hafner, and J. Kallis. 2016. Prospects for using drop nozzles to improve fungicide coverage and control of Sclerotinia head rot. *National Sunflower Association Research Forum*, Fargo, ND. http://www.sunflowernsa.com/research/searchable-database-of-forum-papers/ (accessed September 1, 2016).

Index

A

Acadian flycatcher (*Empidonax virescens*), 86
American bittern (*Botaurus lentiginosus*), 53
American coot (*Fulica americana*), 53
American crow (*Corvus brachyrhynchos*), 123
American goldfinch (*Spinus tristis*), 82
American robin (*Turdus migratorius*), 11, 23, 161
Aminopyridine, or avitrol, 145
Animal Damage Control (ADC) program (Wildlife Services), 125
Anthraquinone, 137–143, 222
Aposematic colors, 169–170
Artificial aural deterrents, 161–162
Auditory frightening devices, 160–163
    artificial aural deterrents, 161–162
    bioacoustics, 160–161
    combatting habituation, 162–163
Avian brood parasite, *see* Brown-headed cowbird (*Molothrus ater*)
Avian salmonellosis, 22

B

Balloons, 164
Barn owl (*Tyto alba*), 53
BBS, *see* Breeding Bird Survey
Bell's vireo (*Vireo bellii pusillus*), 90, 129–130
Bioacoustics, 160–161
Bird Conservation Regions (BCR), 194
Bird-Resistant Synthetic Sunflower Variety 1 (BRS1), 5
Blackbird damage, *see* Damage, strategies for evading; Damage to crops, economic impact of
Black-capped vireo (*Vireo atricapilla*), 90
Blacklegged ticks (*Ixodes scapularis*), 23
Black spiny-tailed iguanas (*Ctenosaura similis*), 10
Black vultures (*Coragyps atratus*), 192
Blue racer (*Coluber constrictor*), 53
Blue spruce (*Picea pungens*), 68
Boat-tailed grackle (*Quiscalus major*), 33, 88
Bohemian waxwings (*Bombycilla garrulus*), 90
Box elder (*Acer negundo*), 68
Breeding Bird Survey (BBS), 72, 86, 120, 194
Brewer's blackbird (*Euphagus cyanocephalus*), 56, 86, 107
Bronzed cowbird (*Molothrus aeneus*), 79
Brown-headed cowbird (*Molothrus ater*), 21, 51, 66, 77–99, 144, 207, 218
    agricultural damages, 88
    behavior, 84–85
    cowbird nestling behavior, 85
    depredation of host nests, 85
    diet, 83–84
    dynamic distribution, 79–80
    feather molt, 84
    female breeding ranges and territoriality, 84–85
    future research, 92
    life history, 80–83
    mating system, 84
    migration, 86
    Molothrus clade, 79
    populations, 86–88
    range expansion and conservation concerns, 80
    subspecies, 79
    taxonomy, 79
BRS1, *see* Bird-Resistant Synthetic Sunflower Variety 1
Bull snake (*Pituophis melanoleucus*), 53
Bulrush (*Scirpus*), 49, 105
Burmese pythons (*Python bivittatus*), 10

C

California Field Station, 12
Canada, blackbird research in, 4–5
Canadian Wildlife Service, 2–3
Cattail (*Typha* spp.), 49, 102, 175, 180
CBC data, *see* Christmas Bird Count data
Cedar waxwings (*Bombycilla cedrorum*), 11, 82
Chemical repellents, 135–158
    aminopyridine, or avitrol, 145
    anthraquinone, 137–143
    avian repellent testing in North America, 136–137
    future of, 222–223
    insect repellents, 152
    methiocarb, 144–145
    methyl anthranilate and dimethyl anthranilate, 145–150
    miscellaneous candidate repellents, 154–155
    miscellaneous plant derivatives, 152–154
    nonlethal chemical repellents, 138–142, 143
    registered fungicides, 151–152
    registered insecticides, 150–151
    suggested future research, 155
Chestnut-capped blackbirds (*Agelaius ruficapillus, Chrysomus ruficapillus*), 153
Chlamydiosis, 22
Christmas Bird Count (CBC) data, 86, 107
Climate change, 112–113
Climate effects, *see* Habitat and climate, effects of
Common cuckoo (*Cuculus canorus*), 78
Common grackle (*Quiscalus quiscula*), 23, 33, 65–75, 102, 160, 207, 218
    agricultural damages, 72–73
    behavior, 71
    breeding range, 67
    brood parasitism, 69
    diet, 70
    eggs, 68
    feather molt and plumages, 70–71
    future research, 73
    historical changes, 68
    incubation, 69
    life history, 68–69
    mating system, 71
    migration, 71–72

nests, 68
nonbreeding, 67
parental care, 69
populations, 72
predatory behavior, 71
subspecies, 67
taxonomy, 66–67
territories, 71
timing, 68
Common starlings (*Sturnus vulgaris*), 104
Conservation Reserve Program (CRP), 111
Corn (*Zea mays*), 124
Corn damage estimates, 209–210

D

Damage, strategies for evading, 175–189
    advancing harvest date, 177
    bird-resistant crops, 183–185
    cultural practices, 176
    glyphosate herbicide, 181
    upland roost sites, management of, 183
    wetland roost sites, management of, 180–183
    wildlife conservation food plots, 177–179
Damage to crops, economic impact of, 207–216
    bird management tools, assessing costs and benefits of, 210–211
    case study (sunflower damage), 212–214
    destruction, 208
    estimates of indirect or secondary damage, 212
    examples of direct or primary damage estimates, 208–210
    IO models, 209
    research needs, 214
Decoy traps, 124–125
Deer mouse (*Peromyscus maniculatus*), 53
Dickcissels (*Spiza americana*), 33, 87
Dimethyl anthranilate, 145–150
Double-crested cormorants (*Phalacrocorax auritus*), 163, 192
DRC-1339 toxicant, 126
Drones, 167
Dynamiting roosts, 123–124

E

Eastern red cedar (*Juniperus virginiana*), 68
Eastern red-winged blackbird (*Agelaius phoeniceus phoeniceus*), 19
Economic impact of damage, *see* Damage to crops, economic impact of
Effigies, 166
Encephalitis, 23
Endangered Species Act, 219
Environmental assessment (EA), 3
Environmental Impact Statement (EIS), 3
EPA, *see* U.S. Environmental Protection Agency
European starlings (*Sturnus vulgaris*), 2, 18, 22, 119

F

Falconry, 167–168
Feral swine (*Sus scrofa*), 10
Floodlight traps, 124
Florida Field Station, 10
Frightening devices, 159–174
    aposematic colors, 169–170
    artificial aural deterrents, 161–162
    auditory frightening devices, 160–163
    balloons, 164
    bioacoustics, 160–161
    combatting habituation, 162–163
    effigies and models, 166
    falconry, 167–168
    flags and streamers, 166
    future of, 223–225
    hawk kites, 164
    hazing with aircraft, 166–167
    lasers, 168–169
    light and color, 168–170
    reflective tape, 165
    remote controlled models and drones, 167
    ultraviolet wavelengths, 169
    visual frightening devices, 163–168
Fungicides (registered), 151–152
FWS, *see* U.S. Fish and Wildlife Service

G

Garter snakes (*Thamnophis* spp.), 53
Genome engineering, 226
Glyphosate herbicide, 181
Golden-cheeked warbler (*Setophaga chrysoparia*), 90
Grackle, *see* Common grackle (*Quiscalus quiscula*)
Gray catbirds (*Dumetella carolinensis*), 90
Great blue heron (*Ardea herodias*), 68, 102
Great horned owl (*Bubo virginianus*), 53
Great tailed grackles (*Quiscalus mexicanus*), 33
Green peach aphids (*Myzus persicae*), 154
Grosbeak (*Coccothraustes vespertinus*), 104
Gulls (*Larus* spp.), 53, 161

H

Habitat and climate, effects of, 101–118
    breeding habitat, 104–105
    breeding season climate and weather associations, 106–107
    climate change, 112–113
    definition of habitat, 102
    energy requirements of birds, 102
    landscape effects, 109–112
    migratory climate and weather associations, 109
    migratory habitat, 108–109
    nesting microhabitat, 105
    winter climate and weather associations, 107–108
    winter habitat, 107
Hawk kites, 164
Hawthorn (*Crataegus rotundifolia*), 68

INDEX

Hazing with aircraft, 166–167
Hellebore (*Veratrum* spp.), 136
Histoplasmosis (*Histoplasma capsulatum*), 23
Honeybees (*Apis mellifera*), 179
House finch (*Haemorhous mexicanus*), 82, 124
House sparrow (*Passer domesticus*), 22, 104, 145
House wren (*Troglodytes aedon*), 23

I

Indigo bunting (*Passerina cyanea*), 86
Insecticides (registered), 150–151
Insect repellents, 152
Intergovernmental Panel on Climate Change (IPCC), 112
IO models, 209
Island scrub jays (*Aphelocoma insularis*), 90

J

Jack pine (*Pinus banksiana*), 90, 127
Johne's disease (*Mycobacterium avium paratuberculosis*), 22

K

Kentucky Field Station, 11–12
Kirtland's warbler (*Setophaga kirtlandii*), 9, 90, 127

L

Lasers, 168–169
Laws, *see* Regulations, policy, and research (history of)
Loblolly pine (*Pinus taeda*), 68
Lyme disease, 23

M

Magpies (*Pica* spp.), 53
Marsh wren (*Cistothorus palustris*), 53, 105, 182
Methiocarb, 144–145
Methyl anthranilate, 145–150
Migratory Bird Treaty Act (MBTA), 2, 191
Mink (*Mustela vison*), 53
Monk parakeets (*Myiopsitta monachus*), 10

N

National Environmental Policy Act (NEPA), 3
National Environmental Protection Act, 219
National Wildlife Research Center (NWRC), 3, 6, 7
North American Breeding Bird Survey (NABBS), 16, 23–24
North Atlantic Oscillation, 109, 113
North Dakota Field Station, 10–11
Northern cardinal (*Cardinalis cardinalis*), 81, 86
Northern Great Plains, allowable take of red-winged blackbirds in, 191–206
  adult and juvenile survival rates, 195–196
  allowable take of breeding females, 202–203
  allowable take of other sex and age classes, 203
  first and last ages of reproduction, 195
  intrinsic growth rate, estimate of, 196–198
  legal context for allowable take, 192
  population size in the northern Great Plains, 198–201
  quantitative framework for allowable take, 192–194
  reproductive rate, 196
  territoriality and the units of take, 202
  yield curve, 193
Northern harriers (*Circus cyaneus*), 53
Northern white cedar (*Thuja occidentalis*), 68
Norway rats (*Rattus norvegicus*), 153
NWRC, *see* National Wildlife Research Center

O

Ohio Field Station, 8–9
Osprey (*Pandion haliaetus*), 68

P

PA-14, winter roost spraying with, 125–126
Passenger pigeon (*Ectopistes migratorius*), 119
Pheasant (*Phasianus colchicus*), 144
Phragmites (*Phragmites australis*), 180
Poison baits, 124
Policy, *see* Regulations, policy, and research (history of)
Population dynamics and management, 119–133
  annual cycle of red-winged blackbird population, 121–123
  decoy traps, 124–125
  dynamiting roosts, 123–124
  floodlight traps, 124
  major organized population control efforts, 1970s through 2000s, 125–130
  poison baits, 124
  population control efforts, 1950s through 1960s, 123–125
  ripening sunflowers in the Dakotas, 126–127
  sprouting rice in Louisiana, 127
  trapping cowbirds to reduce nest parasitism, 127–130
  winter roost spraying with surfactant PA-14, 125–126
  zoonotic disease threats, 120
PRISM Climate Group, 112
Prothonotary warbler (*Protonotaria citrea*), 85
Purple loosestrife (*Lythrum salicaria*), 104

Q

Quaking aspen (*Populus tremuloides*), 48
Quelea (*Quelea quelea*), 119

R

Raccoon (*Procyon lotor*), 53
Red-billed quelea (*Quelea quelea*), 166
Red fox (*Vulpes vulpes*), 53
Red maple (*Acer rubrum*), 109

Red-winged blackbirds (Agelaius phoeniceus), 17–41, 102, 191, 207, 218; *see also* Northern Great Plains, allowable take of red-winged blackbirds in
    aerial mass color-marking, 27
    avian salmonellosis, 22
    breeding biology, 20–22
    brood parasitism, 21–22
    chlamydiosis, 22
    corn, damage to, 33–34
    crop damage, 32–35
    disease transmission, 22–26
    distribution and populations, 23–26
    encephalitis, 23
    fall migration and annual feather replacement, 29
    food habits, 30–31
    histoplasmosis, 23
    Johne's disease, 22
    Lyme disease, 23
    nest predation, 21
    polygyny and territoriality, 20
    rice, damage to, 33
    shiga toxin-producing *Escherichia coli*, 22–23
    spring migration, 27–29
    sunflower, damage to, 34–35
    sweet corn, damage to, 34
    taxonomy, 19–20
    West Nile virus, 23
    winter location, 26–27
Reed (*Phragmites*), 49, 105
Reflective tape, 165
Regulations, policy, and research (history of), 1–15
    California Field Station, 12
    Canada, blackbird research in, 4–5
    Canadian Wildlife Service, 2–3
    Florida Field Station, 10
    Kentucky Field Station, 11–12
    Migratory Bird Treaty Act, 2
    National Environmental Policy Act, 3–4
    North Dakota Field Station, 10–11
    NWRC headquarters, 6
    Ohio Field Station, 8–9
    United States, blackbird research in, 6–12
    U.S. Department of Agriculture, 3
    U.S. Depredation Order for Blackbirds, 2
    WS decision model, 4, 5
Remote controlled models, 167
Research (blackbird management), future of, 217–234; *see also* Regulations, policy, and research (history of)
    blackbird ecology, 219–221
    chemical repellents, 222–223
    economics and human dimensions, 227–228
    evading strategies, 225–227
    frightening devices, 223–225
    lethal control, 221
    management tools, 221–227
Rice (*Oryza sativa*), 123
Rice damage estimates, 208
Rice stinkbug (*Oebalus pugnax*), 83
Rice water weevil (*Lissorhoptrus oryzophilus*), 83
Rock pigeons (*Columba livia*), 145
Rook (*Corvus frugilegus*), 119
Rusty blackbird (*Euphagus carolinus*), 2, 86, 107

S

Scaly-naped pigeons (*Patagioenas squamosa*), 192
Scotch pine (*Pinus sylvestris*), 68
Screaming cowbird (*Molothrus rufoaxillaris*), 79
Sedge (*Carex*), 105
Shiga toxin-producing *Escherichia coli* (STEC), 22–23
Siberian elm (*Ulmus pumila*), 68
Siberian peashrub (*Caragana arborescens*), 68
Snow geese (*Chen caerulescens*), 166
Spikerush (*Eleocharis* spp.), 49
Streamers, 166
Striped skunk (*Mephitis mephitis*), 53
Sunflower (*Helianthus annuus*), 25, 126, 226
Sunflower damage estimates, 208–209
Sweetgum (*Liquidambar styraciflua*), 109

T

Tricolored blackbirds (*Agelaius tricolor*), 2, 144

U

UAS, *see* Unmanned aircraft systems
Ultraviolet wavelengths, 169
United States, blackbird research in, 6–12
    blackbird research field stations, 8–12
    blackbird research headquarters, 8
    California Field Station, 12
    Florida Field Station, 10
    Kentucky Field Station, 11–12
    North Dakota Field Station, 10–11
    NWRC headquarters, 6
    Ohio Field Station, 8–9
Unmanned aircraft systems (UAS), 224
Upland roost sites, management of, 183
U.S. Department of Agriculture (USDA), 3, 179
U.S. Department of the Interior (USDI), 8, 125
U.S. Depredation Order for Blackbirds, 2
U.S. Environmental Protection Agency (EPA), 125
U.S. Fish and Wildlife Service (FWS, USFWS), 2, 192

V

Visual frightening devices, 163–168
    balloons, 164
    effigies and models, 166
    falconry, 167–168
    flags and streamers, 166
    hawk kites, 164
    hazing with aircraft, 166–167
    reflective tape, 165
    remote controlled models and drones, 167

INDEX

W

WCFP, *see* Wildlife conservation food plots
West Nile virus (WNV), 23
Wetland roost sites, management of, 180–183
White pine (*Pinus strobus*), 137
Wildlife conservation food plots (WCFP), 175, 177–179, 211
Wildlife Services (WS), 3
Wild rice (*Zizania aquatica*), 30
Wild turkeys (*Meleagris gallopavo*), 169
Willow (*Salix*), 105
Willow flycatcher (*Empidonax traillii extimus*), 90
WNV, *see* West Nile virus

Y

Yellow-headed blackbirds (*Xanthocephalus xanthocephalus*), 21, 43–63, 66, 102, 160, 207, 218
    agricultural damages, 58–59
    behavior, 54–57
    breeding range, 47–48
    definitive basic plumage, 44–46
    description, 44–47
    diet, 53–54
    displays, 55–56
    eggs, 50–51
    feather molt, 54
    formative plumage, 46
    habitat, 48
    incubation, 51
    juvenile plumage, 46–47
    life history, 48–54
    migration, 56–57
    movements, 55
    nestlings and fledglings, 51–52
    nests, 49–50
    populations, 57–58
    reproductive success, 52
    second-year birds, 53
    socialization, 56
    survival, 52–53
    vocalizations, 54–55
    winter, 48
Yield curve, 193

Z

Zoonotic disease
    reservoirs for, 3
    threats, 120
    transmission, 23